SCHAUM'S OUTLINE OF

THEORY AND PROBLEMS

of

MODERN PHYSICS

·

by

RONALD GAUTREAU, Ph.D.
Associate Professor of Physics
New Jersey Institute of Technology

and

WILLIAM SAVIN, Ph.D.
Associate Professor of Physics
New Jersey Institute of Technology

SCHAUM'S OUTLINE SERIES
McGRAW-HILL BOOK COMPANY

New York St. Louis San Francisco Auckland Bogotá Hamburg London
Madrid Mexico Milan Montreal New Delhi Panama Paris
São Paulo Singapore Sydney Tokyo Toronto

Dedicated to the memory of
Professor Marcus M. Mainardi

0-07-023062-5

10 11 12 13 14 15 SH SH 8 7

Library of Congress Cataloging in Publication Data

Gautreau, Ronald.
   Schaum's outline of theory and problems of modern physics.

   (Schaum's outline series)
   Includes index.
   1. Physics—Problems, exercises, etc.   I. Savin, William, joint author.   II. Title.
QC32.G34   530′.076   78-8752
ISBN 0-07-023062-5

Preface

The area of Modern Physics embraces topics that have evolved since roughly the turn of this century. These developments can be mind-boggling, as with the effects on time predicted by Einstein's Special Theory of Relativity, or quite practical, like the many devices based upon semiconductors, whose explanation lies in the band theory of solids.

The scope of the present book may be gauged from the Table of Contents. Each chapter consists of a succinct presentation of the principles and "meat" of a particular subject, followed by a large number of completely solved problems that naturally develop the subject and illustrate the principles. It is the authors' conviction that these solved problems are a valuable learning tool. The solved problems have been made short and to the point, and have been ordered in terms of difficulty. They are followed by unsolved supplementary problems, with answers, which allow the reader to check his grasp of the material.

It has been assumed that the reader has had the standard introductory courses in general physics, and the book is geared primarily at the sophomore or junior level, although we have also included problems of a more advanced nature. While it will certainly serve as a supplement to any standard Modern Physics text, this book is sufficiently comprehensive and self-contained to be used by itself to learn the principles of Modern Physics.

We extend special thanks to David Beckwith for meticulous editing and for input that improved the final version of the book. Any mistakes are ours, of course, and we would appreciate having these pointed out to us. Finally, we are indebted to our families for their enormous patience with us throughout the long preparation of this work.

RONALD GAUTREAU
WILLIAM SAVIN

New Jersey Institute of Technology

Contents

PART I   **THE SPECIAL THEORY OF RELATIVITY**   1

Chapter 1 **Galilean Transformations**   1

    1.1 Events and Coordinates   1
    1.2 Galilean Coordinate Transformations   1
    1.3 Galilean Velocity Transformations   1
    1.4 Galilean Acceleration Transformations   2
    1.5 Invariance of an Equation   2

Chapter 2 **The Postulates of Einstein**   7

    2.1 Absolute Space and the Ether   7
    2.2 The Michelson-Morley Experiment   7
    2.3 Length and Time Measurements—A Question of Principle   7
    2.4 The Postulates of Einstein   7

Chapter 3 **The Lorentz Coordinate Transformations**   12

    3.1 The Constancy of the Speed of Light   12
    3.2 The Invariance of Maxwell's Equations   12
    3.3 General Considerations in Solving Problems Involving Lorentz Transformations   13
    3.4 Simultaneity   13

Chapter 4 **Relativistic Length Measurements**   17

    4.1 The Definition of Length   17

Chapter 5 **Relativistic Time Measurements**   20

    5.1 Proper Time   20
    5.2 Time Dilation   20

Chapter 6 **Relativistic Space-Time Measurements**   24

Chapter 7 **Relativistic Velocity Transformations**   34

    7.1 The Lorentz Velocity Transformations and the Speed of Light   34
    7.2 General Considerations in Solving Velocity Problems   34

Chapter 8 **Mass, Energy and Momentum in Relativity**   39

    8.1 The Need to Redefine Classical Momentum   39
    8.2 The Variation of Mass with Velocity   39
    8.3 Newton's Second Law in Relativity   39

8.4 Mass and Energy Relationship: $E = mc^2$ 40

8.5 Momentum and Energy Relationship   40

8.6 Units for Energy and Momentum   40

8.7 General Considerations in Solving Mass-Energy Problems   41

Chapter 9 The Relativistic Doppler Effect   50

PART II THE QUANTUM THEORY OF ELECTROMAGNETIC RADIATION   53

Chapter 10 The Theory of Photons   53

Chapter 11 The Photoelectric Effect   56

11.1 Experimental Results   56

11.2 Theory of the Photoelectric Effect   56

Chapter 12 The Compton Effect   62

Chapter 13 Pair Production and Annihilation   68

13.1 Pair Production   68

13.2 Pair Annihilation   68

Chapter 14 Absorption of Photons   74

PART III MATTER WAVES   77

Chapter 15 De Broglie Waves   77

15.1 The Wave-Particle Duality of Electromagnetic Radiation   77

15.2 The Wave-Particle Duality of Matter   77

Chapter 16 Experimental Verification of De Broglie's Hypothesis   82

16.1 The Bragg Law of Diffraction   82

16.2 Electron Diffraction Experiments   83

Chapter 17 The Probability Interpretation of De Broglie Waves   88

17.1 A Probability Interpretation for Electromagnetic Radiation   88

17.2 A Probability Interpretation of Matter   88

Chapter 18 The Heisenberg Uncertainty Principle   91

18.1 Measurements and Uncertainties   91

18.2 The Uncertainty Relation for Position and Momentum   92

18.3 The Uncertainty Relation for Energy and Time   92

18.4 The Principle of Complementarity   92

CONTENTS

PART IV   HYDROGENLIKE ATOMS   99

Chapter 19 The Bohr Atom   99
19.1 The Hydrogen Spectrum   99
19.2 The Bohr Theory of the Hydrogen Atom   99
19.3 Emission of Radiation in Bohr's Theory   100
19.4 Energy Level Diagrams   101
19.5 Hydrogenic Atoms   103
19.6 μ-Mesic and π-Mesic Atoms   103

Chapter 20 Electron Orbital Motion and the Zeeman Effect   112
20.1 Orbital Angular Momentum from a Classical Viewpoint   112
20.2 Classical Magnetic Dipole Moment   112
20.3 Classical Energy of a Magnetic Dipole Moment in an External Magnetic Field   113
20.4 The Zeeman Experiment   113
20.5 Quantization of the Magnitude of the Orbital Angular Momentum   114
20.6 Quantization of the Direction of the Orbital Angular Momentum   114
20.7 Explanation of the Zeeman Effect   114

Chapter 21 The Stern–Gerlach Experiment and Electron Spin   120
21.1 The Stern–Gerlach Experiment   120
21.2 Electron Spin   120

Chapter 22 Electron Spin and Fine Structure   124
22.1 Spin-Orbit Coupling   124
22.2 Fine Structure   124
22.3 Total Angular Momentum (the Vector Model)   124

PART V   MANY-ELECTRON ATOMS   128

Chapter 23 The Pauli Exclusion Principle   128
23.1 Introduction   128
23.2 The Pauli Exclusion Principle   128
23.3 A Single Particle in a One-Dimensional Box   128
23.4 Many Particles in a One-Dimensional Box   128

Chapter 24 Many-Electron Atoms and the Periodic Table   133
24.1 Spectroscopic Notation for Electron Configurations in Atoms   133
24.2 The Periodic Table and an Atomic Shell Model   133
24.3 Spectroscopic Notation for Atomic States   134
24.4 Atomic Excited States and LS Coupling   135
24.5 The Anomalous Zeeman Effect   135

Chapter 25 Inner-Electron Transitions: X-Rays   149
25.1 X-Ray Apparatus   149
25.2 Production of Bremsstrahlung   149

CONTENTS

25.3 Production of Characteristic X-Ray Spectra   150
25.4 The Moseley Relation   150
25.5 X-Ray Absorption Edges   152
25.6 Auger Effect   152
25.7 X-Ray Fluorescence   153

PART VI   NUCLEAR PHYSICS   163

Chapter 26   Nucleon and Deuteron Properties   163

26.1 The Nucleons   163
26.2 Nucleon Forces   163
26.3 The Deuteron   164

Chapter 27   Properties of Nuclei   168

27.1 Designation of Nuclei   168
27.2 Relative Number of Protons and Neutrons   168
27.3 The Nucleus as a Sphere   169
27.4 Nuclear Binding Energy   169

Chapter 28   Nuclear Models   173

28.1 Liquid Drop Model   173
28.2 Shell Model   175

Chapter 29   The Decay of Unstable Nuclei   184

29.1 Introduction   184
29.2 The Statistical Radioactive Decay Law   184
29.3 Gamma Decay   185
29.4 Alpha Decay   185
29.5 Beta Decay and the Neutrino   186

Chapter 30   Nuclear Reactions   199

30.1 Introduction   199
30.2 Classification of Nuclear Reactions   199
30.3 Laboratory and Center-of-Mass Systems   200
30.4 Energetics of Nuclear Reactions   201
30.5 Nuclear Cross Sections   201

Chapter 31   Fission and Fusion   209

31.1 Nuclear Fission   209
31.2 Nuclear Fusion   210

Chapter 32   Elementary Particles   215

32.1 Elementary Particle Genealogy   215
32.2 Particle Interactions   216

32.3 Conservation Laws   216
32.4 Conservation of Leptons   217
32.5 Conservation of Baryons   217
32.6 Conservation of Isotopic Spin   217
32.7 Conservation of Strangeness   218
32.8 Conservation of Parity   219
32.9 Short-lived Particles and the Resonances   219

PART VII **ATOMIC SYSTEMS**   227

Chapter 33 **Molecular Bonding**   227

33.1 Ionic Bonding   227
33.2 Covalent Bonding   227
33.3 Other Types of Bonding   228

Chapter 34 **Excitations of Diatomic Molecules**   232

34.1 Molecular Rotations   232
34.2 Molecular Vibrations   232
34.3 Combined Excitations   233

Chapter 35 **Kinetic Theory**   241

35.1 The Ideal Gas Law   241

Chapter 36 **Distribution Functions**   249

36.1 Discrete Distribution Functions   249
36.2 Continuous Distribution Functions   250
36.3 Fundamental Distribution Functions and Density of States 250

Chapter 37 **Classical Statistics: The Maxwell-Boltzmann Distribution** 256

Chapter 38 **Quantum Statistics**   266

38.1 Fermi-Dirac Statistics   266
38.2 Bose-Einstein Statistics   266
38.3 High-Temperature Limit   267
38.4 Two Useful Integrals   267

Chapter 39 **The Band Theory of Solids**   287

Appendix   296

Index   305

Chapter 1

Galilean Transformations

1.1 EVENTS AND COORDINATES

We begin by considering the concept of a physical event. The event might be the striking of a tree by a lightning bolt or the collision of two particles, and happens at a point in space and at an instant in time. The particular event is specified by an observer by assigning to it four coordinates: the three position coordinates x, y, z that measure the distance from the origin of a coordinate system where the observer is located, and the time coordinate t that the observer records with his clock.

Consider now two observers, O and O', where O' travels with a constant velocity v with respect to O along their common x-x' axis (Fig. 1-1). Both observers are equipped with metersticks and clocks so that they can measure coordinates of events. Further, suppose both observers adjust their clocks so that when they pass each other at $x = x' = 0$, the clocks read $t = t' = 0$. Any given event P will have eight numbers associated with it, the four coordinates (x, y, z, t) assigned by O and the four coordinates (x', y', z', t') assigned (to the same event) by O'.

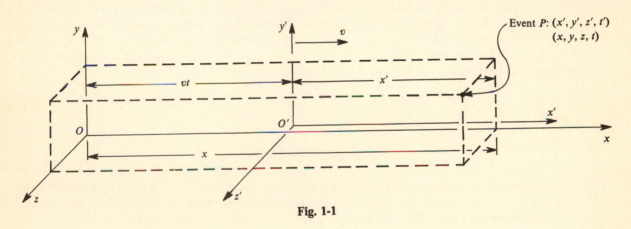

Fig. 1-1

1.2 GALILEAN COORDINATE TRANSFORMATIONS

The relationship between the measurements (x, y, z, t) of O and the measurements (x', y', z', t') of O' for a particular event is obtained by examining Fig. 1-1:

$$x' = x - vt \qquad y' = y \qquad z' = z$$

In addition, in classical physics it is implicitly assumed that

$$t' = t$$

These four equations are called the *Galilean coordinate transformations*.

1.3 GALILEAN VELOCITY TRANSFORMATIONS

In addition to the coordinates of an event, the velocity of a particle is of interest. Observers O and O' will describe the particle's velocity by assigning three components to it, with (u_x, u_y, u_z) being the velocity components as measured by O, and (u'_x, u'_y, u'_z) being the velocity components as measured by O'.

1

The relationship between (u_x, u_y, u_z) and (u'_x, u'_y, u'_z) is obtained from the time differentiation of the Galilean coordinate transformations. Thus, from $x' = x - vt$,

$$u'_x = \frac{dx'}{dt'} = \frac{d}{dt}(x - vt)\frac{dt}{dt'} = \left(\frac{dx}{dt} - v\right)(1) = u_x - v$$

Altogether, the *Galilean velocity transformations* are

$$u'_x = u_x - v \qquad u'_y = u_y \qquad u'_z = u_z$$

1.4 GALILEAN ACCELERATION TRANSFORMATIONS

The acceleration of a particle is the time derivative of its velocity, i.e. $a_x = du_x/dt$, etc. To find the *Galilean acceleration transformations* we differentiate the velocity transformations and use the facts that $t' = t$ and $v = $ constant to obtain

$$a'_x = a_x \qquad a'_y = a_y \qquad a'_z = a_z$$

Thus the measured acceleration components are the same for all observers moving with uniform relative velocity.

1.5 INVARIANCE OF AN EQUATION

By *invariance* of an equation it is meant that the equation will have the same form when determined by two observers. In classical theory it is assumed that space and time measurements of two observers are related by the Galilean transformations. Thus, when a particular form of an equation is determined by one observer, the Galilean transformations can be applied to this form to determine the form for the other observer. If both forms are the same, the equation is invariant under the Galilean transformations. See Problems 1.11 and 1.12.

Solved Problems

1.1. A passenger in a train moving at 30 m/s passes a man standing on a station platform at $t = t' = 0$. Twenty seconds after the train passes him, the man on the platform determines that a bird flying along the tracks in the same direction as the train is 800 m away. What are the coordinates of the bird as determined by the passenger?

The coordinates assigned to the bird by the man on the station platform are

$$(x, y, z, t) = (800 \text{ m}, 0, 0, 20 \text{ s})$$

The passenger measures the distance x' to the bird as

$$x' = x - vt = 800 \text{ m} - (30 \text{ m/s})(20 \text{ s}) = 200 \text{ m}$$

Therefore the bird's coordinates as determined by the passenger are

$$(x', y', z', t') = (200 \text{ m}, 0, 0, 20 \text{ s})$$

1.2. Refer to Problem 1.1. Five seconds after making the first coordinate measurement, the man on the platform determines that the bird is 850 m away. From these data find the velocity of the bird (assumed constant) as determined by the man on the platform and by the passenger on the train.

The coordinates assigned to the bird at the second position by the man on the platform are

$$(x_2, y_2, z_2, t_2) = (850 \text{ m}, 0, 0, 25 \text{ s})$$

Hence, the velocity u_x of the bird as measured by the man on the platform is

$$u_x = \frac{x_2 - x_1}{t_2 - t_1} = \frac{850 \text{ m} - 800 \text{ m}}{25 \text{ s} - 20 \text{ s}} = +10 \text{ m/s}$$

The positive sign indicates the bird is flying in the positive x-direction.   The passenger finds that at the second position the distance x_2' to the bird is

$$x_2' = x_2 - vt_2 = 850 \text{ m} - (30 \text{ m/s})(25 \text{ s}) = 100 \text{ m}$$

Thus, $(x_2', y_2', z_2', t_2') = (100 \text{ m}, 0, 0, 25 \text{ s})$, and the velocity u_x' of the bird as measured by the passenger is

$$u_x' = \frac{x_2' - x_1'}{t_2' - t_1'} = \frac{100 \text{ m} - 200 \text{ m}}{25 \text{ s} - 20 \text{ s}} = -20 \text{ m/s}$$

so that, as measured by the passenger, the bird is moving in the negative x'-direction.   Note that this result is consistent with that obtained from the Galilean velocity transformation:

$$u_x' = u_x - v = 10 \text{ m/s} - 30 \text{ m/s} = -20 \text{ m/s}$$

1.3.   A sample of radioactive material, at rest in the laboratory, ejects two electrons in opposite directions.   One of the electrons has a speed of $0.6c$ and the other has a speed of $0.7c$, as measured by a laboratory observer.   According to classical velocity transformations, what will be the speed of one electron as measured from the other?

Let observer O be at rest with respect to the laboratory and let observer O' be at rest with respect to the particle moving with speed $0.6c$ (taken in the positive direction).   Then, from the Galilean velocity transformation,

$$u_x' = u_x - v = -0.7c - 0.6c = -1.3c$$

This problem demonstrates that velocities greater than the speed of light are possible with the Galilean transformations, a result that is inconsistent with Special Relativity.

1.4.   A train moving with a velocity of 60 mi/hr passes through a railroad station at 12:00.   Twenty seconds later a bolt of lightning strikes the railroad tracks one mile from the station in the same direction that the train is moving.   Find the coordinates of the lightning flash as measured by an observer at the station and by the engineer of the train.

Both observers measure the time coordinate as

$$t = t' = (20 \text{ s})\left(\frac{1 \text{ hr}}{3600 \text{ s}} \right) = \frac{1}{180} \text{ hr}$$

The observer at the station measures the spatial coordinate to be $x = 1$ mi.   The spatial coordinate as determined by the engineer of the train is

$$x' = x - vt = 1 \text{ mi} - (60 \text{ mi/hr})\left(\frac{1}{180} \text{ hr} \right) = \frac{2}{3} \text{ mi}$$

1.5.   A hunter on the ground fires a bullet in the northeast direction which strikes a deer 0.25 miles from the hunter.   The bullet travels with a speed of 1800 mi/hr.   At the instant when the bullet is fired, an airplane is directly over the hunter at an altitude of one mile and is traveling due east with a velocity of 600 mi/hr.   When the bullet strikes the deer, what are the coordinates as determined by an observer in the airplane?

Using the Galilean transformations,

$$t' = t = \frac{0.25 \text{ mi}}{1800 \text{ mi/hr}} = 1.39 \times 10^{-4} \text{ hr}$$

$$x' = x - vt = (0.25 \text{ mi}) \cos 45° - (600 \text{ mi/hr})(1.39 \times 10^{-4} \text{ hr}) = 0.094 \text{ mi}$$

$$y' = y = (0.25 \text{ mi}) \sin 45° = 0.177 \text{ mi}$$

$$z' = z - h = 0 - 1 \text{ mi} = -1 \text{ mi}$$

1.6.   An observer, at rest with respect to the ground, observes the following collision.   A particle of mass $m_1 = 3$ kg moving with velocity $u_1 = 4$ m/s along the x-axis approaches a second particle of mass $m_2 = 1$ kg moving with velocity $u_2 = -3$ m/s along the x-axis.   After a head-on collision the ground observer finds that m_2 has velocity $u_2^* = 3$ m/s along the x-axis.   Find the velocity u_1^* of m_1 after the collision.

$$\text{initial momentum} = \text{final momentum}$$
$$m_1 u_1 + m_2 u_2 = m_1 u_1^* + m_2 u_2^*$$
$$(3 \text{ kg})(4 \text{ m/s}) + (1 \text{ kg})(-3 \text{ m/s}) = (3 \text{ kg})u_1^* + (1 \text{ kg})(3 \text{ m/s})$$
$$9 \text{ kg} \cdot \text{m/s} = (3 \text{ kg})u_1^* + 3 \text{ kg} \cdot \text{m/s}$$

Solving, $u_1^* = 2$ m/s.

1.7.   A second observer, O', who is walking with a velocity of 2 m/s relative to the ground along the x-axis observes the collision described in Problem 1.6.   What are the system momenta before and after the collision as determined by him?

Using the Galilean velocity transformations,

$$u_1' = u_1 - v = 4 \text{ m/s} - 2 \text{ m/s} = 2 \text{ m/s}$$
$$u_2' = u_2 - v = -3 \text{ m/s} - 2 \text{ m/s} = -5 \text{ m/s}$$
$$u_1^{*'} = u_1^* - v = 2 \text{ m/s} - 2 \text{ m/s} = 0$$
$$u_2^{*'} = u_2^* - v = 3 \text{ m/s} - 2 \text{ m/s} = 1 \text{ m/s}$$
$$(\text{initial momentum})' = m_1 u_1' + m_2 u_2' = (3 \text{ kg})(2 \text{ m/s}) + (1 \text{ kg})(-5 \text{ m/s}) = 1 \text{kg} \cdot \text{m/s}$$
$$(\text{final momentum})' = m_1 u_1^{*'} + m_2 u_2^{*'} = (3 \text{ kg})(0) + (1 \text{ kg})(1 \text{ m/s}) = 1 \text{kg} \cdot \text{m/s}$$

Thus, as a result of the Galilean transformations, O' also determines that momentum is conserved (but at a different value from that found by O).

1.8.   An open car traveling at 100 ft/s has a boy in it who throws a ball upward with a velocity of 20 ft/s.   Write the equation of motion (giving position as a function of time) for the ball as seen by (a) the boy, (b) an observer stationary on the road.

(a)   For the boy in the car the ball travels straight up and down, so

$$y' = v_0 t' + \tfrac{1}{2} a t'^2 = (20 \text{ ft/s})t' + \tfrac{1}{2}(-32 \text{ ft/s}^2)t'^2 = 20t' - 16t'^2$$
$$x' = z' = 0$$

(b)   For the stationary observer, one obtains from the Galilean transformations

$$t = t'$$
$$x = x' + vt = 0 + 100t \qquad y = y' = 20t - 16t^2 \qquad z = z' = 0$$

1.9.   Consider a mass attached to a spring and moving on a horizontal, frictionless surface.   Show, from the classical transformation laws, that the equations of motion of the mass are the same as determined by an observer at rest with respect to the surface and by a second observer moving with constant velocity along the direction of the spring.

The equation of motion of the mass, as determined by an observer at rest with respect to the surface, is $F = ma$, or

$$-k(x - x_0) = m \frac{d^2x}{dt^2} \tag{1}$$

To determine the equation of motion as found by the second observer we use the Galilean transformations to obtain

$$x = x' + vt' \qquad x_0 = x_0' + vt' \qquad \frac{d^2x}{dt^2} = \frac{d^2x'}{dt'^2}$$

Substituting these values in (1) gives

$$-k(x' - x_0') = m \frac{d^2x'}{dt'^2} \tag{2}$$

Because (*1*) and (*2*) have the same form, the equation of motion is invariant under the Galilean transformations.

1.10.   Show that the electromagnetic wave equation,

$$\frac{\partial^2 \phi}{\partial x^2} + \frac{\partial^2 \phi}{\partial y^2} + \frac{\partial^2 \phi}{\partial z^2} - \frac{1}{c^2} \frac{\partial^2 \phi}{\partial t^2} = 0$$

is not invariant under the Galilean transformations.

The equation will be invariant if it retains the same form when expressed in terms of the new variables x', y', z', t'. We first find from the Galilean transformations that

$$\frac{\partial x'}{\partial x} = 1 \qquad \frac{\partial x'}{\partial t} = -v \qquad \frac{\partial t'}{\partial t} = \frac{\partial y'}{\partial y} = \frac{\partial z'}{\partial z} = 1$$

$$\frac{\partial x'}{\partial y} = \frac{\partial x'}{\partial z} = \frac{\partial y'}{\partial x} = \frac{\partial t'}{\partial x} = \cdots = 0$$

From the chain rule and using the above results we have

$$\frac{\partial \phi}{\partial x} = \frac{\partial \phi}{\partial x'}\frac{\partial x'}{\partial x} + \frac{\partial \phi}{\partial y'}\frac{\partial y'}{\partial x} + \frac{\partial \phi}{\partial z'}\frac{\partial z'}{\partial x} + \frac{\partial \phi}{\partial t'}\frac{\partial t'}{\partial x} = \frac{\partial \phi}{\partial x'} \qquad \text{and} \qquad \frac{\partial^2 \phi}{\partial x^2} = \frac{\partial^2 \phi}{\partial x'^2}$$

Similarly,

$$\frac{\partial^2 \phi}{\partial y^2} = \frac{\partial^2 \phi}{\partial y'^2} \qquad \frac{\partial^2 \phi}{\partial z^2} = \frac{\partial^2 \phi}{\partial z'^2}$$

Moreover,

$$\frac{\partial \phi}{\partial t} = -v\frac{\partial \phi}{\partial x'} + \frac{\partial \phi}{\partial t'} \qquad \frac{\partial^2 \phi}{\partial t^2} = \frac{\partial^2 \phi}{\partial t'^2} - 2v\frac{\partial^2 \phi}{\partial x'\partial t'} + v^2\frac{\partial^2 \phi}{\partial x'^2}$$

Substituting these expressions in the wave equation gives

$$\frac{\partial^2 \phi}{\partial x'^2} + \frac{\partial^2 \phi}{\partial y'^2} + \frac{\partial^2 \phi}{\partial z'^2} - \frac{1}{c^2}\frac{\partial^2 \phi}{\partial t'^2} + \frac{1}{c^2}\left(2v\frac{\partial^2 \phi}{\partial x'\partial t'} - v^2\frac{\partial^2 \phi}{\partial x'^2}\right) = 0$$

Therefore the wave equation is *not* invariant under the Galilean transformations, for the form of the equation has changed.

The electromagnetic wave equation follows from Maxwell's equations of electromagnetic theory. By applying the procedure described here to Maxwell's equations, one finds that Maxwell's equations also are *not* invariant under Galilean transformations.   Compare with Problem 6.23.

Supplementary Problems

1.11.   A man (O') in the back of a 20-ft flatcar moving at 30 ft/s records that a flashbulb is fired in the front of the flatcar two seconds after he has passed a man (O) on the ground.   Find the coordinates of the event as determined by each observer.   *Ans.* $(x', t') = (20 \text{ ft}, 2 \text{ s})$; $(x, t) = (80 \text{ ft}, 2 \text{ s})$

1.12.   A boy sees a deer run directly away from him.   The deer is running with a speed of 20 mi/hr.   The boy gives chase and runs with a speed of 8 mi/hr.   What is the speed of the deer relative to the boy?
Ans.   12 mi/hr

1.13.   A boy in a train throws a ball in the forward direction with a speed of 20 mi/hr.   If the train is moving with a speed of 80 mi/hr, what is the speed of the ball as measured by a man on the ground?
Ans.   100 mi/hr

1.14.   A passenger walks backward along the aisle of a train with a speed of 2 mi/hr as the train moves along a straight track at a constant speed of 60 mi/hr with respect to the ground. What is the passenger's speed as measured by an observer standing on the ground?   *Ans.*   58 mi/hr

1.15.   A conductor standing on a railroad platform synchronizes his watch with the engineer in the front of a train traveling at 60 mi/hr. The train is 1/4 mile long. Two minutes after the train leaves the platform a brakeman in the caboose lights a cigarette. What are the coordinates of the brakeman, as determined by the engineer and by the conductor, when the cigarette is lit?
Ans. $(x', t') = (-\frac{1}{4} \text{ mi}, 2 \text{ min})$; $(x, t) = (1\frac{3}{4} \text{ mi}, 2 \text{ min})$

1.16.   A man sitting in a train lights two cigarettes, one ten minutes after the other. The train is moving in a straight line with a velocity of 20 m/s. What is the distance separation as measured by a man on the ground?   *Ans.*   12,000 m

1.17.   A one-kilogram ball is constrained to move to the north at 3 m/s. It makes a perfectly elastic collision with an identical second ball which is at rest, and both balls move on a north-south axis after the collision. Compute, in the laboratory system, the total momentum before and after the collision.
Ans.   3 kg · m/s

1.18.   For Problem 1.17 calculate the total energy before and after the collision.   *Ans.*   4.5 J

1.19.   Refer to Problem 1.17. Calculate the total momentum before and after the collision as measured by an observer moving northwards at 1.5 m/s.   *Ans.*   0

1.20.   For the observer in Problem 1.19 calculate the total energy before and after the collision.
Ans.   2.25 J

1.21.   Repeat Problems 1.19 and 1.20 for an observer moving eastwards at 2 m/s.
Ans.   5 kg · m/s 37° north of west; 8.5 J

1.22.   A person is in a boat moving eastwards with a speed of 15 ft/s. At the instant that the boat passes a dock a person on the dock throws a rock northwards. The rock strikes the water 6 s later at a distance of 150 ft from the dock. Find the coordinates of the splash as measured by the person in the boat.
Ans. $(x, y, t) = (-90 \text{ ft}, 150 \text{ ft}, 6 \text{ s})$

1.23.   Consider a one-dimensional, elastic collision that takes place along the x-axis of O. Show, from the classical transformation equations, that kinetic energy will also be conserved as determined by a second observer, O', who moves with constant velocity u along the x-axis of O.

Chapter 2

The Postulates of Einstein

2.1 ABSOLUTE SPACE AND THE ETHER

A consequence of the Galilean velocity transformations is that if a certain observer measures a light signal to travel with the velocity $c = 3 \times 10^8$ m/s, then any other observer moving relative to him will measure the same light signal to travel with a velocity different from c. What determines the particular reference frame such that if an observer is at rest relative to this frame, this privileged observer will measure the value c for the velocity of light signals?

Before Einstein it was generally believed that this privileged observer was the same observer for whom Maxwell's equations were valid. Maxwell's equations describe electromagnetic theory and predict that electromagnetic waves will travel with the speed $c = 1/\sqrt{\epsilon_0 \mu_0} = 3 \times 10^8$ m/s. The space that was at rest with respect to this privileged observer was called "absolute space." Any other observer moving with respect to this absolute space would find the speed of light to be different from c. Since light is an electromagnetic wave, it was felt by 19th century physicists that a medium must exist through which the light propagated. Thus it was postulated that the "ether" permeated all of absolute space.

2.2 THE MICHELSON-MORLEY EXPERIMENT

If an ether exists, then an observer on the earth moving through the ether should notice an "ether wind." An apparatus with the sensitivity to measure the earth's motion through the hypothesized ether was developed by Michelson in 1881, and refined by Michelson and Morley in 1887. The outcome of the experiment was that *no motion through the ether was detected.* See Problems 2.4, 2.5 and 2.6.

2.3 LENGTH AND TIME MEASUREMENTS—
A QUESTION OF PRINCIPLE

The one element common to both the null result of the Michelson-Morley experiment and the fact that Maxwell's equations hold only for a privileged observer is the Galilean transformations. These "obvious" transformations were reexamined by Einstein from what might be termed an "operational" point of view. Einstein took the approach that any quantity relevant to physical theories should, at least in principle, have a well-defined procedure by which it is measured. If such a procedure cannot be formulated, then the quantity should not be employed in physics.

Einstein could find no way to justify operationally the Galilean transformation $t' = t$, i.e. the statement that two observers *can* measure the time of an event to be the same. Consequently, the transformation $t' = t$, and with it the rest of the Galilean transformations, was rejected by Einstein.

2.4 THE POSTULATES OF EINSTEIN

Einstein's guiding idea, which he called the *Principle of Relativity*, was that *all* nonaccelerating observers should be treated equally in all respects, even if they are moving (at constant velocity) relative to each other. This principle can be formalized as follows:

Postulate 1: The laws of physics are the same (invariant) for all inertial (nonaccelerating) observers.

Newton's laws of motion are in accord with the Principle of Relativity, but Maxwell's equations together with the Galilean transformations are in conflict with it.   Einstein could see no reason for a basic difference between dynamical and electromagnetic laws.   Hence his

Postulate 2:   In vacuum the speed of light as measured by all inertial observers is

$$c = 1/\sqrt{\epsilon_0 \mu_0} = 3 \times 10^8 \text{ m/s}$$

independent of the motion of the source.

Solved Problems

2.1.   Suppose that a clock B is located at a distance L from an observer.   Describe how this clock can be synchronized with clock A, which is at the observer's location.

Set the (stopped) clock B to read $t_B = L/c$.   At $t_A = 0$ (as recorded by clock A) send a light signal towards the distant clock B.   Start clock B when the signal reaches it.

2.2.   A flashbulb is located 30 km from an observer.   The bulb is fired and the observer *sees* the flash at 1:00 P.M.   What is the actual time that the bulb is fired?

The time for the light signal to travel 30 km is

$$\Delta t = \frac{\Delta s}{c} = \frac{30 \times 10^3 \text{ m}}{3 \times 10^8 \text{ m/s}} = 1 \times 10^{-4} \text{ s}$$

Therefore, the flashbulb was fired 1×10^{-4} s before 1:00 P.M.

2.3.   A rod is moving from left to right.   When the left end of the rod passes a camera, a picture is taken of the rod together with a stationary calibrated meterstick.   In the developed picture the left end of the rod coincides with the zero mark and the right end coincides with the 0.90-m mark on the meterstick.   If the rod is moving at $0.8c$ with respect to the camera, determine the *actual* length of the rod.

In order that the light signal from the right end of the rod be recorded by the camera, it must have started from the 0.90-m mark at an earlier time given by

$$\Delta t = \frac{\Delta s}{c} = \frac{0.90 \text{ m}}{3 \times 10^8 \text{ m/s}} = 3 \times 10^{-9} \text{ s}$$

During this time interval the left end of the rod will advance through a distance Δs^* given by (see Fig. 2-1)

$$\Delta s^* = v \, \Delta t = (0.8 \times 10^8 \text{ m/s})(3 \times 10^{-9} \text{ s}) = 0.72 \text{ m}$$

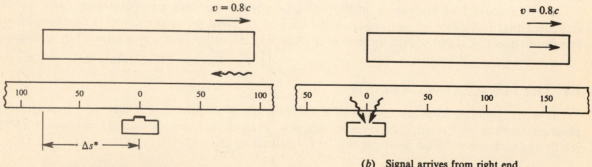

(*a*) Signal starts from right
end; camera shutter closed.

(*b*) Signal arrives from right end
and is recorded by open camera
together with signal from left end.

Fig. 2-1

Therefore, the actual length of the rod is $L = 0.90$ m $+ 0.72$ m $= 1.62$ m.   This result illustrates that photographing a moving rod will *not* give its correct length.

2.4.   Let two events occur at equal distances from an observer.   Suppose the observer adopts the following statement as a definition of simultaneity of equidistant events: "The two events are simultaneous if the light signals emitted from each event reach me at the same time."   Show that, according to this definition, if the observer determines that two events are simultaneous, then another observer, moving relative to him, will in general determine that the two events are *not* simultaneous.

From Fig. 2-2 it is seen that if the two light signals reach the first observer (O) at the same time, they will necessarily reach the second observer (O') at different times.   Since the two signals started out equidistant from O', he will, according to the above definition, determine that the two events did not occur simultaneously, but that event B happened before event A.

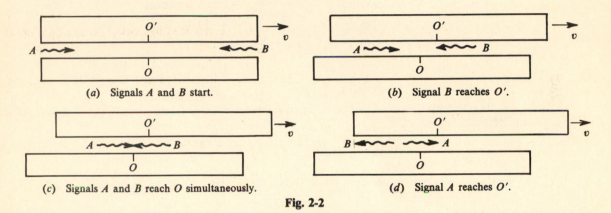

<table>
<tr><td>(a)   Signals A and B start.</td><td>(b)   Signal B reaches O'.</td></tr>
<tr><td>(c)   Signals A and B reach O simultaneously.</td><td>(d)   Signal A reaches O'.</td></tr>
</table>

Fig. 2-2

2.5.   Figure 2-3 diagrams a Michelson-Morley interferometer oriented with one arm (A) parallel to the "ether wind."   Show that if the apparatus is rotated through $90°$, the number of fringes, ΔN, that moves past the telescope crosshairs is, to first order in $(v/c)^2$,

$$\Delta N = \frac{v^2}{\lambda c^2} (l_A + l_B)$$

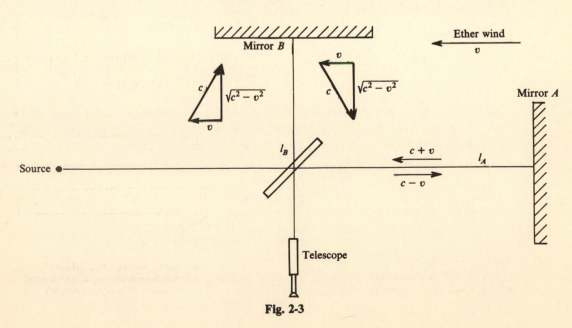

Fig. 2-3

For arm A, the time for light to travel to mirror A is obtained by dividing the path length l_A by the velocity of light, which from the Galilean velocity transformations is $c - v$. On return the path length is still l_A, but now the velocity is $c + v$, so the total time for the round trip is

$$t_A = \frac{l_A}{c - v} + \frac{l_A}{c + v} = \frac{2l_A/c}{1 - (v^2/c^2)}$$

To travel along the other arm a light ray must be aimed such that its resultant velocity vector (velocity with respect to the ether plus velocity of the ether with respect to the interferometer) is perpendicular to arm A. This gives a speed of $\sqrt{c^2 - v^2}$ for both directions along path l_B, so the time for the round trip is

$$t_B = \frac{2l_B}{\sqrt{c^2 - v^2}} = \frac{2l_B/c}{\sqrt{1 - (v^2/v^2)}}$$

If we assume $v/c \ll 1$, the times t_A and t_B can be expanded to first order in $(v/c)^2$ and the time difference taken.

$$t_A \approx \frac{2l_A}{c}\left(1 + \frac{v^2}{c^2}\right) \qquad t_B \approx \frac{2l_B}{c}\left(1 + \frac{v^2}{2c^2}\right)$$

and

$$\delta = t_A - t_B \approx \frac{2(l_A - l_B)}{c} + \frac{2l_A v^2}{c^3} - \frac{l_B v^2}{c^3}$$

Now if the interferometer is rotated 90°, l_A and l_B are interchanged, and there is also a reversal of the time difference. Thus

$$\delta' \approx \frac{2(l_A - l_B)}{c} + \frac{l_A v^2}{c^3} - \frac{2l_B v^2}{c^3}$$

and the interference pattern observed would show a fringe shift of ΔN fringes, where

$$\Delta N = \frac{\delta - \delta'}{T} = \frac{c(\delta - \delta')}{\lambda} = \frac{(l_A + l_B)v^2}{\lambda c^2}$$

Here T and λ are the period and wavelength of the light.

2.6. Assume that the earth's velocity through the ether is the same as its orbital velocity, so that $v = 10^{-4}c$. Consider a Michelson-Morley experiment where the arms of the interferometer are each 10 m long and one arm is in the direction of motion of the earth through the ether. Calculate the difference in time for the two light waves to travel along each of the arms.

Refer to Problem 2.5.

$$\delta \approx \frac{2}{c}(l_A - l_B) + \frac{2v^2}{c^3}\left(l_A - \tfrac{1}{2}l_B\right) = \frac{2(10^{-4}c)^2}{(3 \times 10^8 \text{ m/s})c^2}(5 \text{ m}) = 3.33 \times 10^{-16} \text{ s}$$

2.7. The original Michelson-Morley experiment used an interferometer with arms of 11 m and sodium light of 5900 Å. The experiment would reveal a fringe shift of 0.005 fringes. What upper limit does a null result place on the speed of the earth through the ether?

From Problem 2.5, the number of fringes, ΔN, seen to pass the telescope crosshairs is

$$\Delta N = \frac{v^2}{\lambda c^2}(l_A + l_B) = \frac{2lv^2}{\lambda c^2}$$

$$0.005 = \frac{2(11 \text{ m})v^2}{(5900 \times 10^{-10} \text{ m})(3 \times 10^8 \text{ m/s})^2}$$

Solving, $v = 3.47 \times 10^3$ m/s.

The earth's orbital velocity is 3×10^4 m/s, so the interferometer was sensitive enough to detect this motion. No fringe shift was observed.

Supplementary Problems

2.8.   Repeat Problem 2.3 for the case where the picture is taken when the *right* end of the rod passes the camera.   *Ans.*   0.18 m

2.9.   At the instant that the midpoint of a moving meterstick passes a camera, the camera shutter opens and a picture is taken of the meterstick together with a stationary calibrated rule, as in Problem 2.3.   If its speed relative to the camera is $0.8c$, what will be the length of the moving meterstick as recorded on the film?   *Ans.*   2.778 m

2.10.   Refer to Problem 2.4.   If the two signals reach O' simultaneously, what is their time sequence as determined by O?   *Ans.*   A occurs before B

2.11.   Assume that the oribital speed of the earth, 3×10^4 m/s, is equal to the speed of the earth through the ether.   If light takes t_A seconds to travel through an equal-arm Michelson-Morley apparatus in a direction parallel to this motion, calculate how long it will take light to travel perpendicular to this motion.   *Ans.* $(1 - 0.5 \times 10^{-8})t_A$

Chapter 3

The Lorentz Coordinate Transformations

Postulate 2 of Section 2.4 requires that the Galilean coordinate transformations be replaced by the *Lorentz coordinate transformations*. For the two observers of Fig. 1-1 these are:

$$x' = \frac{x - vt}{\sqrt{1 - (v^2/c^2)}} \qquad t' = \frac{t - \dfrac{v}{c^2} x}{\sqrt{1 - (v^2/c^2)}} \qquad y' = y \qquad z' = z \qquad (3.1)$$

These equations can be inverted to give (see Problem 3.8)

$$x = \frac{x' + vt'}{\sqrt{1 - (v^2/c^2)}} \qquad t = \frac{t' + \dfrac{v}{c^2} x'}{\sqrt{1 - (v^2/c^2)}} \qquad y = y' \qquad z = z' \qquad (3.2)$$

In (3.1) and (3.2), v is the velocity of O' with respect to O along their common axis; v is positive if O' moves in the positive x-direction and negative if O' moves in the negative x-direction It has also been assumed that both origins coincide when the clocks are started, so that $t' = t = 0$ when $x' = x = 0$. Note that the inverse transformations can be obtained from the first set of transformations by interchanging primed and unprimed variables and letting $v \to -v$. This is to be expected from Postulate I, since both observers are completely equivalent and observer O moves with velocity $-v$ with respect to O'.

3.1 THE CONSTANCY OF THE SPEED OF LIGHT

Suppose that at the instant when O and O' pass each other (at $t = t' = 0$), a light signal is sent from their common origin in the positive x-x' direction. If O finds that the signal's spatial and time coordinates are related by $x = ct$, then, according to (3.1), O' will find that

$$x' = \frac{x - vt}{\sqrt{1 - (v^2/c^2)}} = \frac{ct - vt}{\sqrt{1 - (v^2/c^2)}} = \frac{ct[1 - (v/c)]}{\sqrt{[1 - (v/c)][1 + (v/c)]}} = \sqrt{\frac{1 - (v/c)}{1 + (v/c)}} \; ct$$

$$t' = \frac{t - \dfrac{v}{c^2} x}{\sqrt{1 - (v^2/c^2)}} = \frac{t - \dfrac{v}{c^2} ct}{\sqrt{1 - (v^2/c^2)}} = \frac{t[1 - (v/c)]}{\sqrt{[1 - (v/c)][1 + (v/c)]}} = \sqrt{\frac{1 - (v/c)}{1 + (v/c)}} \; t$$

Thus O' will find that $x' = ct'$, in agreement with the second postulate of Einstein. Note also that for this event $t' \neq t$, in definite disagreement with the Galilean assumption.

3.2 THE INVARIANCE OF MAXWELL'S EQUATIONS

As discussed in Chapters 1 and 2, Maxwell's equations of electromagnetic theory are not invariant under Galilean transformations. However, as shown by H. A. Lorentz (before Einstein), they are invariant under Lorentz transformations. See Problems 6.21 through 6.23.

3.3 GENERAL CONSIDERATIONS IN SOLVING PROBLEMS INVOLVING LORENTZ TRANSFORMATIONS

When attacking any space-time problem, the key concept to keep in mind is that of "event." Most problems are concerned with two observers measuring the space and time coordinates of an event (or events). Thus, each event has eight numbers associated with it: (x, y, z, t), as assigned by O, and (x', y', z', t'), as assigned by O'. The Lorentz coordinate transformations express the relationships between these assignments.

Many times problems are concerned with the determination of the spatial interval and/or the time interval between two events. In this case a useful technique is to subtract from each other the appropriate Lorentz transformations describing each event. For example, suppose observer O' measures the time and spatial intervals between two events, A and B, and it is desired to obtain the time interval between these same two events as measured by O. From (3.2) one obtains upon subtracting t_A from t_B

$$t_B - t_A = \frac{(t'_B - t'_A) + (v/c^2)(x'_B - x'_A)}{\sqrt{1 - (v^2/c^2)}} \qquad (3.3)$$

Since all the quantities on the right-hand side of this equation are known, one can determine $t_B - t_A$.

3.4 SIMULTANEITY

Two events are *simultaneous* to an observer if he measures that the two events occur at the same time. With classical physics, when one observer determined that two events were simultaneous, then, since $t' = t$ from the Galilean transformations, every other observer would also find that the two events were simultaneous. In relativistic physics, on the other hand, two events that are simultaneous to one observer will, in general, not be simultaneous to another observer.

Suppose, for example, that events A and B are simultaneous as determined by O', so that $t'_A = t'_B$. According to (3.3), observer O will measure the time separation of these same two events as

$$t_B - t_A = \frac{(v/c^2)(x'_B - x'_A)}{\sqrt{1 - (v^2/c^2)}}$$

If the two events occur at the same spatial location, so that $x'_B = x'_A$, then the two events will also be simultaneous as determined by O. But if $x'_B \neq x'_A$, O will determine that the two events are *not* simultaneous.

Note that if the two events occur at the same spatial location, only *one* clock is needed by each observer to determine if the events are simultaneous. On the other hand, if the two events are separated spatially, then each observer needs *two* clocks, properly synchronized, to determine whether or not the two events are simultaneous.

Solved Problems

3.1. Evaluate $\sqrt{1 - (v^2/c^2)}$ for (a) $v = 10^{-2}c$; (b) $v = 0.9998c$.

In the following we make use of the binomial expansion,

$$(1 + x)^n = 1 + nx + \frac{n(n-1)}{2}x^2 + \cdots$$

(a) Setting $x = -10^{-4}$ and $n = 1/2$ in the binomial expansion, and, because x is so small, keeping only the first two terms of the expansion, we obtain

$$(1 - 10^{-4})^{1/2} \approx 1 + \tfrac{1}{2}(-10^{-4}) = 1 - 0.00005 = 0.99995$$

(b)
$$\sqrt{1 - (v^2/c^2)} = \sqrt{1 - (0.9998)^2} = \sqrt{1 - (1 - 0.0002)^2}$$

To evaluate $(1 - 0.0002)^2$ we employ the binomial expansion to obtain

$$(1 - 0.0002)^2 \approx 1 - 2(0.0002) = 1 - 0.0004$$

Using this in the above expression we obtain

$$\sqrt{1 - (v^2/c^2)} \approx \sqrt{1 - (1 - 0.0004)} = \sqrt{0.0004} = 0.02$$

3.2. As measured by O a flashbulb goes off at $x = 100$ km, $y = 10$ km, $z = 1$ km at $t = 5 \times 10^{-4}$ s. What are the coordinates x', y', z', and t' of this event as determined by a second observer, O', moving relative to O at $-0.8c$ along the common x-x' axis?

From the Lorentz transformations,

$$x' = \frac{x - vt}{\sqrt{1 - (v^2/c^2)}} = \frac{100 \text{ km} - (-0.8 \times 3 \times 10^5 \text{ km/s})(5 \times 10^{-4} \text{ s})}{\sqrt{1 - (0.8)^2}} = 367 \text{ km}$$

$$t' = \frac{t - \dfrac{v}{c^2}x}{\sqrt{1 - (v^2/c^2)}} = \frac{5 \times 10^{-4} \text{ s} - \dfrac{(-0.8)(100 \text{ km})}{3 \times 10^5 \text{ km/s}}}{\sqrt{1 - (0.8)^2}} = 12.8 \times 10^{-4} \text{ s}$$

$$y' = y = 10 \text{ km}$$
$$z' = z = 1 \text{ km}$$

3.3. Suppose that a particle moves relative to O' with a constant velocity of $c/2$ in the $x'y'$-plane such that its trajectory makes an angle of $60°$ with the x'-axis. If the velocity of O' with respect to O is $0.6c$ along the x-x' axis, find the equations of motion of the particle as determined by O.

The equations of motion as determined by O' are

$$x' = u'_x t' = \frac{c}{2}(\cos 60°)t' \qquad y' = u'_y t' = \frac{c}{2}(\sin 60°)t'$$

Substituting from (3.1) in the first expression, we obtain

$$\frac{x - vt}{\sqrt{1 - (v^2/c^2)}} = \frac{c}{2}(\cos 60°)\frac{t - \dfrac{v}{c^2}x}{\sqrt{1 - (v^2/c^2)}}$$

$$x - (0.6c)t = \frac{c}{2}(\cos 60°)\left(t - \frac{0.6}{c}x\right)$$

$$x = (0.74c)t$$

Substituting in the second expression then gives

$$y' = y = \frac{c}{2}(\sin 60°) \frac{t - \frac{v}{c^2}x}{\sqrt{1 - (v^2/c^2)}} = \frac{c}{2}(\sin 60°) \frac{t - (0.6)(0.74\,t)}{\sqrt{1 - (0.6)^2}} = (0.30\,c)t$$

3.4. A train 1/2 mile long (as measured by an observer on the train) is traveling at a speed of 100 mi/hr. Two lightning bolts strike the ends of the train simultaneously as determined by an observer on the ground. What is the time separation as measured by an observer on the train?

We have

$$(100\text{ mi/hr})\left(\frac{1\text{ hr}}{3600\text{ s}}\right) = 2.78 \times 10^{-2}\text{ mi/s}$$

Let events A and B be defined by the striking of each lightning bolt. With O as the ground observer, we have from (3.3)

$$t_B - t_A = \frac{(t'_B - t'_A) + \frac{v}{c^2}(x'_B - x'_A)}{\sqrt{1 - (v^2/c^2)}}$$

$$0 = \frac{(t'_B - t'_A) + \frac{2.78 \times 10^{-2}\text{ mi/s}}{(1.86 \times 10^5\text{ mi/s})^2}(0.5\text{ mi})}{\sqrt{1 - (v^2/c^2)}}$$

Solving, $t'_B - t'_A = -4.02 \times 10^{-3}$ s. The minus sign denotes that event A occurred after event B.

3.5. Observer O notes that two events are separated in space and time by 600 m and 8×10^{-7} s. How fast must an observer O' be moving relative to O in order that the events be simultaneous to O'?

Subtracting two Lorentz transformations, we obtain

$$t'_2 - t'_1 = \frac{(t_2 - t_1) - \frac{v}{c^2}(x_2 - x_1)}{\sqrt{1 - (v^2/c^2)}}$$

$$0 = \frac{8 \times 10^{-7}\text{ s} - \frac{v}{c}\left(\frac{600\text{ m}}{3 \times 10^8\text{ m/s}}\right)}{\sqrt{1 - (v^2/c^2)}}$$

Solving, $v/c = 0.4$.

3.6. The space-time coordinates of two events as measured by O are $x_1 = 6 \times 10^4$ m, $y_1 = z_1 = 0$ m, $t_1 = 2 \times 10^{-4}$ s and $x_2 = 12 \times 10^4$ m, $y_2 = z_2 = 0$ m, $t_2 = 1 \times 10^{-4}$ s. What must be the velocity of O' with respect to O if O' measures the two events to occur simultaneously?

Subtracting two Lorentz transformations:

$$t'_2 - t'_1 = \frac{(t_2 - t_1) - \frac{v}{c^2}(x_2 - x_1)}{\sqrt{1 - (v^2/c^2)}}$$

$$0 = \frac{(1 \times 10^{-4}\text{ s} - 2 \times 10^{-4}\text{ s}) - \frac{v}{c}\left(\frac{12 \times 10^4\text{ m} - 6 \times 10^4\text{ m}}{3 \times 10^8\text{ m/s}}\right)}{\sqrt{1 - (v^2/c^2)}}$$

Solving, $v/c = -1/2$. Therefore v is in the negative x-direction.

3.7.   Refer to Problem 3.6.   What is the spatial separation of the two events as measured by O'?

Subtracting two Lorentz transformations:

$$x_2' - x_1' = \frac{(x_2 - x_1) - v(t_2 - t_1)}{\sqrt{1 - (v^2/c^2)}}$$

From Problem 3.6, $v/c = -1/2$ or $v = -1.5 \times 10^8$ m/s.

$$x_2' - x_1' = \frac{(12 \times 10^4 \text{ m} - 6 \times 10^4 \text{ m}) - (-1.5 \times 10^8 \text{ m/s})(1 \times 10^{-4} \text{ s} - 2 \times 10^{-4} \text{ s})}{\sqrt{1 - (-0.5)^2}} = 5.20 \times 10^4 \text{ m}$$

Supplementary Problems

3.8.   Obtain (3.2) from (3.1).

3.9.   As determined by O' a lightning bolt strikes at $x' = 60$ m, $y' = z' = 0$, $t' = 8 \times 10^{-8}$ s. O' has a velocity of $0.6c$ along the x-axis of O.   What are the space-time coordinates of the strike as determined by O? *Ans.* $(x, y, z, t) = (93 \text{ m}, 0, 0, 2 \times 10^{-7} \text{ s})$

3.10.   Observer O' has a velocity of $0.8c$ relative to O, and clocks are adjusted such that $t = t' = 0$ when $x = x' = 0$.   If O determines that a flashbulb goes off at $x = 50$ m and $t = 2 \times 10^{-7}$ s, what is the time of this event as measured by O'?   *Ans.* 1.11×10^{-7} s

3.11.   Refer to Problem 3.10.   If a second flashbulb flashes at $x' = 10$ m and $t' = 2 \times 10^{-7}$ s as determined by O', what is the time interval between the two events as measured by O?   *Ans.* 1.78×10^{-7} s

3.12.   Refer to Problem 3.11.   What is the spatial separation of the two events as measured by (a) O', (b) O? *Ans.*   (a) 6.67 m; (b) 46.67 m

Chapter 4

Relativistic Length Measurements

4.1 THE DEFINITION OF LENGTH

If a body is at rest with respect to an observer, its length is determined by measuring the difference between the spatial coordinates of the endpoints of the body. Since the body is not moving, these measurements may be made at any time, and the length so determined is called the *rest length* or *proper length* of the body.

For a moving body, however, the procedure is more complicated, since the spatial coordinates of the endpoints of the body *must be measured at the same time*. The difference between these coordinates is then defined to be the length of the body.

Consider now a ruler, oriented along the x-x' direction, that is at rest with respect to observer O'. We wish to determine how the length measurements of O and O' are related to each other when O' is moving relative to O with a velocity v in the x-x' direction. Let the ends of the ruler be designated by A and B. From the inverse Lorentz transformations we obtain

$$x'_B - x'_A = \frac{(x_B - x_A) + v(t_B - t_A)}{\sqrt{1 - (v^2/c^2)}}$$

The difference $x'_B - x'_A = L_0$ is the (proper) length of the ruler as measured by O'. If x_B and x_A are measured by O at the same time, so that $t_B - t_A = 0$, then the difference $x_B - x_A = L$ will be the length of the ruler as measured by O. Thus we have

$$L = L_0\sqrt{1 - (v^2/c^2)}$$

Since $\sqrt{1 - (v^2/c^2)} < 1$ we have $L < L_0$, so that the length of the moving ruler will be measured by O to be contracted. This result is called the *Lorentz-Fitzgerald contraction*.

A Warning!

It is essential to keep clear the distinction between the concepts of "spatial coordinate separation" and "length." A common mistake in solving problems is simply to multiply or divide a given spatial interval by the term $\sqrt{1 - (v^2/c^2)}$. This approach will work if one is concerned with finding the relations between lengths, where the concept of "length" is defined precisely above. However, if one is interested in the spatial interval between two events that do not occur simultaneously, then the answer is obtained from the subtraction technique of Section 3.3; the correct answer will *not* be obtained by multiplying or dividing the original spatial separation by $\sqrt{1 - (v^2/c^2)}$.

Solved Problems

4.1. How fast does a rocket ship have to go for its length to be contracted to 99% of its rest length?

From the expression for length contraction,

$$\frac{L}{L_0} = 0.99 = \sqrt{1 - (v^2/c^2)} \qquad \text{or} \qquad v = 0.141c$$

4.2. Calculate the Lorentz contraction of the earth's diameter (in the plane of the ecliptic) as measured by an observer O' who is stationary with respect to the sun.

Taking the orbital velocity of the earth to be 3×10^4 m/s and the diameter of the earth as 7920 mi, the expression for the Lorentz contraction yields

$$D = D_0\sqrt{1 - (v^2/c^2)} = (7.92 \times 10^3 \text{ mi})\sqrt{1 - \left(\frac{3 \times 10^4 \text{ m/s}}{3 \times 10^8 \text{ m/s}}\right)^2} \approx (7.92 \times 10^3 \text{ mi})(1 - 0.5 \times 10^{-8})$$

Solving, $D_0 - D = 3.96 \times 10^{-5}$ mi $= 2.51$ in. It is seen that relativistic effects are very small at speeds that are normally encountered.

4.3. A meterstick makes an angle of 30° with respect to the x'-axis of O'. What must be the value of v if the meterstick makes an angle of 45° with respect to the x-axis of O?

We have:

$$L_y' = L' \sin \theta' = (1 \text{ m}) \sin 30° = 0.5 \text{ m} \qquad L_x' = L' \cos \theta' = (1 \text{ m}) \cos 30° = 0.866 \text{ m}$$

Since there will be a length contraction only in the x-x' direction,

$$L_y = L_y' = 0.5 \text{ m} \qquad L_x = L_x'\sqrt{1 - (v^2/c^2)} = (0.866 \text{ m})\sqrt{1 - (v^2/c^2)}$$

Since $\tan \theta = L_y/L_x$,

$$\tan 45° = 1 = \frac{0.5 \text{ m}}{(0.866 \text{ m})\sqrt{1 - (v^2/c^2)}}$$

Solving, $v = 0.816c$.

4.4. Refer to Problem 4.3. What is the length of the meterstick as measured by O?

Use the Pythagorean theorem or, more simply,

$$L = \frac{L_y}{\sin 45°} = \frac{0.5 \text{ m}}{\sin 45°} = 0.707 \text{ m}$$

4.5. A cube has a (proper) volume of 1000 cm^3. Find the volume as determined by an observer O' who moves at a velocity of $0.8c$ relative to the cube in a direction parallel to one edge.

The observer measures an edge of the cube parallel to the direction of motion to have the contracted length

$$l_x' = l_x\sqrt{1 - (v^2/c^2)} = (10 \text{ cm})\sqrt{1 - (0.8)^2} = 6 \text{ cm}$$

The lengths of the other edges are unchanged:

$$l_y' = l_y = l_z' = l_z = 10 \text{ cm}$$

Therefore,

$$V' = l_x'l_y'l_z' = (6 \text{ cm})(10 \text{ cm})(10 \text{ cm}) = 600 \text{ cm}^3$$

Supplementary Problems

4.6. An airplane is moving with respect to the earth at a speed of 600 m/s. Its proper length is 50 m. By how much will it appear to be shortened to an observer on earth?   *Ans.* 10^{-10} m

4.7. Compute the contraction in length of a train 1/2 mile long when it is traveling at 100 mi/hr.
Ans. 5.58×10^{-15} mi $= 3.52 \times 10^{-10}$ in.

4.8. At what speed must an observer move past the earth so that the earth appears like an ellipse whose major axis is six times its minor axis?   *Ans.* $0.986c$

4.9. An observer O' holds a 1.00 m stick at an angle of 30° with respect to the positive x'-axis. O' is moving in the positive x-x' direction with a velocity $0.8c$ with respect to observer O. What are the length and angle of the stick as measured by O?   *Ans.* 0.721 m; 43.9°

4.10. A square of area 100 cm^2 is at rest in the reference frame of O. Observer O' moves relative to O at $0.8c$ and parallel to one side of the square. What does O' measure for the area?   *Ans.* 60 cm^2

4.11. For the square of Problem 4.10, find the area measured by O' if O' is moving at a velocity $0.8c$ relative to O and along a diagonal of the square.   *Ans.* 60 cm^2

4.12. Repeat Problem 4.5 if O' moves with the same speed parallel to a diagonal of a face of the cube.
Ans. 600 cm^3

Chapter 5

Relativistic Time Measurements

5.1 PROPER TIME

If an observer, say O, determines that two events A and B occur at the same location, the time interval between these two events can be determined by O with a single clock. This time interval, $t_B - t_A = \Delta t_0$, as measured by O with his single clock, is called the *proper time interval* between the events.

5.2 TIME DILATION

Now consider the same two events A and B as viewed by a second observer, O', moving with a velocity v with respect to O. The second observer will necessarily determine that the two events occur at different locations and will therefore have to use two different, properly synchronized clocks to determine the time separation $t'_B - t'_A = \Delta t'$ between A and B. To find the relationship between the time separations as measured by O and O' we subtract two of the Lorentz time transformations, obtaining

$$\Delta t' = \frac{\Delta t_0 - \frac{v}{c^2}(x_B - x_A)}{\sqrt{1 - (v^2/c^2)}}$$

Because O determines that the two events occur at the same location, $x_B - x_A = 0$. Thus

$$\Delta t' = \frac{\Delta t_0}{\sqrt{1 - (v^2/c^2)}}$$

Since $\sqrt{1 - v^2/c^2} < 1$, $\Delta t' > \Delta t_0$, so that the time interval between the two events as measured by O' is *dilated* (enlarged).

In the above example the single clock was taken to be at rest with respect to O. The same result would obtain, however, if the single clock were taken to be at rest with respect to O'. Thus, in general, suppose a *single* clock advances through a time interval Δt_0. If this clock is moving with a velocity v with respect to an observer, he will determine that his two clocks advance through a time interval Δt given by

$$\Delta t = \frac{\Delta t_0}{\sqrt{1 - (v^2/c^2)}}$$

See Fig. 5-1.

Time dilation is a very real effect. Suppose in Fig. 5-1 cameras are placed at the location of clock 2 and at the location of the single clock, and a picture is taken by each camera when the single clock passes clock 2. When the pictures are developed, each picture will show the same thing—that the single clock has advanced through Δt_0 while clock 2 has advanced through Δt, where Δt and Δt_0 are related by the time dilation expression.

20

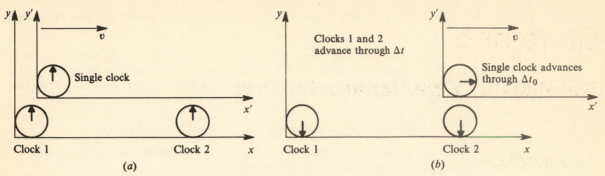

Fig. 5-1. Time Dilation as Viewed by Observer O

A Warning!

It is important to keep clear the distinction between the "time separation" of two events and the "proper time interval" between two events. If observers O and O' measure the time separation between two events that, for both observers, occur at different spatial locations, then these time separations are *not* related by simply multiplying or dividing by $\sqrt{1-(v^2/c^2)}$.

Solved Problems

5.1. The average lifetime of μ-mesons with a speed of $0.95c$ is measured to be 6×10^{-6} s. Compute the average lifetime of μ-mesons in a system in which they are at rest.

The time measured in a system in which the μ-mesons are at rest is the proper time.

$$\Delta t_0 = (\Delta t)\sqrt{1-(v^2/c^2)} = (6 \times 10^{-6} \text{ s})\sqrt{1-(0.95)^2} = 1.87 \times 10^{-6} \text{ s}$$

5.2. An airplane is moving with respect to the earth with a speed of 600 m/s. As determined by earth clocks, how long will it take for the airplane's clock to fall behind by two microseconds?

From the time dilation expression,

$$\Delta t_{\text{earth}} = \frac{\Delta t_{\text{plane}}}{\sqrt{1-\dfrac{v^2}{c^2}}} = \frac{\Delta t_{\text{plane}}}{\sqrt{1-\left(\dfrac{6 \times 10^2 \text{ m/s}}{3 \times 10^8 \text{ m/s}}\right)^2}} \approx \frac{\Delta t_{\text{plane}}}{1-2 \times 10^{-12}}$$

$$(2 \times 10^{-12})\Delta t_{\text{earth}} \approx \Delta t_{\text{earth}} - \Delta t_{\text{plane}} = 2 \times 10^{-6} \text{ s}$$

$$\Delta t_{\text{earth}} \approx 10^6 \text{ s} = 11.6 \text{ days}$$

This result indicates the smallness of relativistic effects at ordinary speeds.

5.3. Observers O and O' approach each other with a relative velocity of $0.6c$. If O measures the initial distance to O' to be 20 m, how much time will it take, as determined by O, before the two observers meet?

We have:

$$\Delta t = \frac{\text{distance}}{\text{velocity}} = \frac{20 \text{ m}}{0.6 \times 3 \times 10^8 \text{ m/sec}} = 11.1 \times 10^{-8} \text{ s}$$

5.4. In Problem 5.3, how much time will it take, as determined by O', before the two observers meet?

The two events under consideration are: (A) the position of O' when O makes his initial measurement, and (B) the coincidence of O and O'. Both of these events occur at the origin of O'. Therefore the time lapse measured by O' is equal to the proper time between the two events. From the time dilation expression,

$$\Delta t_0 = (\Delta t)\sqrt{1 - (v^2/c^2)} = (11.1 \times 10^{-8}\text{ s})\sqrt{1 - (0.6)^2} = 8.89 \times 10^{-8}\text{ s}$$

This problem can also be solved by noting that the initial distance as determined by O' is related to the distance measured by O through the Lorentz contraction:

$$L' = L_0\sqrt{1 - (v^2/c^2)} = (20\text{ m})\sqrt{1 - (0.6)^2} = 16\text{ m}$$

Then

$$\Delta t' = \frac{L'}{v} = \frac{16\text{ m}}{0.6 \times 3 \times 10^8\text{ m/s}} = 8.89 \times 10^{-8}\text{ s}$$

5.5. Pions have a half-life of 1.8×10^{-8} s. A pion beam leaves an accelerator at a speed of $0.8c$. Classically, what is the expected distance over which half the pions should decay?

We have:

$$\text{distance} = v\,\Delta t = (0.8 \times 10^8\text{ m/s})(1.8 \times 10^{-8}\text{ s}) = 4.32\text{ m}$$

5.6. Determine the answer to Problem 5.5 relativistically.

The half-life of 1.8×10^{-8} s is determined by an observer at rest with respect to the pion beam. From the point of view of an observer in the laboratory, the half-life has been increased because of the time dilation, and is given by

$$\Delta t = \frac{\Delta t_0}{\sqrt{1 - (v^2/c^2)}} = \frac{1.8 \times 10^{-8}\text{ s}}{\sqrt{1 - (0.8)^2}} = 3 \times 10^{-8}\text{ s}$$

Therefore, the distance traveled is

$$d = v\,\Delta t = (0.8 \times 3 \times 10^8\text{ m/s})(3 \times 10^{-8}\text{ s}) = 7.20\text{ m}$$

For an observer at rest with respect to the pion beam, the distance d_p the pions have to travel is shorter than the laboratory distance d_l by the Lorentz contraction:

$$d_p = d_l\sqrt{1 - (v^2/c^2)} = d_l\sqrt{1 - (0.8)^2} = 0.6d_l$$

The time elapsed when this distance is covered is

$$\Delta t_0 = \frac{d_p}{v} \qquad \text{or} \qquad 1.8 \times 10^{-8}\text{ s} = \frac{0.6d_l}{0.8 \times 3 \times 10^8\text{ m/s}}$$

Solving, $d_l = 7.20$ m, which agrees with the answer determined from time dilation.

Supplementary Problems

5.7. An atom will decay in 2×10^{-6} s. What will be the decay time as measured by an observer in a laboratory when the atom is moving with a speed of $0.8c$?   *Ans.* 3.33×10^{-6} s

5.8. How fast would a rocket ship have to go if an observer on the rocket ship ages at half the rate of an observer on the earth?   *Ans.* $0.866c$

5.9. A man with 60 years to live wants to visit a distant galaxy which is 160,000 light years away. What must be his constant speed?   *Ans.* $v/c = 1 - (0.703 \times 10^{-7})$

5.10. A particle moving at $0.8c$ in a laboratory decays after traveling 3 m. How long did it exist as measured by an observer in the laboratory?   *Ans.* 1.25×10^{-8} s

5.11. What does an observer moving with the particle of Problem 5.10 measure for the time the particle lived before decaying?   *Ans.* 0.75×10^{-8} s

Chapter 6

Relativistic Space-Time Measurements

In the previous chapters we discussed, more or less separately, relativistic space measurements and relativistic time measurements. There are, however, many types of problems where space and time measurements are intertwined and cannot be treated separately.

Solved Problems

6.1. A meterstick moves with a velocity of $0.6c$ relative to you along the direction of its length. How long will it take for the meterstick to pass you?

The length of the meterstick as measured by you is obtained from the Lorentz contraction:

$$L = L_0\sqrt{1 - (v^2/c^2)} = (1 \text{ m})\sqrt{1 - (0.6)^2} = 0.8 \text{ m}$$

The time for the meterstick to pass you is then found from

$$\text{distance} = \text{velocity} \times \text{time}$$
$$0.8 \text{ m} = (0.6 \times 3 \times 10^8 \text{ m/s}) \times \Delta t$$
$$\Delta t = 4.44 \times 10^{-9} \text{ s}$$

6.2. It takes 10^5 years for light to reach us from the most distant parts of our galaxy. Could a human travel there, at constant speed, in 50 years?

The distance traveled by light in 10^5 years is, according to an observer at rest with respect to the earth,

$$d_0 = c(\Delta t) = 10^5 c$$

where c is expressed in, say, mi/yr. If this observer now moves with constant speed v with respect to the earth, the distance d that he has to travel is shortened according to the Lorentz contraction:

$$d = d_0\sqrt{1 - (v^2/c^2)} = (10^5 c)\sqrt{1 - (v^2/c^2)}$$

The time interval available to travel this distance is 50 years, so that

$$v = \frac{d}{\Delta t} = \frac{10^5 c\sqrt{1 - (v^2/c^2)}}{50}$$

Solving,

$$\frac{v}{c} = \sqrt{1 - 2.5 \times 10^{-7}} \approx 0.999999875$$

Therefore a human traveling at this speed will find that when he completes the trip he has aged 50 years.

6.3. A μ-meson with an average lifetime of 2×10^{-6} s is created in the upper atmosphere at an elevation of 6000 m. When it is created it has a velocity of $0.998c$ in a direction toward the

24

earth. What is the average distance that it will travel before decaying, as determined by an observer on the earth? (Classically, this distance is

$$d = v\,\Delta t = (0.998 \times 3 \times 10^8 \text{ m/s})(2 \times 10^{-6} \text{ s}) = 599 \text{ m}$$

so that μ-mesons would not, on the average, reach the earth.)

As determined by an observer on the earth, the lifetime is increased because of time dilation:

$$\Delta t_{\text{earth}} = \frac{\Delta t}{\sqrt{1 - (v^2/c^2)}} = \frac{2 \times 10^{-6} \text{ s}}{\sqrt{1 - (0.998)^2}} = 31.6 \times 10^{-6} \text{ s}$$

The average distance traveled, as determined by an earth observer, is

$$d = v\,\Delta t_{\text{earth}} = (0.998 \times 3 \times 10^8 \text{ m/s})(31.6 \times 10^{-6} \text{ s}) = 9470 \text{ m}$$

Thus, an observer on the earth determines that, on the average, a μ-meson will reach the earth.

6.4. Consider an observer at rest with respect to the μ-meson of Problem 6.3. How far will he measure the earth to approach him before the μ-meson disintegrates? Compare this distance with the distance he measures from the point of creation of the μ-meson to the earth.

As determined by an observer at rest with respect to the μ-meson, the distance traveled by the earth is

$$d = v\,\Delta t_0 = (0.998 \times 3 \times 10^8 \text{ m/s})(2 \times 10^{-6} \text{ s}) = 599 \text{ m}$$

The initial distance, L, to the earth, however, is shortened because of the Lorentz contraction:

$$L = L_0\sqrt{1 - (v^2/c^2)} = (6 \times 10^3 \text{ m})\sqrt{1 - (0.998)^2} = 379 \text{ m}$$

Thus, an observer on the μ-meson determines that, on the average, it will reach the earth, in agreement with the result of Problem 6.3.

6.5. A pilot in a rocket ship traveling with a velocity of $0.6c$ passes the earth and adjusts his clock so that it coincides with 12:00 P.M. on earth. At 12:30 P.M., as determined by the pilot, the rocket ship passes a space station that is stationary with respect to the earth. What time is it at the station when the rocket passes?

From the time dilation expression,

$$\Delta t_{\text{station}} = \frac{\Delta t_{\text{rocket}}}{\sqrt{1 - (v^2/c^2)}} = \frac{30 \text{ min}}{\sqrt{1 - (0.6)^2}} = 37.5 \text{ min}$$

Therefore, the time at the space station is 12:37.5 P.M.

6.6. In Problem 6.5, what is the distance from the earth to the space station as determined (*a*) by the pilot? (*b*) by an observer on the earth?

(*a*)   distance = velocity × time = $(0.6 \times 3 \times 10^8 \text{ m/s})(30 \text{ min} \times 60 \text{ s/min})$

                                = 3.24×10^{11} m

(*b*)   distance = velocity × time = $(0.6 \times 3 \times 10^8 \text{ m/s})(37.5 \text{ min} \times 60 \text{ s/min})$

                                = 4.05×10^{11} m

6.7. Refer to Problems 6.5 and 6.6. When the rocket ship passes the space station the pilot reports to earth by radio. When does the earth receive the signal, (*a*) by earth time? (*b*) by rocket time?

(a) According to an earth observer,

$$\text{time} = \frac{\text{distance}}{\text{velocity}} = \frac{4.05 \times 10^{11} \text{ m}}{3 \times 10^8 \text{ m/s}} \times \frac{1 \text{ min}}{60 \text{ s}} = 22.5 \text{ min}$$

Thus the signal arrives, according to an observer on the earth, at

$$12:37.5 \text{ P.M.} + 22.5 \text{ min} = 1:00 \text{ P.M.}$$

(b) According to the pilot,

$$\text{time} = \frac{\text{distance}}{\text{velocity}} = \frac{3.24 \times 10^{11} \text{ m}}{3 \times 10^8 \text{ m/s}} \times \frac{1 \text{ min}}{60 \text{ s}} = 18 \text{ min}$$

Thus, according to the pilot, the signal arrives at the earth at

$$12:30 \text{ P.M.} + 18 \text{ min} = 12:48 \text{ P.M.}$$

6.8.   Suppose an observer O determines that two events are separated by 3.6×10^8 m and occur 2 s apart.   What is the proper time interval between the occurrence of these two events?

There exists a second observer, O', moving relative to the first observer who will determine that the two events occur at the same spatial location.   The proper time interval between the two events is the time interval measured by this observer.   Denoting the two events by A and B, we obtain upon subtracting two Lorentz transformations

$$x'_B - x'_A = \frac{(x_B - x_A) - v(t_B - t_A)}{\sqrt{1 - (v^2/c^2)}}$$

$$0 = \frac{3.6 \times 10^8 \text{ m} - v(2 \text{ s})}{\sqrt{1 - (v^2/c^2)}}$$

$$v = 1.8 \times 10^8 \text{ m/s} = 0.6c$$

Again subtracting two Lorentz transformations, we obtain the proper time interval as

$$t'_B - t'_A = \frac{(t_B - t_A) - \dfrac{v}{c^2}(x_B - x_A)}{\sqrt{1 - (v^2/c^2)}} = \frac{2 \text{ s} - \dfrac{0.6 \times 3.6 \times 10^8 \text{ m/s}}{3 \times 10^8 \text{ m/s}}}{\sqrt{1 - (0.6)^2}} = 1.6 \text{ s}$$

Another way to solve this problem is to use v and the time dilation expression:

$$\Delta t_0 = (\Delta t)\sqrt{1 - (v^2/c^2)} = (2 \text{ s})\sqrt{1 - (0.6)^2} = 1.6 \text{ s}$$

6.9.   For observer O, two events are simultaneous and occur 600 km apart.   What is the time difference between these two events as determined by O', who measures their spatial separation to be 1200 km?

Let A and B designate the two events.   Subtracting two Lorentz transformations, we obtain

$$x'_B - x'_A = \frac{(x_B - x_A) - v(t_B - t_A)}{\sqrt{1 - (v^2/c^2)}}$$

$$12 \times 10^5 \text{ m} = \frac{6 \times 10^5 \text{ m} - v(0)}{\sqrt{1 - (v^2/c^2)}}$$

$$\frac{v}{c} = 0.866$$

Again subtracting two Lorentz transformations:

$$t'_B - t'_A = \frac{(t_B - t_A) - \frac{v}{c^2}(x_B - x_A)}{\sqrt{1 - (v^2/c^2)}} = \frac{0 - \frac{0.866(6 \times 10^5 \text{ m})}{3 \times 10^8 \text{ m/s}}}{\sqrt{1 - (0.866)^2}} = -3.46 \times 10^{-3} \text{ s}$$

The minus sign denotes that event A occurred after event B as determined by O'.

6.10. Observer O', moving with a speed of $0.8c$ relative to a space platform, travels to α-Centauri, which, at a distance of 4 light years, is the nearest star to the platform. When he reaches the star he immediately turns around and returns to the platform at the same speed. When O' reaches the space platform, compare his age with that of his twin brother O, who has stayed on the platform.

According to O the time elapsed during the trip from the space platform to α-Centauri is

$$\Delta t = \frac{\text{distance}}{\text{velocity}} = \frac{4 \text{ yr} \times (\text{distance traveled by light/yr})}{0.8 \times (\text{distance traveled by light/yr})} = 5 \text{ yr}$$

Since the return trip takes place with the same speed, the total time elapsed, as measured by the platform observer O, is

$$\Delta t_{\text{round trip}} = 10 \text{ yr}$$

O' measures the proper time interval between the departure from the platform and the arrival at the star. Hence, from the time dilation expression,

$$\Delta t_0 = (\Delta t)\sqrt{1 - (v^2/c^2)} = (5 \text{ yr})\sqrt{1 - (0.8)^2} = 3 \text{ yr}$$

and the total time elapsed, as measured by O', is

$$\Delta t_{\text{travel}} = 6 \text{ yr}$$

Therefore, O' is 4 years younger than O when they meet. This result illustrates the famous "twin effect" in Special Relativity. Note that the motion of the twins is definitely *not* symmetrical. In order to get back home the traveling twin must turn around. This turning around is real (O' experiences measurable accelerations), in contrast to the apparent turning around that O' observes of O (who experiences no acceleration during his entire history). Thus the motion of O' is equivalent to that of two different inertial observers, one moving with $v = +0.8c$ and the other moving with $v = -0.8c$. Twin O, on the other hand, is equivalent to one single inertial observer.

6.11. Refer to Problem 6.10. Suppose that every year (as determined by O), O sends a light signal to O'. How many signals are received by O' on each leg of his journey? (In other words, what would twin O' actually *see* if he looked at his brother O through a telescope?)

As determined by O, brother O' reaches α-Centauri at $t = 5$ yr. In order for a light signal to reach α-Centauri simultaneously with O', it must have been sent by O at an earlier time, determined by

$$\text{time} = \frac{\text{distance}}{\text{velocity}} = \frac{4 \text{ yr} \times (\text{distance traveled by light/yr})}{\text{distance traveled by light/yr}} = 4 \text{ yr}$$

Therefore, a signal sent by O at $t = 1$ yr reaches α-Centauri simultaneously with O'. Since O sends a total of 10 signals, the remaining signals all reach O' on the return journey.

6.12. Refer to Problems 6.10 and 6.11. Suppose that every year (as determined by O'), O' sends a light signal to O. Consider the signal sent by O' just as he reaches α-Centauri. What is the time, as determined by O, when this signal is received. (That is, what would twin O *see* if he looked at his brother O' through a telescope?)

As determined by O, brother O' reaches α-Centauri at $t = 5$ yr. A light signal sent by O' from α-Centauri will reach O in a time interval (as determined by O) of

$$\Delta t = \frac{\text{distance}}{\text{velocity}} = \frac{4 \text{ yr} \times (\text{distance traveled by light/yr})}{\text{distance traveled by light/yr}} = 4 \text{ yr}$$

Therefore, this signal reaches O at $t = 5$ yr $+ 4$ yr $= 9$ yr. Hence, of the six signals sent by O', three of them are received by O during the first nine years (one every three years) and the remaining three are received by O during the last year.

6.13. A man in the back of a rocket shoots a high-speed bullet towards a target in the front of the rocket. The rocket is 60 m long and the bullet's speed is $0.8c$, both as measured by the man. Find the time that the bullet is in flight as measured by the man.

We have:

$$\Delta t = \frac{\text{distance}}{\text{velocity}} = \frac{60 \text{ m}}{0.8 \times 3 \times 10^8 \text{ m/s}} = 2.50 \times 10^{-7} \text{ s}$$

6.14. Refer to Problem 6.13. If the rocket moves with a speed of $0.6c$ relative to the earth, find the time that the bullet is in flight as measured by an observer on the earth.

Subtracting two inverse Lorentz transformations:

$$t_B - t_A = \frac{(t'_B - t'_A) + \dfrac{v}{c^2}(x'_B - x'_A)}{\sqrt{1 - (v^2/c^2)}} = \frac{2.5 \times 10^{-7} \text{ s} + \dfrac{(0.6)(60 \text{ m})}{3 \times 10^8 \text{ m/s}}}{\sqrt{1 - (0.6)^2}} = 4.63 \times 10^{-7} \text{ s}$$

6.15. The rest lengths of spaceships A and B are 90 m and 200 m, respectively. As they travel in opposite directions a pilot in spaceship A determines that the nose of spaceship B requires 5×10^{-7} s to traverse the length of A. What is the relative velocity of the two spaceships?

As determined by pilot A,

$$v = \frac{d}{\Delta t} = \frac{90 \text{ m}}{5 \times 10^{-7} \text{ s}} = 1.8 \times 10^8 \text{ m/s} = 0.6c$$

6.16. In Problem 6.15, what is the time interval, as determined by a pilot in the nose of B, between passing the front and rear ends of A?

The relative velocities are the same as determined by each observer. The pilot B measures the length of spaceship A to be contracted according to

$$L = L_0\sqrt{1 - (v^2/c^2)} = (90 \text{ m})\sqrt{1 - (0.6)^2} = 72 \text{ m}$$

The time interval as measured by B is then

$$\Delta t_B = \frac{L}{v} = \frac{72 \text{ m}}{0.6 \times 3 \times 10^8 \text{ m/s}} = 4 \times 10^{-7} \text{ s}$$

6.17. A rocket ship 90 m long travels at a constant velocity of $0.8c$ relative to the ground. As the nose of the rocket ship passes a ground observer, the pilot in the nose of the ship shines a flashlight toward the tail of the ship. What time does the signal reach the tail of the ship as recorded by (a) the pilot, (b) the ground observer?

(a) Let events A and B be defined by the emission of the light signal and the light signal's striking the tail of the rocket, respectively. Since the signal travels at speed c in the negative direction,

$$t'_B - t'_A = \frac{x'_B - x'_A}{-c} = \frac{-90 \text{ m}}{-3 \times 10^8 \text{ m/s}} = 3 \times 10^{-7} \text{ s}$$

(b) Subtraction of two inverse Lorentz transformations gives

$$t_B - t_A = \frac{(t'_B - t'_A) + \frac{v}{c^2}(x'_B - x'_A)}{\sqrt{1 - (v^2/c^2)}} = \frac{3 \times 10^{-7} \text{ s} + (0.8)\dfrac{(-90 \text{ m})}{3 \times 10^8 \text{ m/s}}}{\sqrt{1 - (0.8)^2}} = 1 \times 10^{-7} \text{ s}$$

6.18. Refer to Problem 6.17. When does the tail of the rocket pass the ground observer, (a) according to the ground observer? (b) according to the pilot?

(a) As determined by the ground observer, the length, L, of the rocket is

$$L = L_0\sqrt{1 - (v^2/c^2)} = (90 \text{ m})\sqrt{1 - (0.8)^2} = 54 \text{ m}$$

Then

$$\Delta t = \frac{L}{v} = \frac{54 \text{ m}}{0.8 \times 3 \times 10^8 \text{ m/s}} = 2.25 \times 10^{-7} \text{ s}$$

(b)
$$\Delta t' = \frac{L'}{v} = \frac{90 \text{ m}}{0.8 \times 3 \times 10^8 \text{ m/s}} = 3.75 \times 10^{-7} \text{ s}$$

6.19. The speed of a rocket with respect to a space station is 2.4×10^8 m/s, and observers O' and O in the rocket and the space station, respectively, synchronize their clocks in the usual fashion (i.e. $t = t' = 0$ when $x = x' = 0$). Suppose that O looks at O''s clock through a telescope. What time does he see on O''s clock when his own clock reads 30 s?

Let events A and B be defined, respectively, by the emission of the light signal from O' and the reception of the same signal by O. Our problem is to find t'_A. Applying the inverse Lorentz transformations to event A, we obtain

$$t_A = \frac{t'_A + (v/c^2)x'_A}{\sqrt{1 - (v^2/c^2)}} = \frac{t'_A + (v/c^2)(0)}{\sqrt{1 - (0.8)^2}} = \frac{t'_A}{0.6}$$

$$x_A = \frac{x'_A + vt'_A}{\sqrt{1 - (v^2/c^2)}} = \frac{0 + (0.8 \times 3 \times 10^8 \text{ m/s})t'_A}{\sqrt{1 - (0.8)^2}} = (4.0 \times 10^8 \text{ m/s})t'_A$$

The light signal travels in the negative direction at speed c, so that

$$x_B - x_A = -c(t_B - t_A)$$

Substituting from above,

$$0 - (4.0 \times 10^8 \text{ m/s})t'_A = \left(-3 \times 10^8 \frac{\text{m}}{\text{s}}\right)\left(30 \text{ s} - \frac{t'_A}{0.6}\right)$$

Solving, $t'_A = 10.0$ s.

This result, and that of Problem 6.20, point out the distinction between *seeing* an event and measuring the coordinates of the same event.

6.20. Refer to Problem 6.19. If O' looks at O's clock through a telescope, what time does his own clock read when he sees O's clock reading 30 s?

Let events A and B be defined by the emission of the light signal from O and the reception of the same signal by O', respectively. Our problem is to find t'_B. Applying the Lorentz transformations to event A gives

$$x'_A = \frac{x_A - vt_A}{\sqrt{1 - (v^2/c^2)}} = \frac{0 - (3 \times 10^8 \text{ m/s})(30 \text{ s})}{\sqrt{1 - (0.8)^2}} = -150 \times 10^8 \text{ m}$$

$$t'_A = \frac{t_A - (v/c^2)x_A}{\sqrt{1 - (v^2/c^2)}} = \frac{30 \text{ s} - (v/c^2)(0)}{\sqrt{1 - (0.8)^2}} = 50 \text{ s}$$

As measured by O', the light signal travels in the positive direction with speed c, so that

$$x'_B - x'_A = c(t'_B - t'_A)$$

Substituting from above,

$$0 - (-150 \times 10^8 \text{ m}) = (3 \times 10^8 \text{ m/s})(t'_B - 50 \text{ s})$$

Solving, $t'_B = 100$ s.

6.21. The equation for a spherical pulse of light starting from the origin at $t = t' = 0$ is

$$x^2 + y^2 + z^2 - c^2t^2 = 0$$

Show from the Lorentz transformations that O' will also measure this same pulse to be spherical, in accord with Einstein's second postulate stating that the velocity of light is the same for all observers.

From the inverse Lorentz transformations

$$x^2 = \left[\frac{x' + vt'}{\sqrt{1 - (v^2/c^2)}} \right]^2 = \frac{1}{1 - (v^2/c^2)} (x'^2 + v^2t'^2 + 2vx't')$$

$$t^2 = \left[\frac{t' + (v/c^2)x'}{\sqrt{1 - (v^2/c^2)}} \right]^2 = \frac{1}{1 - (v^2/c^2)} \left(\frac{v^2}{c^4} x'^2 + t'^2 + \frac{2v}{c^2} x't' \right)$$

$$y^2 = y'^2 \qquad z^2 = z'^2$$

Substituting, one finds that

$$x^2 + y^2 + z^2 - c^2t^2 = x'^2 + y'^2 + z'^2 - c^2t'^2$$

Therefore, since $x^2 + y^2 + z^2 - c^2t^2 = 0$, we also have

$$x'^2 + y'^2 + z'^2 - c^2t'^2 = 0$$

so that the pulse as determined by O' is also spherical.

6.22. Show that the differential expression

$$dx^2 + dy^2 + dz^2 - c^2dt^2$$

is invariant under a Lorentz transformation.

If the expression is invariant, it will retain the same form when expressed in terms of the primed coordinates. From the inverse Lorentz transformations one finds

$$dx^2 = \left[\frac{dx' + v\, dt'}{\sqrt{1 - (v^2/c^2)}} \right]^2 = \frac{1}{1 - (v^2/c^2)} (dx'^2 + v^2\, dt'^2 + 2v\, dx'\, dt')$$

$$dt^2 = \left[\frac{dt' + (v/c^2)dx'}{\sqrt{1 - (v^2/c^2)}} \right]^2 = \frac{1}{1 - (v^2/c^2)} \left(\frac{v^2}{c^4} dx'^2 + dt'^2 + \frac{2v}{c^2} dx'\, dt' \right)$$

$$dy^2 = dy'^2 \qquad dz^2 = dz'^2$$

Substituting these expressions one finds

$$dx^2 + dy^2 + dz^2 - c^2 dt^2 = dx'^2 + dy'^2 + dz'^2 - c^2 dt'^2$$

6.23. Show that the electromagnetic wave equation,

$$\frac{\partial^2 \phi}{\partial x^2} + \frac{\partial^2 \phi}{\partial y^2} + \frac{\partial^2 \phi}{\partial z^2} - \frac{1}{c^2} \frac{\partial^2 \phi}{\partial t^2} = 0$$

is invariant under a Lorentz transformation.

The equation will be invariant if it retains the same form when expressed in terms of the new variables x', y', z', t'. To express the wave equation in terms of the primed variables we first find from the Lorentz transformations that

$$\frac{\partial x'}{\partial x} = \frac{1}{\sqrt{1 - (v^2/c^2)}} \qquad \frac{\partial x'}{\partial t} = -\frac{v}{\sqrt{1 - (v^2/c^2)}}$$

$$\frac{\partial t'}{\partial x} = -\frac{v/c^2}{\sqrt{1 - (v^2/c^2)}} \qquad \frac{\partial t'}{\partial t} = \frac{1}{\sqrt{1 - (v^2/c^2)}}$$

$$\frac{\partial y'}{\partial y} = \frac{\partial z'}{\partial z} = 1 \qquad \frac{\partial x'}{\partial y} = \frac{\partial x'}{\partial z} = \frac{\partial y'}{\partial x} = \cdots = 0$$

From the chain rule, and using the above results, we have

$$\frac{\partial \phi}{\partial x} = \frac{\partial \phi}{\partial x'} \frac{\partial x'}{\partial x} + \frac{\partial \phi}{\partial y'} \frac{\partial y'}{\partial x} + \frac{\partial \phi}{\partial z'} \frac{\partial z'}{\partial x} + \frac{\partial \phi}{\partial t'} \frac{\partial t'}{\partial x} = \frac{1}{\sqrt{1 - (v^2/c^2)}} \frac{\partial \phi}{\partial x'} + \frac{-v/c^2}{\sqrt{1 - (v^2/c^2)}} \frac{\partial \phi}{\partial t'}$$

Differentiating again with respect to x, we have

$$\frac{\partial^2 \phi}{\partial x^2} = \frac{1}{1 - (v^2/c^2)} \left(\frac{\partial^2 \phi}{\partial x'^2} + \frac{v^2}{c^4} \frac{\partial^2 \phi}{\partial t'^2} - \frac{2v}{c^2} \frac{\partial^2 \phi}{\partial x' \partial t'} \right)$$

Similarly we have

$$\frac{\partial \phi}{\partial t} = \frac{-v}{\sqrt{1 - (v^2/c^2)}} \frac{\partial \phi}{\partial x'} + \frac{1}{\sqrt{1 - (v^2/c^2)}} \frac{\partial \phi}{\partial t'}$$

$$\frac{\partial^2 \phi}{\partial t^2} = \frac{1}{1 - (v^2/c^2)} \left(v^2 \frac{\partial^2 \phi}{\partial x'^2} + \frac{\partial^2 \phi}{\partial t'^2} - 2v \frac{\partial^2 \phi}{\partial x' \partial t'} \right)$$

$$\frac{\partial^2 \phi}{\partial y^2} = \frac{\partial^2 \phi}{\partial y'^2} \qquad \frac{\partial^2 \phi}{\partial z^2} = \frac{\partial^2 \phi}{\partial z'^2}$$

Substituting these in the wave equation, we obtain

$$\frac{\partial^2 \phi}{\partial x^2} + \frac{\partial^2 \phi}{\partial y^2} + \frac{\partial^2 \phi}{\partial z^2} - \frac{1}{c^2} \frac{\partial^2 \phi}{\partial t^2} = \frac{\partial^2 \phi}{\partial x'^2} + \frac{\partial^2 \phi}{\partial y'^2} + \frac{\partial^2 \phi}{\partial z'^2} - \frac{1}{c^2} \frac{\partial^2 \phi}{\partial t'^2}$$

so that the equation is invariant under Lorentz transformations. Recall that the wave equation is not invariant under Galilean transformations (Problem 1.10).

Supplementary Problems

6.24. An unstable particle with a mean lifetime of 4 μs is formed by a high-energy accelerator and projected through a laboratory with a speed of 0.6c. (*a*) What is the mean lifetime of the particle as determined by an observer in the laboratory? (*b*) What will be the average distance that the particle will travel in the laboratory before disintegrating? (*c*) How far would an observer at rest with respect to the particle determine that he has traveled before the particle disintegrates? *Ans.* (*a*) 5 μs; (*b*) 900 m; (*c*) 720 m

6.25. A μ-meson with a lifetime of 8×10^{-6} s is formed 10,000 m high in the upper atmosphere and is moving directly toward the earth. If the μ-meson decays just as it reaches the earth's surface, what is its speed relative to the earth? *Ans.* 0.972c

6.26. A meterstick moves along the x-axis with a velocity of 0.6c. The midpoint of the meterstick passes O at $t = 0$. As determined by O, where are the ends of the meterstick at $t = 0$?
Ans. 40 cm and −40 cm

6.27. Observer O measures the area of a circle at rest in his xy-plane to be 12 cm². An observer O' moving relative to O at 0.8c also observes the figure. What area does O' measure? *Ans.* 7.2 cm²

6.28. As determined by observer O a red light flashes, and, 10^{-6} s later, a blue light flashes 600 m farther out on the x-axis. What are the magnitude and direction of the velocity of a second observer, O', if he measures the red and blue flashes to occur simultaneously? *Ans.* +0.5c

6.29. Refer to Problem 6.28. What is the spatial separation of the red and blue flashes as determined by O'? *Ans.* 519.6 m

6.30. A rocket ship 150 m long travels at a speed of 0.6c. As the tail of the rocket passes by a man on a stationary space platform, he shines a flashlight in the direction of the nose. (*a*) How far from the platform is the nose when the light reaches it? (*b*) As measured by the observer on the space platform, how much time elapses between the emission and arrival of the light signal? (*c*) What is the time interval between emission and reception of the signal as determined by an observer in the nose of the rocket? *Ans.* (*a*) 300 m; (*b*) 10^{-6} s; (*c*) 0.5×10^{-6} s

6.31. Two events occur at the same place and are separated by a 4 s time interval as determined by one observer. If a second observer measures the time separation between these two events to be 5 s, what is his determination of their spatial separation? *Ans.* 9×10^8 m

6.32. An observer sets off two flashbulbs that are on his x-axis. He records that the first bulb is set off at his origin at 1 o'clock and the second bulb is set off 20 s later at $x = 9 \times 10^8$ m. A second observer is moving along the common x-x' axis with a speed of $-0.6c$ with respect to the first observer. What are the time and spatial separations between the two flashes as measured by the second observer?
Ans. 27.3 s; 56.3×10^8 m

6.33. The relative speed of O and O' is 0.8c. At $t' = 2 \times 10^{-7}$ s, a super bullet is fired from $x' = 100$ m. Traveling in the negative x'-direction with a constant speed, it strikes a target at the origin of O' at $t' = 6 \times 10^{-7}$ s. As determined by O, what is the velocity of the bullet and how far did it travel? *Ans.* -3×10^7 m/s; −6.67 m

6.34. A ground observer determines that it takes 5×10^{-7} s for a rocket to travel between two markers in the ground that are 90 m apart. What is the speed of the rocket as determined by the ground observer? *Ans.* 0.6c

6.35.   Refer to Problem 6.34.   As determined by an observer in the rocket, what is the distance between the two markers and the time interval between passing the two markers?   *Ans.*   72 m; 4×10^{-7} s

6.36.   A laser beam is rotated at 150 rev/min and throws a beam on a screen 50,000 miles away.   What is the sweep speed of the beam across the screen?   *Ans.* 7.85×10^5 mi/s (note: $c = 1.86 \times 10^5$ mi/s)

6.37.   Show that the expressions $x^2 + y^2 + z^2 - c^2t^2$ and $dx^2 + dy^2 + dz^2 - c^2dt^2$ are not invariant under Galilean transformations.

Chapter 7

Relativistic Velocity Transformations

To find the velocity transformations we consider an arrangement identical to that for the Lorentz coordinate transformations (Fig. 1-1). One observer, O', moves along the common x-x' axis at a constant velocity v with respect to a second observer, O. Each observer measures the velocity of a single particle, with O recording (u_x, u_y, u_z) and O' recording (u'_x, u'_y, u'_z) for the components of the particle's velocity. From the Lorentz coordinate transformations one finds the following *Lorentz velocity transformations* (see Problem 7.1):

$$u'_x = \frac{u_x - v}{1 - (v/c^2)u_x} \qquad u'_y = \frac{u_y\sqrt{1 - (v^2/c^2)}}{1 - (v/c^2)u_x} \qquad u'_z = \frac{u_z\sqrt{1 - (v^2/c^2)}}{1 - (v/c^2)u_x} \tag{7.1}$$

As before, the velocity v is positive if O' moves in the positive x-direction and negative if O' moves in the negative x-direction. When these equations are inverted, one obtains

$$u_x = \frac{u'_x + v}{1 + (v/c^2)u'_x} \qquad u_y = \frac{u'_y\sqrt{1 - (v^2/c^2)}}{1 + (v/c^2)u'_x} \qquad u_z = \frac{u'_z\sqrt{1 - (v^2/c^2)}}{1 + (v/c^2)u'_x} \tag{7.2}$$

Note that, as with the Lorentz coordinate transformations, the inverse velocity transformations can be obtained from the velocity transformations (7.1) by interchanging primed and unprimed variables and letting $v \to -v$. This is to be expected from symmetry, since from Postulate 1 of Section 2.4 both observers are completely equivalent, and observer O moves with a velocity of $-v$ with respect to O'.

7.1 THE LORENTZ VELOCITY TRANSFORMATIONS AND THE SPEED OF LIGHT

Consider now the experiment discussed in Section 3.1 where a light signal is sent in the x-x' direction from the common origin when O and O' pass each other at $t = t' = 0$. If O measures the signal's velocity components to be $u_x = c$, $u_y = u_z = 0$, then, by (7.1), O' will measure

$$u'_x = \frac{u_x - v}{1 - (v/c^2)u_x} = \frac{c - v}{1 - (v/c^2)c} = c \qquad u'_y = u'_z = 0$$

Thus O' also determines that the light signal travels with speed c, in accord with the second postulate of Einstein.

7.2 GENERAL CONSIDERATIONS IN SOLVING VELOCITY PROBLEMS

In velocity problems there are three objects involved: two observers, O and O', and a particle, P. The particle P has two velocities (and, hence, six numbers) associated with it: its velocity with respect to O, (u_x, u_y, u_z), and its velocity with respect to O', (u'_x, u'_y, u'_z). The quantity v appearing in the velocity transformations is the velocity of O' with respect to O.

When attacking a velocity problem, one should first determine which objects in the problem are to be identified with O, O', and P. Sometimes this identification is dictated; other times the identification can be made arbitrarily (see, for example, Problem 7.3). Once the identification has been made, one then uses the appropriate Lorentz velocity transformations to achieve the answer.

In dealing with velocity problems the best way to avoid mistakes is not to forget the phrase "with respect to." The phrase "velocity of an object" is meaningless (both classically and relativistically) because a velocity is always measured with respect to something.

Solved Problems

7.1.   Derive the Lorentz velocity transformation for the x-direction.

Taking the differentials of the Lorentz coordinate transformations (3.1), one finds

$$dx' = \frac{dx - v\,dt}{\sqrt{1 - (v^2/c^2)}} \qquad dt' = \frac{dt - \left(\dfrac{v}{c^2}\right)dx}{\sqrt{1 - (v^2/c^2)}}$$

Dividing dx' by dt' gives

$$u'_x = \frac{dx'}{dt'} = \frac{dx - v\,dt}{dt - \left(\dfrac{v}{c^2}\right)dx} = \frac{\dfrac{dx}{dt} - v}{1 - \dfrac{v}{c^2}\dfrac{dx}{dt}} = \frac{u_x - v}{1 - (v/c^2)u_x}$$

7.2.   At what speeds will the Galilean and Lorentz expressions for u'_x differ by 2%?

The Galilean transformation is $u'_{xG} = u_x - v$ and the Lorentz transformation is

$$u'_{xR} = \frac{u_x - v}{1 - (v/c^2)u_x} = \frac{u'_{xG}}{1 - (v/c^2)u_x}$$

Rearranging,

$$\frac{u'_{xR} - u'_{xG}}{u'_{xR}} = \frac{vu_x}{c^2}$$

Thus, if the product vu_x exceeds $0.02\,c^2$, the error in using the Galilean transformation instead of the Lorentz transformation will exceed 2%.

7.3.   Rocket A travels to the right and rocket B travels to the left, with velocities $0.8c$ and $0.6c$, respectively, relative to the earth. What is the velocity of rocket A measured from rocket B?

Let observers O, O' and the particle be associated with the earth, rocket B, and rocket A, respectively. Then

$$u'_x = \frac{u_x - v}{1 - (v/c^2)u_x} = \frac{0.8c - (-0.6c)}{1 - \dfrac{(-0.6c)(0.8c)}{c^2}} = 0.946c$$

The problem can also be solved with other associations. For example, let observers O, O' and the particle be associated with rocket A, rocket B, and the earth, respectively. Then

$$u'_x = \frac{u_x - v}{1 - (v/c^2)u_x} \qquad \text{or} \qquad 0.6c = \frac{-0.8c - v}{1 - (v/c^2)(-0.8c)}$$

Solving, $v = -0.946c$, which agrees with the above answer. (The minus sign appears because v is the velocity of O' with respect to O, which, with the present association, is the velocity of rocket B with respect to rocket A.)

7.4.   Repeat Problem 7.3 if rocket A travels with a velocity of $0.8c$ in the $+y$-direction relative to the earth.   (Rocket B still travels in the $-x$-direction.)

Let observers O, O' and the particle be associated with the earth, rocket B, and rocket A, respectively.   Then

$$u_x' = \frac{u_x - v}{1 - (v/c^2)u_x} = \frac{0 - (-0.6c)}{1 - 0} = 0.6c$$

$$u_y' = \frac{u_y\sqrt{1 - (v^2/c^2)}}{1 - (v/c^2)u_x} = \frac{(0.8c)\sqrt{1 - (0.6)^2}}{1 - 0} = 0.64c$$

which give the magnitude and direction of the desired velocity as

$$u' = \sqrt{u_x'^2 + u_y'^2} = \sqrt{(0.6c)^2 + (0.64c)^2} = 0.88c$$

and

$$\tan\phi' = \frac{u_y'}{u_x'} = \frac{0.64c}{0.60c} = 1.07 \quad \text{or} \quad \phi' = 46.8°$$

7.5.   A particle moves with a speed of $0.8c$ at an angle of $30°$ to the x-axis, as determined by O. What is the velocity of the particle as determined by a second observer, O', moving with a speed of $-0.6c$ along the common x-x' axis?

For observer O we have

$$u_x = (0.8c)\cos 30° = 0.693c \qquad u_y = (0.8c)\sin 30° = 0.400c$$

Using the Lorentz velocity transformations, we have for observer O'

$$u_x' = \frac{u_x - v}{1 - (v/c^2)u_x} = \frac{0.693c - (-0.6c)}{1 - \dfrac{(-0.6c)}{c^2}(0.693c)} = 0.913c$$

$$u_y' = \frac{u_y\sqrt{1 - (v^2/c^2)}}{1 - (v/c^2)u_x} = \frac{(0.4c)\sqrt{1 - (0.6)^2}}{1 - \dfrac{(-0.6c)}{c^2}(0.693c)} = 0.226c$$

The speed measured by observer O' is

$$u' = \sqrt{u_x'^2 + u_y'^2} = \sqrt{(0.913c)^2 + (0.226c)^2} = 0.941c$$

and the angle ϕ' the velocity makes with the x'-axis is

$$\tan\phi' = \frac{u_y'}{u_x'} = \frac{0.226c}{0.913c} = 0.248 \quad \text{or} \quad \phi' = 13.9°$$

7.6.   Consider a radioactive nucleus that moves with a constant speed of $0.5c$ relative to the laboratory.   The nucleus decays and emits an electron with a speed of $0.9c$ relative to the nucleus along the direction of motion.   Find the velocity of the electron in the laboratory frame.

Let the laboratory observer, the radioactive nucleus and the electron be respectively associated with O, O' and the particle.   Then

$$u_x = \frac{u_x' + v}{1 + (v/c^2)u_x'} = \frac{0.9c + 0.5c}{1 + \dfrac{(0.5c)(0.9c)}{c^2}} = 0.966c$$

7.7. Refer to Problem 7.6. Suppose that the nucleus decays by emitting an electron with a speed of $0.9c$ in a direction perpendicular to the direction of (the laboratory's) motion as determined by an observer at rest with respect to the nucleus. Find the velocity of the electron as measured by an observer in the laboratory frame.

With the same association as in Problem 7.6, one has

$$u_x = \frac{u'_x + v}{1 + (v/c^2)u'_x} = \frac{0 + 0.5c}{1 + 0} = 0.5c$$

$$u_y = \frac{u'_y\sqrt{1 - (v^2/c^2)}}{1 + (v/c^2)u'_x} = \frac{(0.9c)\sqrt{1 - (0.5)^2}}{1 + 0} = 0.779c$$

whence

$$u = \sqrt{u_x^2 + u_y^2} = \sqrt{(0.5c)^2 + (0.779c)^2} = 0.926c$$

and

$$\tan \phi = \frac{u_y}{u_x} = \frac{0.779c}{0.5c} = 1.56 \quad \text{or} \quad \phi = 57.3°$$

7.8. At $t = 0$ observer O emits a photon traveling in a direction of $60°$ with the x-axis. A second observer, O', travels with a speed of $0.6c$ along the common x-x' axis. What angle does the photon make with the x'-axis of O'?

We have:

$$u_x = c \cos 60° = 0.500c \qquad u_y = c \sin 60° = 0.866c$$

$$u'_x = \frac{u_x - v}{1 - (v/c^2)u_x} = \frac{0.5c - 0.6c}{1 - \dfrac{(0.6c)(0.5c)}{c^2}} = -0.143c$$

$$u'_y = \frac{u_y\sqrt{1 - (v^2/c^2)}}{1 - (v/c^2)u_x} = \frac{(0.866c)\sqrt{1 - (0.6)^2}}{1 - \dfrac{(0.6c)(0.5c)}{c^2}} = 0.990c$$

Thus

$$\tan \phi' = \frac{u'_y}{u'_x} = \frac{0.990c}{-0.143c} = -6.92$$

and $\phi' = 81.8°$ above the negative x'-axis. The magnitude of the velocity as measured by O' is

$$u' = \sqrt{u_x'^2 + u_y'^2} = \sqrt{(-0.143c)^2 + (0.990c)^2} = c$$

as is necessary.

7.9. The speed of light in still water is c/n, where the index of refraction for water is approximately $n = 4/3$. Fizeau, in 1851, found that the speed (relative to the laboratory) of light in water moving with a speed V (relative to the laboratory) could be expressed as

$$u = \frac{c}{n} + kV$$

where the "dragging coefficient" was measured by him to be $k \approx 0.44$. Determine the value of k predicted by the Lorentz velocity transformations.

An observer at rest relative to the water will measure the speed of light to be $u'_x = c/n$. Treating the light as a particle, the laboratory observer will find its speed to be

$$u_x = \frac{u'_x + V}{1 + \dfrac{V}{c^2}u'_x} = \frac{\dfrac{c}{n} + V}{1 + \dfrac{V}{c^2}\dfrac{c}{n}} = \left(\frac{c}{n} + V\right)\left(1 + \frac{V}{nc}\right)^{-1}$$

For small values of V the approximation

$$\left(1 + \frac{V}{nc}\right)^{-1} \approx 1 - \frac{V}{nc}$$

yields

$$u_x \approx \left(\frac{c}{n} + V\right)\left(1 - \frac{V}{nc}\right) \approx \frac{c}{n} + \left[1 - \frac{1}{n^2}\right]V$$

where terms of order V^2/c have been neglected.   Thus

$$k \approx 1 - \frac{1}{n^2} = 1 - \frac{1}{(4/3)^2} = 0.438$$

which agrees with Fizeau's experimental result.

Supplementary Problems

7.10.   A rocket moves with a velocity of $c/3$ with respect to a man holding a lantern.   The pilot of the rocket measures the speed of light reaching him from the lantern.   Determine this speed from the Lorentz velocity transformations.   *Ans.* c

7.11.   The pilot of a rocket moving at a velocity of $0.8c$ relative to the earth observes a second rocket approaching in the opposite direction at a velocity of $0.7c$.   What does an observer on earth measure for the second rocket's velocity?   *Ans.* $0.227c$

7.12.   An observer in rocket A finds that rockets C and B are moving away from him in opposite directions at speeds of $0.6c$ and $0.8c$, respectively.   What is the speed of C as measured by B?   *Ans.* $0.946c$ (classically, $1.4c$)

7.13.   An observer O' is moving along the x-x' axis with a speed of $c/2$ with respect to another observer, O.   Observer O measures a particle moving in the positive y-direction with a speed $c/\sqrt{3}$.   Calculate the velocity of the particle as measured by O'.   *Ans.* $c/\sqrt{2}$, $135°$

7.14.   A man standing on the platform of a space station observes two rocket ships approaching him from opposite directions at speeds of $0.9c$ and $0.8c$.   At what speed does one rocket ship move with respect to the other?   *Ans.* $0.988c$

7.15.   Derive the Lorentz velocity transformations for the y- and z-directions.

7.16.   Starting from the Lorentz velocity transformations, (7.1), obtain the inverse Lorentz velocity transformations, (7.2).

7.17.   A K^0-meson, at rest, decays into a π^+-meson and a π^--meson, each with a speed of $0.827c$.   When a K^0-meson traveling at a speed of $0.6c$ decays, what is the greatest speed that one of the π-mesons can have?   *Ans.* $0.954c$

Chapter 8

Mass, Energy and Momentum in Relativity

8.1 THE NEED TO REDEFINE CLASSICAL MOMENTUM

One of the major developments to come out of the Special Theory of Relativity is that the mass of a body will vary with its velocity. A heuristic argument for this variation can be given as follows.

Consider a ballistics experiment where an observer, say O', fires a bullet in the y'-direction into a block that cannot move relative to him. It is reasonable to suppose that the amount the bullet penetrates into the block is determined by the y'-component of the momentum of the bullet, given by $p'_y = m'u'_y$, where m' is the mass of the bullet as measured by O'.

Now consider the same experiment from the point of view of observer O who sees observer O' moving in the common x-x' direction with a velocity v. Since the tunnel left by the bullet is at right angles to the direction of relative motion, O will agree with O' as to the distance that the bullet penetrates into the block, and therefore would expect to find the same value as O' for the y-component of the bullet's momentum.

As determined by O, $p_y = mu_y$, where m is the mass of the bullet as measured by O. From the Lorentz velocity transformations we find, since $u'_x = 0$, that

$$u_y = \frac{u'_y\sqrt{1 - (v^2/c^2)}}{1 + (v/c^2)u'_x} = u'_y\sqrt{1 - (v^2/c^2)}$$

so that $p_y = mu'_y\sqrt{1 - (v^2/c^2)}$. Since from above $p'_y = m'u'_y$, it is seen that if both observers assign the same mass to the bullet, so that $m' = m$, they will find $p'_y \neq p_y$, contrary to what is expected.

8.2 THE VARIATION OF MASS WITH VELOCITY

At this point we have two choices. We can assume that momentum principles—in particular, conservation of momentum—do not apply at large velocities. Or, we can look for a way to redefine the momentum of a body in order to make momentum principles applicable to Special Relativity. The latter alternative was chosen by Einstein. He showed that all observers will find classical momentum principles to hold if the mass m of a body varies with its speed u according to

$$m = \frac{m_0}{\sqrt{1 - (u^2/c^2)}}$$

where m_0, the *rest mass*, is the mass of the body measured when it is at rest with respect to the observer. See Problem 8.1.

8.3 NEWTON'S SECOND LAW IN RELATIVITY

The classical expression of Newton's second law is that the net force on a body is equal to the rate of change of the body's momentum. To include relativistic effects, allowance must be made for the fact that the mass of a body varies with its velocity. Thus the relativistic generalization of Newton's second law is

$$\mathbf{F} = \frac{d\mathbf{p}}{dt} = \frac{d}{dt}\left[\frac{m_0\mathbf{u}}{\sqrt{1-(u^2/c^2)}}\right] = \frac{d}{dt}(m\mathbf{u})$$

8.4   MASS AND ENERGY RELATIONSHIP: $E = mc^2$

In relativistic mechanics, as in classical mechanics, the kinetic energy, K, of a body is equal to the work done by an external force in increasing the speed of the body from zero to some value u, i.e.

$$K = \int_{u=0}^{u=u} \mathbf{F} \cdot d\mathbf{s}$$

Using Newton's second law, $\mathbf{F} = d(m\mathbf{u})/dt$, one finds (Problem 8.21) that this expression reduces to

$$K = mc^2 - m_0c^2$$

The kinetic energy, K, represents the difference between the *total energy*, E, of the moving particle and the *rest energy*, E_0, of the particle when at rest, so that

$$E - E_0 = mc^2 - m_0c^2$$

If the rest energy is chosen so that $E_0 = m_0c^2$, we obtain Einstein's famous relation

$$E = mc^2$$

which shows the equivalence of mass and energy. Thus, even when a body is at rest it still has an energy content given by $E_0 = m_0c^2$, so that in principle a massive body can be completely converted into another, more familiar, form of energy.

8.5   MOMENTUM AND ENERGY RELATIONSHIP

Since momentum is conserved, but not velocity, it is often useful to express the energy of a body in terms of its momentum rather than its velocity. To this end, if the expression

$$m = \frac{m_0}{\sqrt{1-(u^2/c^2)}}$$

is squared and both sides are multiplied by $c^4[1 - (u^2/c^2)]$, one obtains

$$m^2c^4 - m^2u^2c^2 = m_0^2c^4$$

Using the results $E = mc^2$, $E_0 = m_0c^2$, and $|\mathbf{p}| = mu$, we find the desired relationship between E and p to be

$$E^2 = (pc)^2 + E_0^2 \quad \text{or} \quad (K + m_0c^2)^2 = (pc)^2 + (m_0c^2)^2$$

8.6   UNITS FOR ENERGY AND MOMENTUM

The *electron volt* (eV) is the kinetic energy of a body whose charge equals the charge of an electron, after it moves through a potential difference of one volt.

$$1 \text{ eV} = (1.602 \times 10^{-19} \text{ C})(1 \text{ V}) = 1.602 \times 10^{-19} \text{ J}$$

$$1 \text{ MeV} = 10^6 \text{ eV} \qquad 1 \text{ GeV} = 10^9 \text{ eV}$$

The relationship 1.602×10^{-19} J = 1 eV can be looked at as a conversion factor between two different units of energy.

The standard units for momentum are kg·m/s. In relativistic calculations, however, units of MeV/c are frequently used for momentum.   These units arise from the energy-momentum expression

$$p = \frac{\sqrt{E^2 - E_0^2}}{c}$$

The conversion factor is

$$1 \frac{\text{MeV}}{c} = 0.534 \times 10^{-21} \frac{\text{kg} \cdot \text{m}}{\text{s}}$$

8.7   GENERAL CONSIDERATIONS IN SOLVING MASS-ENERGY PROBLEMS

A common mistake in solving mass-energy problems is to use the wrong expression for the kinetic energy.   Thus

$$K \neq \tfrac{1}{2} m_0 u^2 \quad \text{and} \quad K \neq \tfrac{1}{2} m u^2$$

The correct expression for the kinetic energy is

$$K = (m - m_0) c^2$$

Likewise, concerning the momentum, we note that

$$p \neq m_0 u$$

Solved Problems

8.1.   Show how Einstein's mass-speed relation resolves the difficulty in the ballistics experiment of Section 8.1.

As determined by O', the mass of the bullet will be, since $u'_x = 0$,

$$m' = \frac{m_0}{\sqrt{1 - \dfrac{u'^2}{c^2}}} = \frac{m_0}{\sqrt{1 - \dfrac{u_x'^2 + u_y'^2}{c^2}}} = \frac{m_0}{\sqrt{1 - \dfrac{u_y'^2}{c^2}}}$$

while the mass of the bullet as measured by O is, since $u_x = v$,

$$m = \frac{m_0}{\sqrt{1 - \dfrac{u^2}{c^2}}} = \frac{m_0}{\sqrt{1 - \dfrac{u_x^2 + u_y^2}{c^2}}} = \frac{m_0}{\sqrt{1 - \dfrac{v^2 + u_y^2}{c^2}}}$$

If we now apply the Lorentz transformation to the quantity inside the last square root, we find

$$1 - \frac{v^2}{c^2} - \frac{u_y^2}{c^2} = 1 - \frac{v^2}{c^2} - \frac{1}{c^2}\left(u_y'\sqrt{1 - \frac{v^2}{c^2}} \right)^2 = \left(1 - \frac{v^2}{c^2} \right)\left(1 - \frac{u_y'^2}{c^2} \right)$$

so that

$$m = \frac{m_0}{\sqrt{1 - (v^2/c^2)}\,\sqrt{1 - (u_y'^2/c^2)}} = \frac{m'}{\sqrt{1 - (v^2/c^2)}}$$

Hence

$$p_y = m u_y'\sqrt{1 - (v^2/c^2)} = \left(\frac{m'}{\sqrt{1 - (v^2/c^2)}} \right) u_y'\sqrt{1 - (v^2/c^2)} = m' u_y' = p_y'$$

8.2. From the rest masses listed in the Appendix calculate the rest energy of an electron in joules and electron-volts.

We have: $E_0 = m_0 c^2 = (9.109 \times 10^{-31} \text{ kg})(2.998 \times 10^8 \text{ m/s})^2 = 8.187 \times 10^{-14}$ J, and

$$(8.187 \times 10^{-14} \text{ J})\left(\frac{1 \text{ eV}}{1.602 \times 10^{-19} \text{ J}} \right)\left(\frac{1 \text{ MeV}}{10^6 \text{ eV}} \right) = 0.511 \text{ MeV}$$

8.3. A body at rest spontaneously breaks up into two parts which move in opposite directions. The parts have rest masses of 3 kg and 5.33 kg and respective speeds of $0.8c$ and $0.6c$. Find the rest mass of the original body.

Since $E_{\text{initial}} = E_{\text{final}}$,

$$m_0 c^2 = \frac{m_{01} c^2}{\sqrt{1 - (v_1^2/c^2)}} + \frac{m_{02} c^2}{\sqrt{1 - (v_2^2/c^2)}} = \frac{(3 \text{ kg}) c^2}{\sqrt{1 - (0.8)^2}} + \frac{(5.33 \text{ kg}) c^2}{\sqrt{1 - (0.6)^2}}$$

$$m_0 = 11.66 \text{ kg}$$

Observe that rest mass is not conserved (see also Problem 8.26).

8.4. What is the speed of an electron that is accelerated through a potential difference of 10^5 V?

Since $K = e \, \Delta V = 10^5 \text{ eV} = 0.1 \text{ MeV}$, we have

$$0.1 \text{ MeV} = K = \frac{m_0 c^2}{\sqrt{1 - (v^2/c^2)}} - m_0 c^2$$

Substituting $m_0 c^2 = 0.511 \text{ MeV}$ (Problem 8.2) and solving, we find $v = 0.548c$.

8.5. Calculate the momentum of a 1 MeV electron.

$$E^2 = (pc)^2 + E_0^2$$

$$(1 \text{ MeV} + 0.511 \text{ MeV})^2 = (pc)^2 + (0.511 \text{ MeV})^2$$

$$p = 1.42 \text{ MeV}/c$$

8.6. Calculate the kinetic energy of an electron whose momentum is 2 MeV/c.

$$E^2 = (pc)^2 + E_0^2$$

$$(K + 0.511 \text{ MeV})^2 = \left(\frac{2 \text{ MeV}}{c} \times c \right)^2 + (0.511 \text{ MeV})^2$$

$$K = 1.55 \text{ MeV}$$

8.7. Calculate the velocity of an electron whose kinetic energy is 2 MeV.

$$K = \frac{m_0 c^2}{\sqrt{1 - (v^2/c^2)}} - m_0 c^2$$

$$2 \text{ MeV} = \frac{0.511 \text{ MeV}}{\sqrt{1 - (v^2/c^2)}} - 0.511 \text{ MeV}$$

$$v = 0.98c$$

8.8. Calculate the momentum of an electron whose velocity is $0.8c$.

$$p = mv = \frac{m_0 c^2}{\sqrt{1-(v^2/c^2)}}\left(\frac{v}{c^2}\right) = \frac{0.511 \text{ MeV}}{\sqrt{1-(0.8)^2}}\left(\frac{0.8}{c}\right) = 0.681 \frac{\text{MeV}}{c}$$

8.9. The rest mass of a μ-meson is $207 m_{0e}$ and its average lifetime when at rest is 2×10^{-6} s. What is the mass of a μ-meson if its average lifetime in the laboratory is 7×10^{-6} s?

From the time dilation expression,

$$\frac{1}{\sqrt{1-(v^2/c^2)}} = \frac{\Delta t}{\Delta t_0} = \frac{7}{2}$$

and so

$$m = \frac{m_0}{\sqrt{1-(v^2/c^2)}} = (207 m_{0e})\left(\frac{7}{2}\right) = 725 m_{0e}$$

8.10. Compute the effective mass of a 5000 Å photon.

$$m_{\text{eff}}c^2 = E_{\text{photon}} = h\nu = hc/\lambda$$

(See Chapter 10.)   Hence

$$m_{\text{eff}} = \frac{6.63 \times 10^{-34} \text{ J} \cdot \text{s}}{(5 \times 10^{-7} \text{ m})(3 \times 10^8 \text{ m/s})} = 4.42 \times 10^{-36} \text{ kg}$$

8.11. An electron is accelerated to an energy of 2 GeV by an electron synchrotron. What is the ratio of the electron's mass to its rest mass?

From $mc^2 = K + m_0 c^2$,

$$\frac{m}{m_0} = \frac{mc^2}{m_0 c^2} = \frac{K + m_0 c^2}{m_0 c^2} = \frac{2000 \text{ MeV} + 0.511 \text{ MeV}}{0.511 \text{ MeV}} = 3915$$

8.12. A ^{235}U nucleus, when it fissions, releases 200 MeV of energy. What percent of the total energy available is this?

The rest mass of a ^{235}U atom is, in terms of the unified atomic mass unit (u), 235 u. Using the conversion 1 u = 931.5 MeV, we have:

$$\text{total available energy} = \text{rest energy of } ^{235}\text{U} = (235 \text{ u})\frac{931.5 \text{ MeV}}{\text{u}} = 219 \times 10^3 \text{ MeV}$$

$$\% \text{ total energy} = \frac{200 \text{ MeV}}{219 \times 10^3 \text{ MeV}} \times 100\% = 0.0913\%$$

8.13. An electron is accelerated from rest to a velocity of $0.5c$. Calculate its change in energy.

$$\text{change in energy} = \frac{m_0 c^2}{\sqrt{1-(v^2/c^2)}} - m_0 c^2 = \frac{0.511 \text{ MeV}}{\sqrt{1-(0.5)^2}} - 0.511 \text{ MeV} = 0.079 \text{ MeV}$$

8.14. At what fraction of the speed of light must a particle move so that its kinetic energy is double its rest energy?

$$K = \frac{m_0 c^2}{\sqrt{1-(v^2/c^2)}} - m_0 c^2 = 2 m_0 c^2 \quad \text{or} \quad \frac{1}{\sqrt{1-(v^2/c^2)}} = 3$$

Solving, $v = 0.943c$.

8.15. An electron's velocity is 5×10^7 m/s. How much energy is needed to double the speed?

$$\text{initial energy} = \frac{m_0 c^2}{\sqrt{1 - \dfrac{v^2}{c^2}}} = \frac{0.511 \text{ MeV}}{\sqrt{1 - \left(\dfrac{0.5 \times 10^8 \text{ m/s}}{3 \times 10^8 \text{ m/s}}\right)^2}} = 0.518 \text{ MeV}$$

$$\text{final energy} = \frac{m_0 c^2}{\sqrt{1 - \dfrac{(2v)^2}{c^2}}} = \frac{0.511 \text{ MeV}}{\sqrt{1 - \left(\dfrac{1 \times 10^8 \text{ m/s}}{3 \times 10^8 \text{ m/s}}\right)^2}} = 0.542 \text{ Mev}$$

$$\text{change in energy} = 0.024 \text{ MeV}$$

8.16. A 1 MeV photon collides with a stationary electron in the vicinity of a heavy nucleus and is absorbed. (A free electron cannot capture a photon.) If the recoil energy of the nucleus can be neglected, what is the velocity of the electron after the collision?

From $E_{\text{initial}} = E_{\text{final}}$,

$$E_p + m_{0e} c^2 + m_{0n} c^2 = \frac{m_{0e} c^2}{\sqrt{1 - (v^2/c^2)}} + m_{0n} c^2 \qquad \text{or} \qquad 1 \text{ MeV} + 0.511 \text{ MeV} = \frac{0.511 \text{ MeV}}{\sqrt{1 - (v^2/c^2)}}$$

Solving, $v = 0.941 c$.

8.17. An electron moves in the laboratory with a speed of $0.6c$. An observer moves with a velocity of $0.8c$ along the direction of motion of the electron. What is the energy of the electron as determined by the observer?

From the Lorentz velocity transformations,

$$u'_x = \frac{u_x - v}{1 - (v/c^2)u_x} = \frac{0.6c - 0.8c}{1 - (0.8)(0.6)} = 0.385c$$

and

$$K' = \frac{m_0 c^2}{\sqrt{1 - (u'_x/c)^2}} - m_0 c^2 = \frac{0.511 \text{ MeV}}{\sqrt{1 - (-0.385)^2}} - 0.511 \text{ MeV} = 0.043 \text{ MeV}$$

8.18. A particle has a total energy of 6×10^3 MeV and a momentum of 3×10^3 MeV/c. What is its rest mass?

Using $E^2 = (pc)^2 + E_0^2$,

$$(6 \times 10^3 \text{ MeV})^2 = \left[(3 \times 10^3 \text{ MeV}/c)c\right]^2 + E_0^2$$

Solving, $E_0 = 5.2 \times 10^3$ MeV, and (see Problem 8.12)

$$m_0 = (5.2 \times 10^3 \text{ MeV})\left(\frac{1 \text{ u}}{931.5 \text{ MeV}}\right) = 5.58 \text{ u}$$

8.19. Refer to Problem 8.18. What is the energy of the particle in a frame where its momentum is 5×10^3 MeV/c?

$$E^2 = (pc)^2 + E_0^2 = \left[(5 \times 10^3 \text{ MeV}/c)c\right]^2 + (5.2 \times 10^3 \text{ MeV})^2$$

Solving, $E = 7.2 \times 10^3$ MeV.

8.20. The K^0-mesor decays at rest into two π^0-mesons. If the rest energy of the K^0 is 498 MeV and of the π^0 is 135 MeV, what is the kinetic energy of each π^0?

Since the initial and final momenta must be equal in the lab frame, the π^0's move off in opposite directions with equal amounts of kinetic energy.

$$E_{\text{initial}} = E_{\text{final}}$$

$$498 \text{ MeV} = 2(135 \text{ MeV}) + 2K$$

$$K = 114 \text{ MeV}$$

8.21. For one-dimensional motion show that

$$K = \int_{u=0}^{u=u} \mathbf{F} \cdot d\mathbf{s} = mc^2 - m_0 c^2$$

For one-dimensional motion,

$$K = \int_{u=0}^{u=u} F \, dx = \int_{u=0}^{u=u} \frac{d}{dt}(mu) \, dx = \int_{u=0}^{u=u} d(mu) \, \frac{dx}{dt}$$

$$= \int_{u=0}^{u=u} (m \, du + u \, dm)u = \int_{u=0}^{u=u} (mu \, du + u^2 \, dm) \qquad (1)$$

From the expression for the variation of mass with velocity we have

$$m = \frac{m_0}{\sqrt{1 - (u^2/c^2)}} \qquad \text{or} \qquad m^2 c^2 - m^2 u^2 = m_0^2 c^2$$

Taking differentials of both sides of this expression, we obtain

$$2mc^2 \, dm - m^2 2u \, du - u^2 2m \, dm = 0$$

which can be rewritten as

$$mu \, du + u^2 \, dm = c^2 \, dm \qquad (2)$$

The left-hand side of (2) is exactly the integrand of (1), so we obtain

$$K = \int_{m=m_0}^{m=m} c^2 \, dm = c^2(m - m_0)$$

8.22. Show from the binomial expansion that $E - E_0$ reduces to $\frac{1}{2} m_0 u^2$ when $u/c \ll 1$.

$$E - E_0 = \frac{m_0 c^2}{\sqrt{1 - (u^2/c^2)}} - m_0 c^2 = m_0 c^2 \left[\left(1 - \frac{u^2}{c^2}\right)^{-1/2} - 1 \right]$$

$$= m_0 c^2 \left[\left(1 + \frac{1}{2} \frac{u^2}{c^2} + \cdots \right) - 1 \right] \approx \frac{1}{2} m_0 u^2$$

8.23. What is the maximum speed that a particle can have such that its kinetic energy can be written as $\frac{1}{2} m_0 v^2$ with an error no greater than 0.5%?

At the maximum speed,

$$\frac{K - \frac{1}{2} m_0 u^2}{K} = 0.005 \qquad \text{or} \qquad K = \frac{\frac{1}{2} m_0 u^2}{0.995}$$

But, as in Problem 8.22,

$$K = m_0 c^2 \left[\left(1 - \frac{u^2}{c^2}\right)^{-1/2} - 1 \right] = m_0 c^2 \left[\left(1 + \frac{1}{2} \frac{u^2}{c^2} + \frac{3}{8} \frac{u^4}{c^4} + \cdots \right) - 1 \right]$$

$$= \frac{1}{2} m_0 u^2 + \frac{3}{8} m_0 u^2 \left(\frac{u^2}{c^2}\right) + \cdots$$

Hence

$$\frac{\frac{1}{2} m_0 u^2}{0.995} \approx \frac{1}{2} m_0 u^2 + \frac{3}{8} m_0 u^2 \left(\frac{u^2}{c^2} \right)$$

Solving, $v \approx 0.082 c$.

8.24. Suppose that a force F acts on a particle in the same direction as its velocity. Find the corresponding expression for Newton's second law.

The force F is the time derivative of the momentum.

$$F = \frac{d}{dt} \left[\frac{m_0 u}{\sqrt{1 - (u^2/c^2)}} \right] = \frac{m_0}{\sqrt{1 - (u^2/c^2)}} \frac{du}{dt} + \frac{m_0 u}{\left[1 - (u^2/c^2) \right]^{3/2}} \frac{u}{c^2} \frac{du}{dt}$$

$$= \frac{m_0 \frac{du}{dt}}{\left[1 - (u^2/c^2) \right]^{3/2}} \left(1 - \frac{u^2}{c^2} + \frac{u^2}{c^2} \right) = \frac{m_0 \frac{du}{dt}}{\left[1 - (u^2/c^2) \right]^{3/2}}$$

8.25. Using Newton's second law, find an expression for the relativistic velocity of a particle of charge q moving in a circle of radius R at right angles to a magnetic field B.

In vector form Newton's second law is

$$\mathbf{F} = \frac{d}{dt} (m\mathbf{u}) = \frac{d}{dt} \left[\frac{m_0 \mathbf{u}}{\sqrt{1 - (u^2/c^2)}} \right] = \frac{d}{dt} \left[\frac{m_0 \mathbf{u}}{\sqrt{1 - (\mathbf{u} \cdot \mathbf{u}/c^2)}} \right]$$

Performing the differentiation by means of the chain rule, we obtain

$$\mathbf{F} = \frac{m_0}{\sqrt{1 - (\mathbf{u} \cdot \mathbf{u}/c^2)}} \frac{d\mathbf{u}}{dt} + \frac{m_0}{\left[1 - (\mathbf{u} \cdot \mathbf{u}/c^2) \right]^{3/2}} \frac{\mathbf{u} \cdot \frac{d\mathbf{u}}{dt}}{c^2} \mathbf{u}$$

In a magnetic field the velocity and acceleration are perpendicular, so that

$$\mathbf{u} \cdot \frac{d\mathbf{u}}{dt} = 0$$

Furthermore,

$$F_r = quB \quad \text{and} \quad \left| \frac{d\mathbf{u}}{dt} \right| = \frac{u^2}{R}$$

Thus

$$quB = \frac{m_0}{\sqrt{1 - (u^2/c^2)}} \frac{u^2}{R} \quad \text{or} \quad u = \frac{qBR/m_0}{\sqrt{1 + (qBR/m_0 c)^2}}$$

The classical velocity is obtained by letting $c \to \infty$ in the above expression.

8.26. Two identical bodies, each with rest mass m_0, approach each other with equal velocities u, collide, and stick together in a perfectly inelastic collision. Determine the rest mass of the composite body.

Since the initial velocities are equal in magnitude, and the final momentum must be zero,

$$E_{\text{initial}} = E_{\text{final}}$$

$$\frac{2 m_0 c^2}{\sqrt{1 - (u^2/c^2)}} = M_0 c^2$$

$$M_0 = \frac{2 m_0}{\sqrt{1 - (u^2/c^2)}} > 2 m_0$$

8.27. What is the rest mass of the composite body in Problem 8.26 as determined by an observer who is at rest with respect to one of the initial bodies?

Consider the body, A, that moves in the $+x$-direction. The velocity v of the observer O' at rest with respect to A is equal to A's velocity, $v = u$. The second body, B, has a velocity $u_B = -u$ as measured by O. Its velocity as measured by O', u'_B, is obtained from the Lorentz velocity transformation:

$$u'_B = \frac{u_B - v}{1 - \dfrac{u_B v}{c^2}} = -\frac{2u}{1 + \dfrac{u^2}{c^2}}$$

Since the composite body, C, is at rest with respect to the laboratory (observer O), its velocity with respect to O' is $u'_C = -u$. From conservation of momentum, as determined by O',

$$\frac{m_0 u'_A}{\sqrt{1 - (u'^2_A/c^2)}} + \frac{m_0 u'_B}{\sqrt{1 - (u'^2_B/c^2)}} = \frac{M_0 u'_C}{\sqrt{1 - (u'^2_C/c^2)}}$$

But $u'_A = 0$.

$$\frac{m_0 \dfrac{-2u}{1 + (u^2/c^2)}}{\sqrt{1 - \left[\dfrac{2u/c}{1 + (u^2/c^2)}\right]^2}} = \frac{M_0(-u)}{\sqrt{1 - (u^2/c^2)}}$$

$$M_0 = \frac{2m_0}{\sqrt{1 - (u^2/c^2)}}$$

in agreement with the value found from energy considerations by observer O (Problem 8.26).

8.28. A particle of rest mass m_0 moving with a speed of $0.8c$ makes a completely inelastic collision with a particle of rest mass $3m_0$ that is initially at rest. What is the rest mass of the resulting single body?

From $p_{\text{final}} = p_{\text{initial}}$,

$$\frac{M_0 u_f}{\sqrt{1 - (u^2_f/c^2)}} = \frac{m_0 u_i}{\sqrt{1 - (u^2_i/c^2)}} = \frac{m_0(0.8c)}{\sqrt{1 - (0.8)^2}} = \frac{4}{3} m_0 c$$

From $E_{\text{final}} = E_{\text{initial}}$,

$$\frac{M_0 c^2}{\sqrt{1 - (u^2_f/c^2)}} = \frac{m_0 c^2}{\sqrt{1 - (u^2_i/c^2)}} + 3m_0 c^2 = \frac{m_0 c^2}{\sqrt{1 - (0.8)^2}} + 3m_0 c^2 = 4.67 m_0 c^2$$

Solving these two equations simultaneously, we get

$$u_f = 0.286c \qquad M_0 = 4.47 m_0$$

8.29. Find the increase in mass of 100 kg of copper if its temperature is increased 100 °C. (For copper the specific heat is $\mathcal{C} = 93$ cal/kg $\cdot$ °C.)

The energy added to the copper block is

$$\Delta E = m\mathcal{C}(\Delta T) = (100 \text{ kg})(93 \text{ cal/kg} \cdot °C)(100 \text{ °C})(4.184 \text{ J/cal}) = 39 \times 10^5 \text{ J}$$

If this energy appears as an increase in mass, then

$$\Delta m = \frac{\Delta E}{c^2} = \frac{39 \times 10^5 \text{ J}}{(3 \times 10^8 \text{ m/s})^2} = 4.33 \times 10^{-11} \text{ kg}$$

This increase is far too small to be measured.

Supplementary Problems

8.30. From the rest masses given in the Appendix calculate the rest mass of one atomic mass unit in joules. *Ans.* 1.49×10^{-10} J

8.31. Calculate the kinetic energy of a proton whose velocity is $0.8c$. *Ans.* 625.5 MeV

8.32. Calculate the momentum of a proton whose kinetic energy is 200 MeV. *Ans.* 644.5 MeV/c

8.33. Calculate the kinetic energy of a neutron whose momentum is 200 MeV/c. *Ans.* 21.0 MeV

8.34. Calculate the velocity of a proton whose kinetic energy is 200 MeV. *Ans.* $0.566c$

8.35. What is the mass of a proton whose kinetic energy is 1 GeV? *Ans.* $m = 2.07 m_{0p}$

8.36. At what velocity must a particle move such that its kinetic energy equals its rest energy? *Ans.* $0.866c$

8.37. Suppose that the relativistic mass of a particle is 5% larger than its rest mass. What is its velocity? *Ans.* $0.305c$

8.38. What is the ratio of the relativistic mass to the rest mass for (*a*) an electron, (*b*) a proton, when it accelerates from rest through a potential difference of 15 megavolts? *Ans.* (*a*) 30.35; (*b*) 1.015

8.39. What is the mass of an electron if it moves through a potential difference that would, according to classical physics, accelerate the electron to the speed of light? *Ans.* $\frac{3}{2} m_0$

8.40. Refer to Problem 8.20. What are the velocity and momentum of each π^0? *Ans.* $0.84c$; 209 MeV/c

8.41. Suppose that electrons in a uniform magnetic field of flux density 0.03 T move in a circle of radius 0.2 m. What are the velocity and kinetic energy of the electrons? *Ans.* $0.962c$; 1.36 MeV

8.42. What is the minimum energy required to accelerate a rocket ship to a speed of $0.8c$ if its final payload rest mass is 5000 kg? *Ans.* 3×10^{20} J

8.43. A 0.8 MeV electron moves in a magnetic field in a circular path with a radius of 5 cm. What is the magnetic induction? *Ans.* 8.07×10^{-2} T

8.44. Compute the radius of a 20 MeV electron moving at right angles to a uniform magnetic field of flux density 5 T. *Ans.* 1.37 cm

8.45. A particle of rest mass m_0 moving with a speed of $0.6c$ collides and sticks to a similar particle initially at rest. What are the rest mass and velocity of the composite particle? *Ans.* $2.12 m_0$; $0.333c$

8.46.   A particle with a rest mass m_0 and kinetic energy $3m_0c^2$ makes a completely inelastic collision with a stationary particle of rest mass $2m_0$.   What are the velocity and rest mass of the composite particle? *Ans.* $0.645c$; $4.58m_0$

8.47.   A π^+-meson whose rest energy is 140 MeV is created 100 km above sea level in the earth's atmosphere. The π^+-meson has a total energy of 1.5×10^5 MeV and is moving vertically downward. If it disintegrates 2×10^{-8} s after its creation, as determined in its own frame of reference, at what altitude above sea level does the disintegration occur?   *Ans.*   93.6 km

Chapter 9

The Relativistic Doppler Effect

Consider a source that emits electromagnetic radiation with a frequency ν_0 as measured by an observer who is at rest with respect to the source. Suppose this same source is in motion with respect to another observer, who measures the frequency ν of the radiation received from the source. With the angle θ and velocity v of the source as defined in Fig. 9-1, the frequency ν as measured by observer O is given by the *Doppler equation*

$$\nu = \nu_0 \frac{\sqrt{1 - (v^2/c^2)}}{1 - (v/c)\cos\theta}$$

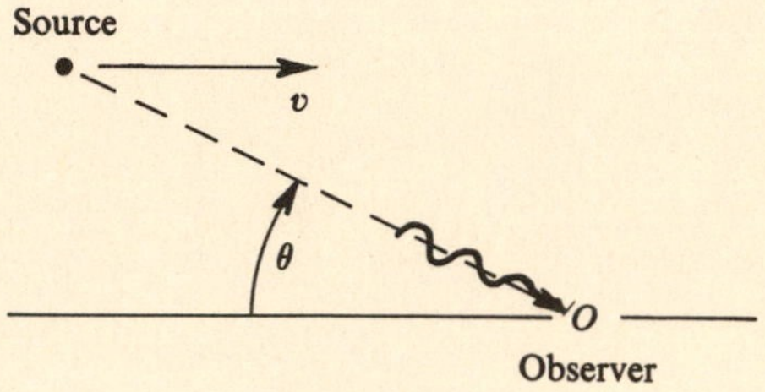

Fig. 9-1

If the source and observer are moving *toward* each other, $\theta = 0$ and we have

$$\nu = \nu_0 \sqrt{\frac{c + v}{c - v}}$$

In this case, $\nu > \nu_0$.

If the source and observer are moving *away from* each other, $\theta = 180°$ and we have

$$\nu = \nu_0 \sqrt{\frac{c - v}{c + v}}$$

In this case, $\nu < \nu_0$.

If the radiation is observed *transverse* to the direction of motion, $\theta = 90°$ and we have

$$\nu = \nu_0 \sqrt{1 - (v^2/c^2)}$$

Thus, $\nu < \nu_0$.

Because all observers measure the speed of light as c, the above equations also allow the change in wavelength to be obtained via $\lambda = c/\nu$.

Solved Problems

9.1.   Evaluate the Doppler equation to first order in v/c when the source and observer are receding from each other.

$$\nu = \nu_0 \sqrt{\frac{1 - \frac{v}{c}}{1 + \frac{v}{c}}} \times \frac{\sqrt{1 + \frac{v}{c}}}{\sqrt{1 + \frac{v}{c}}} = \nu_0 \frac{\sqrt{1 - \frac{v^2}{c^2}}}{1 + \frac{v}{c}} \approx \nu_0 \frac{c}{c + v}$$

which is the classical expression for the Doppler effect when the receiver is stationary with respect to the medium.

9.2.   A car is approaching a radar speed trap at 80 mph.   If the radar set works at a frequency of 20×10^9 Hz, what frequency shift is observed by the patrolman at the radar set?

To first order in v/c, the frequency received by the car is

$$\nu' = \nu_0 \sqrt{\frac{1 + \frac{v}{c}}{1 - \frac{v}{c}}} \approx \nu_0 \sqrt{\left(1 + \frac{v}{c}\right)\left(1 + \frac{v}{c}\right)} = \nu_0\left(1 + \frac{v}{c}\right)$$

The car then acts as a moving source with this frequency.   The frequency received back at the radar set is

$$\nu'' \approx \nu'\left(1 + \frac{v}{c}\right) \approx \nu_0\left(1 + \frac{v}{c}\right)^2 \approx \nu_0\left(1 + \frac{2v}{c}\right)$$

from which (80 mi/hr = 35 m/s)

$$\nu'' - \nu_0 \approx 2\frac{v}{c}\nu_0 = \frac{2 \times 35 \text{ m/s}}{3 \times 10^8 \text{ m/s}} \times 20 \times 10^9 \text{ Hz} = 4.67 \times 10^3 \text{ Hz}$$

9.3.   A star is receding from the earth at a speed of $5 \times 10^{-3}c$.   What is the wavelength shift for the sodium D_2 line (5890 Å)?

The Doppler equation gives

$$\frac{c}{\lambda} = \frac{c}{\lambda_0}\sqrt{\frac{c - v}{c + v}} \qquad \text{or} \qquad \lambda = \lambda_0\sqrt{\frac{1 + (v/c)}{1 - (v/c)}} = (5890 \text{ Å})\sqrt{\frac{1 + 0.005}{1 - 0.005}} = 5920 \text{ Å}$$

Hence, $\Delta\lambda = 5920$ Å $- 5890$ Å $= 30$ Å.   The shift is to a greater wavelength (*red shift*).

9.4.   Suppose that the Doppler shift in the sodium D_2 line (5890 Å) is 100 Å when the light is observed from a distant star.   Determine the star's velocity of recession.

$$\lambda = \lambda_0\sqrt{\frac{1 + (v/c)}{1 - (v/c)}} \qquad \text{or} \qquad 5990 \text{ Å} = (5890 \text{ Å})\sqrt{\frac{1 + (v/c)}{1 - (v/c)}}$$

Solving, $v = 0.017c$.

9.5.   A man in a rocket ship moving with a speed of $0.6c$ away from a space platform shines a light of wavelength 5000 Å toward the platform.   What is the frequency of the light as seen by an observer on the platform?

$$\nu = \nu_0\sqrt{\frac{1 - (v/c)}{1 + (v/c)}} = \frac{3 \times 10^8 \text{ m/s}}{5 \times 10^{-7} \text{ m}}\sqrt{\frac{1 - 0.6}{1 + 0.6}} = 3 \times 10^{14} \text{ Hz}$$

9.6. Refer to Problem 9.5.   What is the frequency of the light as seen by a passenger in a second rocket ship that moves in the opposite direction with a speed of $0.8c$ relative to the space platform?

The velocity of the first rocket relative to the second is found from the Lorentz velocity transformation.

$$u_x' = \frac{u_x - v}{1 - \frac{v}{c^2}u_x} = \frac{0.6c - (-0.8c)}{1 - \frac{(-0.8c)(0.6c)}{c^2}} = 0.946c$$

The frequency observed by the second rocket is then given by

$$\nu = \nu_0\sqrt{\frac{1 - (u_x'/c)}{1 + (u_x'/c)}} = \frac{3 \times 10^8 \text{ m/s}}{5 \times 10^{-7} \text{ m}}\sqrt{\frac{1 - 0.946}{1 + 0.946}} = 1.0 \times 10^{14} \text{ Hz}$$

Supplementary Problems

9.7. What is the Doppler shift in 5500 Å light if the source approaches the observer with velocity of $0.8c$?
Ans. -3667 Å

9.8. Suppose that the largest wavelength visible to the eye is 6500 Å.   How fast must a rocket move in order that a green light ($\lambda = 5000$ Å) on the rocket shall be invisible to an observer on the earth?
Ans. $0.257c$ away from the observer

9.9. How fast must a star recede from the earth in order that a given wavelength shall be shifted by 0.5%?
Ans. $4.99 \times 10^{-3}c$

Chapter 10

The Theory of Photons

The basic postulate of the quantum interpretation is that electromagnetic radiation consists of particle-like discrete bundles of energy called *photons* or *quanta*. Each photon has an energy E that depends only on the frequency v of the radiation and is given by

$$E = hv = h\frac{c}{\lambda}$$

where $h = 6.626 \times 10^{-34}$ J·s is *Planck's constant*. Each photon interacts in an all-or-nothing manner; it either gives up all its energy or none of it.

Since photons travel at the speed of light, they must, according to relativity theory, have zero rest mass; hence, their energy is entirely kinetic. If a photon exists, then it moves at the speed of light, c; if it ceases to move with speed c, it ceases to exist. For $m_0 = 0$, the relativistic momentum-energy relation (Section 8.5) becomes $E = pc$. Thus, each photon has a momentum of

$$p = \frac{E}{c} = \frac{hv}{c} = \frac{h}{\lambda}$$

From the quantum point of view, a beam of electromagnetic energy is composed of photons traveling at the speed c. The intensity of the beam will be proportional to the number of photons crossing a unit area per unit time. Hence, if the beam is monochromatic (one frequency), the intensity I will be given by

$$I = \text{(energy of one photon)} \times \frac{\text{number of photons}}{\text{area} \times \text{time}}$$

Finally, we note for convenience in calculations the following expressions in nonstandard units:

$$h = 4.136 \times 10^{-15} \text{ eV·s}$$
$$hc = 12.4 \text{ keV·Å}$$

where $1 \text{ eV} = 10^{-3} \text{ keV} = 1.602 \times 10^{-19}$ J and $1 \text{ Å} = 10^{-10}$ m.

Solved Problems

10.1. Find the wavelength and frequency of a 1.0 keV photon.

$$\lambda = \frac{hc}{E} = \frac{12.4 \text{ keV·Å}}{1.0 \text{ keV}} = 12.4 \text{ Å}$$

$$v = \frac{c}{\lambda} = \frac{3 \times 10^8 \text{ m/s}}{12.4 \times 10^{-10} \text{ m}} = 2.42 \times 10^{17} \text{ Hz}$$

10.2. Find the momentum of a 12.0 MeV photon.

$$p = \frac{E}{c} = 12 \text{ MeV}/c$$

10.3. Calculate the frequency of the photon produced when an electron of 20 keV is brought to rest in one collision with a heavy nucleus.

Assuming all the kinetic energy of the electron is used to produce the photon, we have

$$E_{initial} = E_{final}$$
$$K + \cancel{m_0 c^2} = h\nu + \cancel{m_0 c^2}$$
$$20 \times 10^3 \text{ eV} = (4.136 \times 10^{-15} \text{ eV} \cdot \text{s})\nu$$
$$\nu = 4.84 \times 10^{18} \text{ Hz}$$

10.4. Show that momentum is not conserved in Problem 10.3.

The initial momentum of the electron is found from

$$(K + E_0)^2 = (p_e c)^2 + E_0^2 \quad \text{or} \quad (0.02 \text{ MeV} + 0.511 \text{ MeV})^2 = (p_e c)^2 + (0.511 \text{ MeV})^2$$

whence $p_e = 0.144 \text{ MeV}/c$.   But

$$p_{final} = p_{photon} = \frac{E_{photon}}{c} = \frac{0.02 \text{ MeV}}{c}$$

The excess momentum is absorbed by the nucleus that stops the electron.   Because the nucleus is so much more massive than the electron, its change in energy could be neglected in Problem 10.3.

10.5. Find the maximum wavelength of the photon that will separate a molecule whose binding energy is 15 eV.

From $E = hc/\lambda$,

$$15 \text{ eV} = \frac{12.4 \times 10^3 \text{ eV} \cdot \text{Å}}{\lambda} \quad \text{or} \quad \lambda = 827 \text{ Å}$$

10.6. What energy does a photon have if its momentum is equal to that of a 3 MeV electron?

The momentum and energy of an electron are related by

$$E^2 = (p_e c)^2 + E_0^2 \quad \text{or} \quad (3 \text{ MeV} + 0.511 \text{ MeV})^2 = (p_e c)^2 + (0.511 \text{ MeV})^2$$

whence $p_e = 3.47 \text{ MeV}/c$.   The energy of the photon is

$$E = pc = p_e c = (3.47 \text{ MeV}/c)c = 3.47 \text{ MeV}$$

10.7. Monochromatic light of wavelength 3000 Å is incident normally on a surface of area 4 cm². If the intensity of the light is $15 \times 10^{-2} \text{ W/m}^2$, determine the rate at which photons strike the surface.

The energy per photon is

$$E = \frac{hc}{\lambda} = \frac{(6.63 \times 10^{-34} \text{ J} \cdot \text{s})(3 \times 10^8 \text{ m/s})}{3 \times 10^{-7} \text{ m}} = 6.63 \times 10^{-19} \text{ J}$$

The total energy flux is

$$IA = (15 \times 10^{-2} \text{ W/m}^2)(4 \times 10^{-4} \text{ m}^2) = 6 \times 10^{-5} \text{ W} = 6 \times 10^{-5} \text{ J/s}$$

Hence, the rate at which photons strike the surface is

$$\frac{6 \times 10^{-5} \text{ J/s}}{6.63 \times 10^{-19} \text{ J/photon}} = 9.05 \times 10^{13} \text{ photons/s}$$

10.8. A radio station operates at a frequency of 103.7 MHz with a power output of 200 kW. Determine the rate of emission of quanta from the station.

The energy of each quantum is

$$E = h\nu = (6.63 \times 10^{-34} \text{ J} \cdot \text{s})(103.7 \times 10^6 \text{ s}^{-1}) = 6.88 \times 10^{-26} \text{ J}$$

so

$$\frac{\text{number of quanta}}{\text{time}} = 200 \times 10^3 \frac{\text{J}}{\text{s}} \times \frac{1 \text{ quantum}}{6.88 \times 10^{-26} \text{ J}} = 2.91 \times 10^{30} \frac{\text{quanta}}{\text{s}}$$

Supplementary Problems

10.9. Find the wavelength and frequency of a 1 MeV photon.   *Ans.* 1.24×10^{-2} Å; 2.42×10^{20} Hz

10.10. Find the wavelength and frequency of a photon whose momentum is 0.02 MeV/c.
Ans. 6.20×10^{-1} Å; 4.84×10^{18} Hz

10.11. Find the momentum of a 4 keV photon.   *Ans.*   4 keV/c

10.12. Find the energy of a photon whose momentum is 10 MeV/c.   *Ans.*   10 MeV

10.13. Find the energy of a photon whose wavelength is 4000 Å.   *Ans.*   3.1 eV

10.14. Find the energy and momentum of a photon whose frequency is 10^6 Hz.
Ans. 4.14×10^3 MeV; 4.14×10^3 MeV/c

10.15. Find the momentum of a photon whose wavelength is 10 Å   *Ans.*   1.24 MeV/c

10.16. A 1 MeV electron is brought to rest by a single collision and produces one photon. Find the wavelength of the photon.   *Ans.* 12.4×10^{-3} Å

10.17. If the maximum wavelength of a photon needed to separate a diatomic molecule is 3000 Å, what is its binding energy?   *Ans.*   4.13 eV

10.18. What is the momentum of a photon if it has the same energy as a 10 MeV alpha particle?
Ans.   10 MeV/c

10.19. A radio station has a power output of 150 kW at 101.1 MHz. Find the number of photons crossing a unit area per unit time one mile from the radio station. Assume the radio station radiates uniformly in all directions.   *Ans.* 6.39×10^{21} photons/ft$^2 \cdot$s

10.20. A plane, 300 MHz electromagnetic wave is incident normally on a surface of area 50 cm^2. If the intensity of the wave is 9×10^{-5} W/m^2, determine the rate at which photons strike the surface.
Ans. 2.26×10^{18} photons/s

10.21. A light source of frequency 6×10^{14} Hz produces 10 W. How many photons are produced in 1 second?
Ans. 2.52×10^{19} photons

10.22. Refer to Problem 10.8. Treating the radio station as a point source radiating uniformly in all directions, find the number of photons inside a cubical radio 20 cm on a side located 15 km from the radio station.
Ans. 2.75×10^{10} photons

Chapter 11

The Photoelectric Effect

11.1 EXPERIMENTAL RESULTS

In a photoelectric experiment light shines on a metal surface in an evacuated tube and electrons are emitted from this surface, as shown in Fig. 11-1. The frequency v and intensity I of the light, the retarding voltage V, and the material of the emitter can be varied. If the electrons are sufficiently energetic they will be able to overcome the retarding potential V and will reach the collector and be recorded as current i in the ammeter A. In order to be able to reach the collector the electrons must have a kinetic energy equal to or greater than the electrical potential energy that they must gain in going between emitter and collector, i.e.

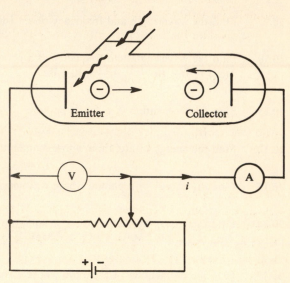

Fig. 11-1

$$\frac{1}{2}\, m_e v^2 \geq eV$$

If their energy is less than this value, they will be turned back before reaching the collector and will not be recorded as current.

The experimental results are:
(1) The current begins almost instantaneously, even for light of very low intensity. The delay between when the incident light strikes the surface and when the electrons are observed is of the order of 10^{-9} s and is independent of the intensity.
(2) When the frequency and retarding potential are held fixed, the current is directly proportional to the intensity of the incident light.
(3) When the frequency and light intensity are held fixed, the current decreases as the retarding voltage is increased, reaching zero for a certain *stopping voltage*, V_s. This stopping voltage is independent of the intensity.
(4) For a given emitter material the stopping voltage varies linearly with the frequency according to the relation

$$eV_s = hv - eW_0$$

The value of the constant term, eW_0, varies from material to material, but the slope h remains the same for all materials, being numerically equal to Planck's constant (see Problem 11.2).
(5) For a given material there exists a *threshold frequency*, v_{th}, below which no electrons will be emitted, no matter how great the light intensity.

11.2 THEORY OF THE PHOTOELECTRIC EFFECT

A wave picture of light can explain only result (2), the increase of current with intensity, since the more intense the light, the more energy transmitted by the wave, and the more electrons that should be emitted. The other results, however, are completely inexplicable in terms of a wave picture (see Problem 11.1).

The quantum interpretation of light is capable of explaining all the experimental results. In the quantum picture the energy carried by a photon is absorbed by a single electron. If the electron is ejected from the material, the difference between the energy absorbed by the electron and the energy with which the electron was bound to the surface appears as kinetic energy of the electron. The electrons are bound to the surface with varying energies, but the binding energy of the least tightly bound electrons depends on the material of the emitter. The energy required to remove these least tightly bound electrons is called the *work function*, ϕ, of the material. Hence, the electrons will be ejected with various kinetic energies ranging from zero to a maximum value given by

maximum kinetic energy of emitted electron =

(energy carried by photon) − (binding energy of the least tightly bound electron)

thereby explaining experimental result (3). Since $K_{max} = eV_s$, the maximum-energy relation becomes

$$eV_s = h\nu - \phi$$

where $\phi = eW_0$. Hence the linear relation of result (4) is explained, along with the existence of a threshold frequency (result (5)) given by

$$h\nu_{th} = eW_0$$

Below this threshold frequency the incident photons will not have sufficient energy to release even the least tightly bound electrons, no matter how intense the light. The short delay time of experimental result (1) is also explained, because the absorption of a photon occurs almost instantaneously. Finally, the more intense the light, the larger the photon density, and hence the more electrons that will be ejected, thereby explaining result (2).

Solved Problems

11.1. Consider a potassium surface that is 75 cm away from a 100-watt bulb. Suppose that the energy radiated by the bulb is 5% of the input power. Treating each potassium atom as a circular disc of diameter 1 Å, determine the time required for each atom to absorb an amount of energy equal to its work function of 2.0 eV, according to the wave interpretation of light.

Treating the bulb as a point source, the intensity at the location of the potassium surface is

$$\text{intensity} = \frac{\text{power}}{\text{area of sphere}} = \frac{100 \text{ W} \times 0.05}{4\pi(0.75 \text{ m})^2} = 0.707 \text{ W/m}^2$$

The power incident on each potassium atom is

power per atom = intensity × (area per atom)

$$= \left(0.707 \frac{\text{W}}{\text{m}^2}\right) \frac{\pi(1 \times 10^{-10} \text{ m})^2}{4} = 5.56 \times 10^{-21} \text{ W}$$

The time interval to absorb 2.0 eV of energy is then found from

$$\text{power} = \frac{\text{energy}}{\text{time}} \quad \text{or} \quad \text{time} = \frac{\text{energy}}{\text{power}} = \frac{(2.0 \text{ eV})(1.6 \times 10^{-19} \text{ J/eV})}{5.56 \times 10^{-21} \text{ J/s}} = 57.6 \text{ s}$$

In this calculation it has been assumed that all the incident energy has been absorbed. Since, with a wave picture, much of the incident energy would be reflected, the actual calculated time would be in excess of 57.6 s. Thus a wave picture of electromagnetic radiation predicts an emission time much larger than the experimentally observed time of less than 10^{-9} s.

11.2. When a photoelectric experiment is performed using calcium as the emitter, the following stopping potentials are found:

λ, Å	2536	3132	3650	4047
v, Hz $\times 10^{15}$	1.18	0.958	0.822	0.741
V_s, V	1.95	0.98	0.50	0.14

Find Planck's constant from these data.

The data are graphed in Fig. 11-2.   From the photoelectric equation, the slope of the graph is h/e, so

$$h = e(\text{slope}) = (1.6 \times 10^{-19}\,\text{C})\,\frac{1.66\,\text{V}}{0.40 \times 10^{15}\,\text{s}^{-1}} = 6.6 \times 10^{-34}\,\text{J} \cdot \text{s}$$

$(1\,\text{C} \cdot \text{V} = 1\,\text{J})$.

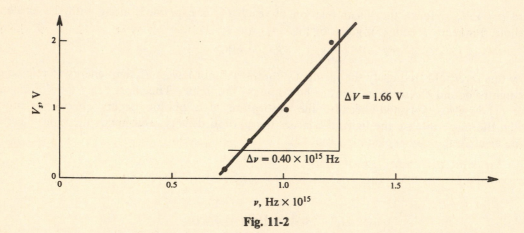

Fig. 11-2

11.3. The kinetic energies of photoelectrons range from zero to 4.0×10^{-19} J when light of wavelength 3000 Å falls on a surface.   What is the stopping potential for this light?

$$K_{max} = 4.0 \times 10^{-19}\,\text{J} \times \frac{1\,\text{eV}}{1.6 \times 10^{-19}\,\text{J}} = 2.5\,\text{eV}$$

Then, from $eV_s = K_{max}$, $V_s = 2.5$ V.

11.4. What is the threshold wavelength for the material in Problem 11.3?

$$eV_s = hv - eW_0 = \frac{hc}{\lambda} - \frac{hc}{\lambda_{th}} \quad \text{or} \quad 2.5\,\text{eV} = \frac{12.4 \times 10^3\,\text{eV} \cdot \text{Å}}{3000\,\text{Å}} - \frac{12.4 \times 10^3\,\text{eV} \cdot \text{Å}}{\lambda_{th}}$$

Solving, $\lambda_{th} = 7590$ Å.

11.5. The emitter in a photoelectric tube has a threshold wavelength of 6000 Å.   Determine the wavelength of the light incident on the tube if the stopping potential for this light is 2.5 V.

The work function is

$$eW_0 = hv_{th} = \frac{hc}{\lambda_{th}} = \frac{12.4 \times 10^3\,\text{eV} \cdot \text{Å}}{6000\,\text{Å}} = 2.07\,\text{eV}$$

The photoelectric equation then gives

$$eV_s = h\nu - eW_0 = \frac{hc}{\lambda} - eW_0 \quad \text{or} \quad 2.5 \text{ eV} = \frac{12.4 \times 10^3 \text{ eV} \cdot \text{Å}}{\lambda} - 2.07 \text{ eV}$$

Solving, $\lambda = 2713$ Å.

11.6. Find the work function for potassium if the largest wavelength for electron emission in a photoelectric experiment is 5620 Å.

$$\phi = eW_0 = \frac{hc}{\lambda_{th}} = \frac{12.4 \times 10^3 \text{ eV} \cdot \text{Å}}{5620 \text{ Å}} = 2.21 \text{ eV}$$

11.7. Potassium is illuminated with ultraviolet light of wavelength 2500 Å. If the work function of potassium is 2.21 eV, what is the maximum kinetic energy of the emitted electrons?

$$K_{max} = h\frac{c}{\lambda} - eW_0 = \frac{12.4 \times 10^3 \text{ eV} \cdot \text{Å}}{2500 \text{ Å}} - 2.21 \text{ eV} = 2.75 \text{ eV}$$

11.8. In Problem 11.7 the ultraviolet light has an intensity of 2 W/m². Calculate the rate of electron emission per unit area.

In Problem 11.7 each photon has an energy of 4.96 eV $= 7.94 \times 10^{-19}$ J. Assuming each photon liberates one electron, we have

$$\frac{\text{number of electrons}}{\text{m}^2 \cdot \text{s}} = \frac{\text{number of photons}}{\text{m}^2 \cdot \text{s}} = \frac{2 \text{ J/m}^2 \cdot \text{s}}{7.94 \times 10^{-19} \text{ J/photon}} = 2.52 \times 10^{18} \frac{\text{photons}}{\text{m}^2 \cdot \text{s}}$$

11.9. Suppose the wavelength of the incident light in a photoelectric experiment is increased from 3000 Å to 3010 Å. Find the corresponding change in the stopping potential.

$$eV_s = \frac{hc}{\lambda} - eW_0$$

Treating the change in the wavelength as a differential and remembering that W_0 is a constant, one finds

$$e(dV_s) = -\frac{hc}{\lambda^2} d\lambda = -\frac{12.4 \times 10^3 \text{ eV} \cdot \text{Å}}{(3000 \text{ Å})^2} (10 \text{ Å}) = -1.38 \times 10^{-2} \text{ eV}$$

Solving, $dV_s = -1.38 \times 10^{-2}$ V.

11.10. Find the strength of the transverse magnetic field required to bend all the photoelectrons within a circle of radius 20 cm when light of wavelength 4000 Å is incident on a barium emitter. The work function of barium is 2.5 eV.

$$\frac{1}{2} mv_{max}^2 = h\nu - eW_0 = \frac{hc}{\lambda} - \phi$$

$$\frac{1}{2} (9.11 \times 10^{-31} \text{ kg})v_{max}^2 = \frac{(6.63 \times 10^{-34} \text{ J} \cdot \text{s})(3 \times 10^8 \text{ m/s})}{4 \times 10^{-7} \text{ m}} - (2.5 \text{ eV})\left(1.6 \times 10^{-19} \frac{\text{J}}{\text{eV}}\right)$$

and solving, $v_{max} = 4.62 \times 10^5$ m/s. When these electrons enter the magnetic field, we have

force of magnetic field = mass × radial acceleration

$$ev_{max}B = m\frac{(v_{max})^2}{R_{max}}$$

$$B = \frac{mv_{max}}{eR_{max}} = \frac{(9.11 \times 10^{-31} \text{ kg})(4.62 \times 10^5 \text{ m/s})}{(1.6 \times 10^{-19} \text{ C})(0.20 \text{ m})} = 1.32 \times 10^{-5} \text{ T}$$

It should be noted that this field is comparable to the earth's magnetic field, which is approximately 5.8×10^{-5} T.

11.11. Prove that the photoelectric effect cannot occur for free electrons.

In Fig. 11-3 we look at the hypothetical process in the center-of-mass system, which is defined as that system in which the initial momentum is zero. From the conservation of energy,

$$E_{\text{initial}=} = E_{\text{final}} \quad \text{or} \quad h\nu + mc^2 = m_0 c^2$$

which implies $m_0 > m$. Since this cannot be true, so the process cannot occur.

The electrons participating in the photoelectric effect are not free. The heavy matter present takes off momentum but absorbs a negligible amount of energy. See Problem 8.16.

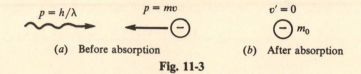

(a) Before absorption   (b) After absorption

Fig. 11-3

Supplementary Problems

11.12. The threshold wavelength for a material is 5000 Å. Find the work function.   *Ans.*   2.48 eV

11.13. For the material in Problem 11.12, what is the stopping potential for 3500 Å photons?   *Ans.*   1.06 V

11.14. The maximum energy of the emitted electrons when a material is illuminated with 3000 Å light is 1.2 eV. Find the work function.   *Ans.*   2.93 eV

11.15. In Problem 11.14 the light has an intensity of 3 W/m^2. What is the rate of electron emission per m^2 if it is 50% efficient?   *Ans.* 2.27×10^{18} electrons/s·m^2

11.16. Find the maximum kinetic energy of electrons emitted from a surface with a threshold wavelength of 6000 Å when light of 4000 Å strikes the surface.   *Ans.*   1.03 eV

11.17. Determine the maximum wavelength of light that will cause emission of electrons from a material whose work function is 3.0 eV   *Ans.*   4133 Å

11.18. Find the energy of the fastest electrons that are emitted when light of wavelength 5000 Å is incident on lithium (work function = 2.13 eV).   *Ans.*   0.35 eV

11.19. When light of wavelength 4500 Å falls on a surface, the stopping potential for the emitted electrons is found to be 0.75 V. What is the stopping potential for the photoelectrons if light of wavelength 3000 Å falls on the surface?   *Ans.*   2.13 V

11.20. The most energetic electrons emitted from a surface by 3500 Å photons are found to be bent in a 18 cm circle by a magnetic field of 1.5×10^{-5} T. Find the work function for the material.   *Ans.*   2.90 eV

11.21. Light of wavelength 4500 Å is incident on two photoelectric tubes. The emitter in the first tube has a threshold wavelength of 6000 Å and the emitter in the second tube has a work function twice as large as in the first tube. Find the stopping potential in each of the tubes.
Ans. $V_{s1} = 0.69$ V; there is no photoelectric emission from the second tube

11.22. Suppose a photon of wavelength 600 Å is absorbed by a hydrogen atom whose ionization energy is 13.6 eV. What is the kinetic energy of the ejected electron?   *Ans.*   7.1 eV

Chapter 12

The Compton Effect

The wave interpretation predicts that when electromagnetic radiation is scattered from a charged particle the scattered radiation will have the same frequency as the incident radiation in all directions. Arthur H. Compton, in 1922, demonstrated that if the quantum interpretation of electromagnetic radiation (Chapter 10) is accepted, then the scattered radiation will have a frequency that is smaller than the incident radiation's and that also depends on the angle of scattering.

Compton's analysis involved, in effect, viewing the scattering of electromagnetic radiation from a charged particle as a perfectly elastic, billiard ball type of collision between a photon and the effectively free charged particle, as shown in Fig. 12-1. Even though the details of the interaction are not known, conservation of energy and momentum can be applied. It is found that the scattered photon undergoes a shift in wavelength, $\Delta\lambda$, given by

$$\Delta\lambda = \lambda' - \lambda = \frac{h}{m_0 c}(1 - \cos\theta)$$

(see Problem 12.10). The quantity $h/m_0 c$ is usually called the *Compton wavelength*; its value, for an electron, is 0.0243 Å. Note that the shift in the wavelength depends only on the scattering angle θ and is independent of the incident photon's energy.

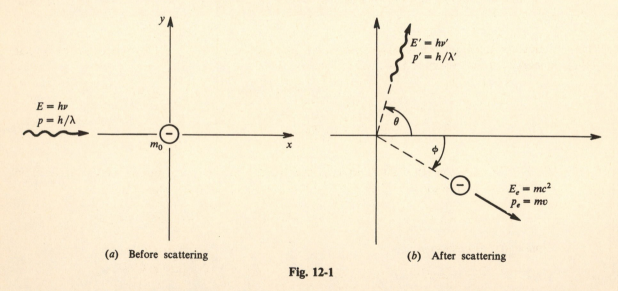

(a) Before scattering   (b) After scattering

Fig. 12-1

Compton verified his theoretical relationship experimentally by scattering X-rays ($\lambda = 0.7$ Å) from graphite. The energy of the X-rays (1.8×10^4 eV) is several orders of magnitude larger than the binding energy of the outer carbon electrons, so treating these electrons as free particles is a good approximation.

Solved Problems

12.1. A 0.3 MeV X-ray photon makes a "head-on" collision with an electron initially at rest. Using conservation of energy and momentum, find the recoil velocity of the electron.

In the notation of Fig. 12-1, the conservation of energy is expressed by

$$E + m_0 c^2 = E' + \frac{m_0 c^2}{\sqrt{1 - (v^2/c^2)}} \quad \text{or} \quad 0.3 \text{ MeV} + 0.511 \text{ MeV} = E' + \frac{0.511 \text{ MeV}}{\sqrt{1 - (v^2/c^2)}}$$

The momentum of a photon is $h\nu/c = E/c$, so the conservation of momentum is ($\theta = 180°$, $\phi = 0$)

$$\frac{E}{c} + 0 = -\frac{E'}{c} + \frac{m_0 v}{\sqrt{1 - (v^2/c^2)}} \quad \text{or} \quad \frac{0.3 \text{ MeV}}{c} = \frac{-E'}{c} + \frac{0.511 \text{ MeV}}{\sqrt{1 - (v^2/c^2)}} \frac{v}{c^2}$$

Simultaneous solution of the energy and momentum equations gives $v = 0.65c$.

12.2. In Problem 12.1 check that the velocity agrees with the value determined from the Compton equation.

$$\lambda' - \lambda = \frac{h}{m_0 c}(1 - \cos\theta) = \frac{h}{m_0 c}(1 - \cos 180°) = \frac{2h}{m_0 c} \quad \text{or} \quad \lambda' = \lambda + \frac{2h}{m_0 c}$$

Multiplying this result by $1/hc$, we obtain

$$\frac{\lambda'}{hc} = \frac{\lambda}{hc} + \frac{2}{m_0 c^2} = \frac{1}{h\nu} + \frac{2}{m_0 c^2} = \frac{1}{0.3 \text{ MeV}} + \frac{2}{0.511 \text{ MeV}} = 7.24 \frac{1}{\text{MeV}}$$

Substituting $E' = (1/7.24)$ MeV into the energy equation of Problem 12.1 and solving for v, we again obtain $v = 0.65c$.

12.3. Calculate the fractional change in the wavelength of an X-ray of wavelength 0.400 Å that undergoes a 90° Compton scattering from an electron.

$$\lambda' - \lambda = \frac{h}{m_0 c}(1 - \cos\theta) = (0.0243 \text{ Å})(1 - \cos 90°) = 0.0243 \text{ Å}$$

$$\frac{\lambda' - \lambda}{\lambda} = \frac{0.0243 \text{ Å}}{0.400 \text{ Å}} = 0.0608$$

12.4. An X-ray of wavelength 0.300 Å undergoes a 60° Compton scattering. Find the wavelength of the scattered photon and the energy of the electron after the scattering.

$$\lambda' = \lambda + \frac{h}{m_0 c}(1 - \cos\theta) = 0.30 \text{ Å} + (0.0243 \text{ Å})(1 - \cos 60°) = 0.312 \text{ Å}$$

From energy conservation,

$$\frac{hc}{\lambda} + m_0 c^2 = \frac{hc}{\lambda'} + K_e + m_0 c^2 \quad \text{or} \quad \frac{12.4 \text{ keV} \cdot \text{Å}}{0.3 \text{ Å}} = \frac{12.4 \text{ keV} \cdot \text{Å}}{0.312 \text{ Å}} + K_e$$

Solving, $K_e = 1.59$ keV.

12.5. In a Compton experiment an electron attains a kinetic energy of 0.100 MeV when an X-ray of energy 0.500 MeV strikes it. Determine the wavelength of the scattered photon if the electron is initially at rest.

$$E_{\text{initial}} = E_{\text{final}}$$

$$E + m_0 c^2 = E' + (K_e + m_0 c^2)$$

$$0.500 \text{ MeV} = E' + 0.100 \text{ MeV}$$

$$E' = 0.400 \text{ MeV}$$

whence

$$\lambda' = \frac{hc}{E'} = \frac{12.4 \times 10^{-3} \text{ MeV} \cdot \text{\AA}}{0.400 \text{ MeV}} = 31 \times 10^{-3} \text{ \AA}$$

12.6. In Problem 12.5 find the angle that the scattered photon makes with the incident direction.

The incident wavelength is

$$\lambda = \frac{hc}{E} = \frac{12.4 \times 10^{-3} \text{ MeV} \cdot \text{\AA}}{0.500 \text{ MeV}} = 24.8 \times 10^{-3} \text{ \AA}$$

From the Compton equation,

$$\lambda' - \lambda = \frac{h}{m_0 c} (1 - \cos \theta)$$

$$31 \times 10^{-3} \text{ \AA} - 24.8 \times 10^{-3} \text{ \AA} = (24.3 \times 10^{-3} \text{ \AA})(1 - \cos \theta)$$

Solving, $\theta = 42°$.

12.7. If the maximum energy imparted to an electron in Compton scattering is 45 keV, what is the wavelength of the incident photon?

If the electron is to have its maximum recoil energy, then the photon is back-scattered. By conservation of energy,

$$E + m_0 c^2 = E' + 45 \text{ keV} + m_0 c^2 \qquad \text{or} \qquad E - E' = 45 \text{ keV} \qquad (1)$$

By conservation of momentum,

$$\frac{E}{c} = -\frac{E'}{c} + p_e \qquad (2)$$

Relating the momentum and energy of the electron by $E_e^2 = (p_e c)^2 + E_0^2$, we have

$$(0.511 \text{ MeV} + 0.045 \text{ MeV})^2 = (p_e c)^2 + (0.511 \text{ MeV})^2 \qquad \text{or} \qquad p_e = 0.219 \text{ MeV}/c$$

Putting this in (2), we have

$$E + E' = 219 \text{ keV} \qquad (3)$$

Solving (1) and (3), we obtain $E = 132$ keV, from which

$$\lambda = \frac{hc}{E} = \frac{12.4 \text{ keV} \cdot \text{\AA}}{132 \text{ keV}} = 9.39 \times 10^{-2} \text{ \AA}$$

12.8. Show that a free electron at rest cannot absorb a photon. (Hence Compton scattering *must* occur with free electrons.)

$$p_{\text{photon}} = p_{\text{electron}} \qquad \text{or} \qquad \frac{h\nu}{c} = p_e$$

$$E_{\text{photon}} = E_{\text{electron}} \qquad \text{or} \qquad h\nu = \sqrt{(p_e c)^2 + (m_0 c^2)^2}$$

Dividing the energy expression by c gives

$$\frac{h\nu}{c} = \sqrt{p_e^2 + m_0^2 c^2} > p_e$$

which contradicts the momentum expression.

Essentially the same problem has been solved in a somewhat different manner in connection with the photoelectric effect (Problem 11.11).

12.9. Determine the maximum scattering angle in a Compton experiment for which the scattered photon can produce a positron-electron pair.

The threshold wavelength for positron-electron pair production is (see Problem 13.4)

$$h\frac{c}{\lambda_{th}} = 2m_0 c^2 \qquad \text{or} \qquad \frac{h}{m_0 c} = 2\lambda_{th}$$

Substituting this result in the Compton equation, we find

$$\lambda' = \lambda + 2\lambda_{th}(1 - \cos\theta)$$

The right-hand side of this expression is the sum of two positive-definite terms. Hence, if

$$2\lambda_{th}(1 - \cos\theta) \ge \lambda_{th}$$

then $\lambda' > \lambda_{th}$ and pair production cannot occur. Taking the equality, we find for θ_{th}:

$$2\lambda_{th}(1 - \cos\theta_{th}) = \lambda_{th} \qquad \text{or} \qquad \cos\theta_{th} = 1/2 \qquad \text{or} \qquad \theta_{th} = 60°$$

Note that this result is independent of the energy of the incident photon.

12.10. Derive the Compton equation, $\lambda' - \lambda = (h/m_0 c)(1 - \cos\theta)$.

Refer to Fig. 12-1. The photon is treated as a particle of energy $E = h\nu = hc/\lambda$ and momentum $p = h/\lambda$. From conservation of energy:

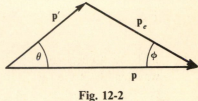

Fig. 12-2

$$\frac{hc}{\lambda} + m_0 c^2 = \frac{hc}{\lambda'} + mc^2$$

Squaring and rearranging we obtain

$$(mc^2)^2 = \frac{h^2 c^2}{\lambda^2 \lambda'^2}(\lambda^2 + \lambda'^2) - \frac{2h^2 c^2}{\lambda\lambda'} + \frac{2hm_0 c^3}{\lambda\lambda'}(\lambda' - \lambda) + (m_0 c^2)^2 \qquad (1)$$

From conservation of momentum we obtain the vector diagram shown in Fig. 12-2. Since $\mathbf{p}_e = \mathbf{p} - \mathbf{p}'$,

$$\mathbf{p}_e \cdot \mathbf{p}_e = p_e^2 = p^2 + p'^2 - 2\mathbf{p}\cdot\mathbf{p}' = \frac{h^2}{\lambda^2\lambda'^2}(\lambda'^2 + \lambda^2 - 2\lambda\lambda'\cos\theta) \qquad (2)$$

Substituting (1) and (2) in the relation $(mc^2)^2 = (p_e c)^2 + (m_0 c^2)^2$, we obtain

$$\frac{h^2 c^2}{\lambda^2\lambda'^2}(\lambda^2 + \lambda'^2) - \frac{2h^2 c^2}{\lambda\lambda'} + \frac{2hm_0 c^3}{\lambda\lambda'}(\lambda' - \lambda) + (m_0 c^2)^2 = \frac{h^2 c^2}{\lambda^2\lambda'^2}(\lambda'^2 + \lambda^2 - 2\lambda\lambda'\cos\theta) + (m_0 c^2)^2$$

Solving, we obtain the Compton relationship

$$\lambda' - \lambda = \Delta\lambda = \frac{h}{m_0 c}(1 - \cos\theta)$$

12.11. In Compton scattering, what is the kinetic energy of the electron scattered at angle ϕ to the incident photon?

$$\text{initial energy} = \text{final energy}$$

$$h\nu + m_0 c^2 = h\nu' + K_e + m_0 c^2$$

or, since $h\nu = pc$,

$$pc = p'c + K_e \qquad (1)$$

From Fig. 12-2, $\mathbf{p}' = \mathbf{p} - \mathbf{p}_e$, and so

$$\mathbf{p}'\cdot\mathbf{p}' = p'^2 = p^2 + p_e^2 - 2pp_e\cos\phi \qquad (2)$$

Using (1) and the relation

$$p_e^2 = \frac{E_e^2 - (m_0 c^2)^2}{c^2} = \frac{1}{c^2}\left[(m_0 c^2 + K_e)^2 - (m_0 c^2)^2\right] = \frac{1}{c^2}(K_e^2 + 2K_e m_0 c^2) \qquad (3)$$

in (2) we obtain

$$K_e\left(m_0 + \frac{p}{c}\right) = pp_e\cos\phi$$

Squaring this and using (3) again, we obtain

$$K_e^2\left(m_0 + \frac{p}{c}\right)^2 = \frac{p^2}{c^2}\left(K_e^2 + 2K_e m_0 c^2\right)\cos^2 \phi$$

Finally, solving for K_e and replacing p by $h\nu/c$, we find

$$K_e = h\nu \, \frac{2\left(\dfrac{h\nu}{m_0 c^2}\right)\cos^2 \phi}{\left(\dfrac{h\nu}{m_0 c^2} + 1\right)^2 - \left(\dfrac{h\nu}{m_0 c^2}\right)^2 \cos^2 \phi}$$

Observe that K_e is maximum for $\phi = 0$.

Supplementary Problems

12.12. Find the Compton wavelength for a proton (rest mass = 938.3 MeV).   *Ans.* 1.32×10^{-5} Å

12.13. Repeat Problem 12.3 for visible light of wavelength 5000 Å.   *Ans.* 4.86×10^{-6}

12.14. A 100 keV photon scatters from a free electron initially at rest.   Find the recoil velocity of the electron if the photon scattering angle is 180°.   (Use energy and momentum conservation.)   *Ans.* $0.319c$

12.15. In Problem 12.14 calculate from the Compton equation the wavelength of the scattered photon.   *Ans.*   0.1726 Å

12.16. If the photon of Problem 12.14 is scattered at an angle of 65° to the incoming beam, find its final wavelength.   *Ans.*   0.138 Å

12.17. For Problem 12.16 calculate the final momentum of the electron.   *Ans.*   102.4 keV/c

12.18. Repeat Problems 12.16 and 12.17 for a scattering angle of 144°.   *Ans.*   0.168 Å; 166 keV/c

12.19. In Compton scattering the scattered photon and electron are detected.   It is found that the electron has a kinetic energy of 75 keV and the photon an energy of 200 keV.   What was the initial wavelength of the photon?   *Ans.*   0.045 Å

12.20. For Problem 12.19 find the scattering angles for the photon and the electron.   *Ans.*   72.5°; 41.7°

12.21. Calculate the percent change in wavelength in a 0.15 Å photon which undergoes a 120° scattering from an electron.   *Ans.*   24.3%

12.22. Find the final wavelength of a scattered photon which undergoes a 90° Compton scattering from a free proton if its original energy is 12 MeV.   (For a proton, $m_0 c^2 = 938.3$ MeV.)   *Ans.* 1.05×10^{-3} Å

12.23. Calculate the maximum energy, in electron volts, that can be transferred to an electron in a Compton experiment when the incident quanta are X-rays of wavelength 0.50 Å.   *Ans.*   4.7 eV

12.24. Repeat Problem 12.23 for visible light photons of wavelength 5000 Å.   *Ans.* 2.41×10^{-5} eV

12.25. For Compton scattering, what is the relation between scattering angles for the photon and the electron?
Ans. $\cot \phi = \left(1 - \dfrac{h\nu}{m_0 c^2}\right) \cot \dfrac{\theta}{2}$

12.26. An electron that undergoes a "head-on" collision with an X-ray photon has a stopping potential of 70 kV. If the electron was initially at rest, what are the wavelengths of the initial and scattered X-ray photons?   *Ans.*   0.0716 Å; 0.1201 Å

Chapter 13

Pair Production and Annihilation

13.1 PAIR PRODUCTION

In the process of *pair production* the energy carried by a photon is completely converted into matter, resulting in the creation of an electron-positron pair, as indicated in Fig. 13-1. (Except for its charge, a positron is identical in all ways to an electron.) Since the charge of the system was initially zero, two oppositely charged particles must be produced in order to conserve charge. In order to produce a pair, the incident photon must have an energy at least equal to the rest energy of the pair; any excess energy of the photon appears as kinetic energy of the particles.

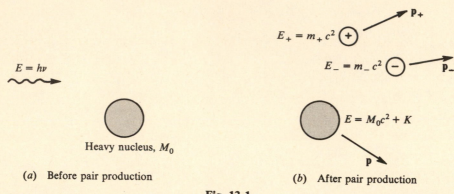

(a) Before pair production (b) After pair production

Fig. 13-1

Pair production cannot occur in empty space (see Problem 13.11). Hence, in Fig. 13-1 the presence of a heavy nucleus is indicated. The nucleus carries away an appreciable amount of the incident photon's momentum, but because of its large mass, its recoil kinetic energy, $K \approx p^2/2M_0$, is usually negligible compared to the kinetic energies of the electron-positron pair. Thus, energy (but not momentum) conservation may be applied with the heavy nucleus ignored, yielding

$$h\nu = m_+ c^2 + m_- c^2 = K_+ + K_- + 2m_0 c^2$$

since the positron and the electron have the same rest mass, $m_0 = 9.11 \times 10^{-31}$ kg.

13.2 PAIR ANNIHILATION

The inverse of pair production can also occur. In *pair annihilation* a positron-electron pair is annihilated, resulting in the creation of two (or more) photons, as shown in Fig. 13-2. At least two photons must be produced in order to conserve energy and momentum. In contrast to pair production, pair annihilation can take place in empty space and both energy and momentum principles are applicable, so that

$$E_{\text{initial}} = E_{\text{final}} \quad \text{or} \quad 2m_0 c^2 + K_+ + K_- = h\nu_1 + h\nu_2$$

$$\mathbf{p}_{\text{initial}} = \mathbf{p}_{\text{final}} \quad \text{or} \quad m_+ \mathbf{v}_+ + m_- \mathbf{v}_- = \frac{h}{2\pi} \mathbf{k}_1 + \frac{h}{2\pi} \mathbf{k}_2$$

where $\mathbf{k}$ is the *propagation vector*, $|\mathbf{k}| = 2\pi/\lambda$.

68

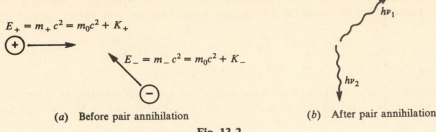

(a) Before pair annihilation   (b) After pair annihilation

Fig. 13-2

Both pair production and annihilation can occur with other particles and antiparticles, such as a proton and an antiproton (see Problem 13.16).

Solved Problems

13.1. A photon of wavelength 0.0030 Å in the vicinity of a heavy nucleus produces an electron-positron pair. Determine the kinetic energy of each of the particles if the kinetic energy of the positron is twice that of the electron.

From $E_{\text{initial}} = E_{\text{final}}$,

$$\frac{hc}{\lambda} = 2m_0c^2 + K_+ + K_- = 2m_0c^2 + 3K_-$$

$$\frac{12.4 \times 10^{-3}\,\text{MeV} \cdot \text{Å}}{0.0030\,\text{Å}} = 2(0.511\,\text{MeV}) + 3K_-$$

$$K_- = 1.04\,\text{MeV}$$

and $K_+ = 2K_- = 2.08$ MeV.

13.2. Find the energies of the two photons that are produced when annihilation occurs between an electron and positron that are initially at rest.

Since the initial momentum of the positron-electron pair is zero, the two photons must travel in opposite directions with equal energies. Applying conservation of energy then yields

$$2m_0c^2 = 2E_\gamma \quad \text{or} \quad E_\gamma = m_0c^2 = 0.511\,\text{MeV}$$

13.3. Pair annihilation takes place when an electron and a positron have a head-on collision, producing two 2.0 MeV photons that travel in opposite directions. Find the kinetic energies of the electron and positron before the collision.

Since the final momentum of the photons is zero, the electron and positron must have had equal kinetic energies before the collision. From energy conservation,

$$2m_0c^2 + 2K = 2E_\gamma$$

$$2(0.511\,\text{MeV}) + 2K = 2(2.0\,\text{MeV})$$

$$K = 1.49\,\text{MeV}$$

13.4. Determine the threshold wavelength for pair production.

The threshold wavelength is that wavelength for which the positron and electron have zero kinetic energy. Conservation of energy for this situation (neglecting the recoil energy of the nucleus) gives

$$\frac{hc}{\lambda_{th}} = 2m_0c^2 \quad \text{or} \quad \lambda_{th} = \frac{hc}{2m_0c^2} = \frac{12.4 \times 10^{-3}\ \text{MeV}\cdot\text{Å}}{2(0.511\ \text{MeV})} = 0.0121\ \text{Å}$$

13.5. Annihilation occurs between an electron and positron at rest, producing three photons. Find the energy of the third photon if the energies of two of the photons are 0.20 MeV and 0.30 MeV.

From conservation of energy,

$$2(0.511\ \text{MeV}) = 0.20\ \text{MeV} + 0.30\ \text{MeV} + E_3 \quad \text{or} \quad E_3 = 0.522\ \text{MeV}$$

13.6. How many positrons can a 200 MeV photon produce?

The energy needed to create an electron-positron pair at rest is twice the rest energy of an electron, or 1.022 MeV. Therefore,

$$\text{maximum number of positrons} = (200\ \text{MeV})\left(\frac{1\ \text{pair}}{1.022\ \text{MeV}}\right)\left(1\ \frac{\text{positron}}{\text{pair}}\right) = 195\ \text{positrons}$$

13.7. A 5 MeV electron undergoes annihilation with a positron that is at rest, producing two photons. One of the photons travels in the direction of the incident electron. Calculate the energy of each photon.

The second photon must travel parallel to ($\epsilon = +1$) or antiparallel to ($\epsilon = -1$) the first photon in order that momentum be conserved in the transverse direction. From conservation of momentum,

$$p_- = \frac{E_1}{c} + \epsilon\frac{E_2}{c} \quad \text{or} \quad E_1 + \epsilon E_2 = p_- c$$

Substituting for $p_- c$ from

$$(K_- + m_0c^2)^2 = (p_-c)^2 + (m_0c^2)^2$$

we obtain

$$E_1 + \epsilon E_2 = \sqrt{(K_- + m_0c^2)^2 - (m_0c^2)^2} = \sqrt{(5.511\ \text{MeV})^2 - (0.511\ \text{MeV})^2} = 5.49\ \text{MeV}$$

Conservation of energy requires

$$E_1 + E_2 = K_- + m_0c^2 + m_0c^2 = 5\ \text{MeV} + 2(0.511\ \text{MeV}) = 6.02\ \text{MeV}$$

Substituting for E_1 in the momentum equation gives

$$-0.53\ \text{MeV} = (\epsilon - 1)E_2$$

Therefore ϵ must be taken equal to -1, so that the second photon travels in the opposite direction from the first. The energies are then found to be

$$E_2 = 0.27\ \text{MeV} \quad E_1 = 5.75\ \text{MeV}$$

13.8. An electron and positron, traveling together as shown in Fig. 13-3, annihilate. Find the wavelengths of the two photons that are produced if they are both to move along the line of motion of the original pair.

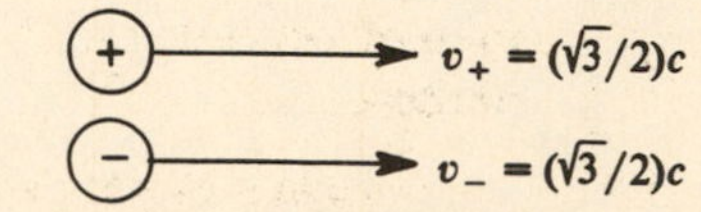

Fig. 13-3

If the process is looked at in the center-of-mass system, the photons move off in opposite directions. Transforming this back to the laboratory frame, one must still find the photons moving in opposite directions, because the laboratory velocity relative to the center of mass is less than c. Conservation of momentum gives

$$2mv = p_1 - p_2 = \frac{h}{\lambda_1} - \frac{h}{\lambda_2}$$

with

$$2mv = \frac{2(m_0 c^2)(v/c)}{c\sqrt{1 - (v^2/c^2)}} = \frac{2(0.511 \text{ MeV})(\sqrt{3}/2)}{c\sqrt{1 - (\sqrt{3}/2)^2}} = 1.770 \ \frac{\text{MeV}}{c}$$

Therefore,

$$\frac{1}{\lambda_1} - \frac{1}{\lambda_2} = \frac{(2mv)c}{hc} = \frac{1.770 \text{ MeV}}{12.4 \times 10^{-3} \text{ MeV} \cdot \text{Å}} = 142.7 \ \text{Å}^{-1} \qquad (1)$$

By conservation of energy,

$$2mc^2 = h\nu_1 + h\nu_2 = \frac{hc}{\lambda_1} + \frac{hc}{\lambda_2}$$

with

$$2mc^2 = \frac{2m_0 c^2}{\sqrt{1 - (v^2/c^2)}} = \frac{2(0.511 \text{ MeV})}{\sqrt{1 - (\sqrt{3}/2)^2}} = 2.044 \text{ MeV}$$

Therefore,

$$\frac{1}{\lambda_1} + \frac{1}{\lambda_2} = \frac{2mc^2}{hc} = \frac{2.044 \text{ MeV}}{12.4 \times 10^{-3} \text{ MeV} \cdot \text{Å}} = 164.8 \ \text{Å}^{-1} \qquad (2)$$

Solving (1) and (2) simultaneously we get

$$\lambda_1 = 6.50 \times 10^{-3} \text{ Å} \qquad \lambda_2 = 9.05 \times 10^{-2} \text{ Å}$$

13.9. An electron and positron moving as in Problem 13.8 annihilate, and the photons produced are observed to have equal scattering angles. Find the energy and scattering angles of the photons.

Since the initial momentum in the transverse direction was zero, the photons must have the same energy E_γ. By conservation of energy and the results of Problem 13.8,

$$2E_\gamma = \frac{2m_0 c^2}{\sqrt{1 - (v^2/c^2)}} = 2.044 \text{ MeV} \quad \text{or} \quad E_\gamma = 1.022 \text{ MeV}$$

By conservation of momentum in the longitudinal direction,

$$2mv = \frac{E_\gamma}{c} \cos \theta + \frac{E_\gamma}{c} \cos (-\theta)$$

$$\frac{2m_0 v}{\sqrt{1 - (v^2/c^2)}} = \frac{2m_0 c}{\sqrt{1 - (v^2/c^2)}} \cos \theta$$

whence $\cos \theta = v/c = \sqrt{3}/2$ and $\theta = 30°$.

13.10. Pair production occurring in a magnetic field of 0.1 T results in a positron and electron having radii of curvature of 120 mm and 40 mm, respectively. Determine the energy of the incident photon.

Applying Newton's second law to a charged particle in a magnetic field (see Problem 8.25), we obtain

$$quB = \frac{m_0}{\sqrt{1 - (u^2/c^2)}} \frac{u^2}{R} \quad \text{or} \quad \frac{1}{\sqrt{1 - (u^2/c^2)}} = \sqrt{1 + \left(\frac{qBR}{m_0 c} \right)^2}$$

Hence the total energy of a charged particle is

$$E = mc^2 = \frac{m_0 c^2}{\sqrt{1 - (u^2/c^2)}} = m_0 c^2 \sqrt{1 + \left(\frac{qBR}{m_0 c}\right)^2}$$

Evaluating the energy of the positron and of the electron:

$$E_+ = (0.511 \text{ MeV}) \sqrt{1 + \left[\frac{(1.6 \times 10^{-19} \text{ C})(0.1 \text{ T})(120 \times 10^{-3} \text{ m})}{(9.11 \times 10^{-31} \text{ kg})(3 \times 10^8 \text{ m/s})}\right]^2} = 3.63 \text{ MeV}$$

$$E_- = (0.511 \text{ MeV}) \sqrt{1 + \left[\frac{(1.6 \times 10^{-19} \text{ C})(0.1 \text{ T})(40 \times 10^{-3} \text{ m})}{(9.11 \times 10^{-31} \text{ kg})(3 \times 10^8 \text{ m/s})}\right]^2} = 1.30 \text{ MeV}$$

Then, by energy conservation (with the heavy nucleus ignored),

$$h\nu = E_+ + E_- = 4.93 \text{ MeV}$$

13.11. Prove that pair production cannot occur in empty space. (Hence, in order for pair production to occur a nucleus must be present.)

The production of a pair of particles is an invariant occurrence—if one observer finds that a pair is produced, then any other observer moving relative to him must also find that a pair is produced. The frequency of a photon, however, will vary from one observer to another because of the Doppler shift (Chapter 9). It is always possible to find an observer moving with a speed such that a given photon's frequency is Doppler-shifted *below* the threshold frequency required for pair production (Problem 13.4). Since this observer will find pair production to be impossible in empty space, it follows that all other observers will also find it impossible to produce a pair in empty space.

Supplementary Problems

13.12. Determine a photon's threshold energy for pair production.   *Ans.*   1.022 MeV

13.13. A 0.0005 Å photon produces an electron-positron pair in the vicinity of a heavy nucleus. If they have the same kinetic energies, find the energy of each particle.   *Ans.*   11.9 MeV

13.14. For Problem 13.13, if the positron's kinetic energy is five times the electron's kinetic energy, find the energy of each particle.   *Ans.*   19.8 MeV; 3.96 MeV

13.15. After pair annihilation two 1 MeV photons are observed moving in opposite directions.   If the electron and positron both had the same kinetic energy, find its value.   *Ans.*   0.49 MeV

13.16. Find the threshold wavelength for proton-antiproton production. The rest mass of a proton (or antiproton) is 938 MeV.   *Ans.* 6.61×10^{-6} Å

13.17. An electron, velocity $0.8c$, annihilates with a positron at rest, producing two photons.   One photon is observed to travel in the direction of the incident electron.   Calculate the energy of each photon.   *Ans.*   1.02 MeV; 0.34 MeV

13.18. If in Problem 13.17 the photon observed is found to move perpendicular to the incident electron, find the energy of each photon.   *Ans.*   0.51 MeV; 0.85 MeV

13.19. Pair production occurs in a magnetic field of 0.05 T and both the electron and positron are observed to have a radius of curvature of 90 mm.   Find the energy of the incident photon.   *Ans.*   2.88 MeV

Chapter 14

Absorption of Photons

The intensity of a beam of radiation will be reduced as it passes through material because photons will be removed or scattered from the forward direction by some combination of the photoelectric effect, the Compton effect, and pair production. The reduction in intensity obeys the exponential attenuation law

$$I = I_0 e^{-\mu x} \qquad (14.1)$$

Here I_0 is the intensity of the radiation incident on the absorber and μ (the *linear absorption coefficient*) is, for a given photon energy, a constant that depends on the particular absorbing material. For any given material μ will vary with the energy (or wavelength) of the radiation because different interactions predominate at different energies.

Solved Problems

14.1. What percentage of incident X-ray radiation passes through 5.0 mm of material whose linear absorption coefficient is 0.07 mm^{-1}?

$$\frac{I}{I_0} = e^{-\mu x} = e^{-(0.07 \text{ mm}^{-1})(5.0 \text{ mm})} = 0.705 = 70.5\%$$

14.2. A monochromatic beam of photons is incident on an absorbing material. If the incident intensity is reduced by a factor of two by 8 mm of material, what is the absorption coefficient?

$$\frac{I_0}{2} = I_0 e^{-\mu(8 \text{ mm})}$$

Solving, $\mu = 0.0866$ mm^{-1}.

14.3. Find the *half-value thickness* of aluminium if $\mu = 0.070$ mm^{-1}.

The half-value thickness is that thickness which reduces the intensity of a photon beam to half its incident value. Thus

$$\frac{1}{2} = e^{-(0.070 \text{ mm}^{-1})x} \quad \text{or} \quad x = 9.9 \text{ mm}$$

14.4. What is the ratio of the intensity of a photon beam to its original intensity after it passes through material whose thickness is equal to two half-thicknesses?

Through each half-thickness the intensity is reduced to one-half its original value. So through two half-thicknesses the incident intensity (I_0) is reduced to one-quarter its initial value ($I_0/4$).

14.5. A beam passes normal to a 20 mm sheet and is attenuated to half its original intensity. The sheet is now rotated through an angle of 40°. Find the intensity of the beam as it now emerges from the sheet.

The linear absorption coefficient is obtained from

$$\frac{1}{2} = e^{-\mu(20 \text{ mm})} \quad \text{or} \quad \mu = 0.0347 \text{ mm}^{-1}$$

When the sheet is rotated through 40°, the new thickness is

$$x_2 = \frac{x_1}{\cos 40°} = \frac{20 \text{ mm}}{\cos 40°} = 26.1 \text{ mm}$$

The new intensity is then found from

$$\frac{I_2}{I_0} = e^{-\mu x_2} = e^{-(0.0347 \text{ mm}^{-1})(26.1 \text{ mm})} = 0.404$$

14.6. What thickness of aluminum ($\mu_a = 0.044$ mm^{-1}) is equivalent to 6.0 mm of lead ($\mu_l = 5.8$ mm^{-1})?

 An equivalent thickness of aluminum will reduce the incident radiation by the same amount that it is reduced in passing through 6.0 mm of lead.

$$\frac{I}{I_0} = e^{-\mu_l x_l} = e^{-\mu_a x_a}$$

from which $\mu_l x_l = \mu_a x_a$. Thus

$$x_a = \frac{\mu_l}{\mu_a} x_l = \frac{5.8 \text{ mm}^{-1}}{0.044 \text{ mm}^{-1}} (6.0 \text{ mm}) = 791 \text{ mm}$$

14.7. Material A has an absorption coefficient of 0.044 mm^{-1} and material B has an absorption coefficient of 0.056 mm^{-1}. If the incident intensity is I_0 and the final intensity is to be $I_0/5$, calculate the thicknesses of A and B, if A is to be twice as thick as B and the beam passes through both materials.

 If the thickness of A is $2x$, the intensity incident on B is $I_0 e^{-\mu_a(2x)}$. Therefore, applying the exponential law to B,

$$I = \frac{I_0}{5} = \left[I_0 e^{-\mu_a(2x)} \right] e^{-\mu_b x} = I_0 e^{-(2\mu_a + \mu_b)x}$$

or

$$5 = e^{(0.144 \text{ mm}^{-1})x}$$

Solving, $x = 11.18$ mm, $2x = 22.36$ mm.

14.8. Derive the formula $I = I_0 e^{-\mu x}$.

 For a given photon energy, the photon flux is reduced in a material because of the photoelectric effect, pair production and Compton scattering. The number of reactions, dN, in the thickness dx is directly proportional to the magnitude of the photon flux, N, and the number of atoms encountered as the photons pass through the small thickness of material. In turn, the number of atoms in dx is proportional to dx. Therefore,

$$-dN = \mu N \, dx$$

Integrating this we get

$$\int_{N_0}^{N} \frac{dN}{N} = -\mu \int_0^x dx$$

$$N = N_0 e^{-\mu x}$$

But the intensity, I, of a monochromatic beam is proportional to N. Hence,

$$I = I_0 e^{-\mu x}$$

Supplementary Problems

14.9. The absorption coefficient for a material is 0.061 mm^{-1}. If the incident intensity is I_0, calculate the thickness of material needed to reduce the beam to $I_0/3$.   *Ans.*   18 mm

14.10. The linear absorption coefficient for a material is 0.055 mm^{-1}. What percentage of a monochromatic beam will pass through 10 mm of the material?   *Ans.*   57.7%

14.11. Through 8.5 mm of material a monochromatic beam is reduced by a factor of three. Find the linear absorption coefficient.   *Ans.*   0.129 mm^{-1}

14.12. For a material $\mu = 0.035$ mm^{-1}. Find the half-value thickness.   *Ans.*   19.8 mm

14.13. What thickness of material A ($\mu_a = 0.060$ mm^{-1}) is equivalent to 8 mm of material B ($\mu_b = 0.131$ mm^{-1})?   *Ans.*   17.5 mm

14.14. The materials of Problem 14.13 are to be of equal thickness and together are to reduce an incident monochromatic beam by a factor of 5. Find their thicknesses.   *Ans.*   8.4 mm

14.15. Radiation of equal intensities of 0.3 Å X-rays ($\mu_a = 0.3$ mm^{-1}) and 0.5 Å X-rays ($\mu_b = 0.72$ mm^{-1}) are incident on a material. Find the thickness of the material if, in the emerging radiation, the 0.3 Å X-rays are twice as intense as the 0.5 Å X-rays.   *Ans.*   1.7 mm

Chapter 15

De Broglie Waves

15.1 THE WAVE-PARTICLE DUALITY OF ELECTROMAGNETIC RADIATION

In Chapters 10 through 13 it was shown that particle characteristics had to be assigned to electromagnetic radiation in order to explain certain experimental observations (photoelectric effect, Compton scattering). It is known from interference and diffraction experiments, however, that electromagnetic radiation also behaves like a wave. Hence electromagnetic radiation exhibits a *wave-particle duality*: in certain circumstances it behaves like a wave, while in other situations it acts like a particle.

It is essential that one clearly understand the distinction between waves and particles, since these are the only two modes of energy transmission. A classical particle is something that has position, momentum, kinetic energy, mass, and electric charge. A classical wave, on the other hand, has the attributes of wavelength, frequency, velocity, amplitude of the disturbance, intensity, energy, and momentum. The most distinctive difference between the two is that a particle can be localized, whereas a wave is spread out and occupies a relatively large portion of space.

15.2 THE WAVE-PARTICLE DUALITY OF MATTER

In 1924, Louis de Broglie proposed that if electromagnetic radiation could behave sometimes like a wave and other times like a particle, then perhaps material objects, like electrons, may at certain times act like waves. In other words, de Broglie proposed that if material objects pass through a slit whose width is comparable to a wavelength associated with them, they will undergo diffraction just as photons do in a single-slit experiment.

For a photon, $v = E/h$ and $\lambda = h/p$. It is seen that the left-hand sides of these equations involve the wave aspects of photons (frequency, wavelength), while on the right-hand sides the particle aspects (energy, momentum) appear. The bridge between the two sides is Planck's constant. Arguing from the symmetry of nature, de Broglie conjectured that wavelengths associated with material bodies would satisfy the same relations that held for photons. He therefore postulated that a material body will have a wavelength given by

$$\lambda = \frac{h}{p} = \frac{h}{mv}$$

There is one important difference between photons and massive objects in the way their wave and particle properties are related. Because $\lambda v = c$ for a photon, only one rule is required to get both wavelength and frequency from a photon's particle properties of energy and momentum. A massive object, on the other hand, requires separate rules for its wavelength ($\lambda = h/p$) and frequency ($v = E/h$).

Solved Problems

15.1. Find the de Broglie wavelength of a 0.01 kg pellet having a velocity of 10 m/s.

$$\lambda = \frac{h}{mv} = \frac{6.63 \times 10^{-34} \text{ J} \cdot \text{s}}{0.01 \text{ kg} \times 10 \text{ m/s}} = 6.63 \times 10^{-33} \text{ m} = 6.63 \times 10^{-23} \text{ Å}$$

In order to observe de Broglie waves, interference or diffraction experiments must be performed using apertures comparable to the de Broglie wavelength. The above de Broglie wavelength of 10^{-23} Å is orders of magnitude smaller than any existing aperture.

15.2. Determine the accelerating potential necessary to give an electron a de Broglie wavelength of 1 Å, which is the size of the interatomic spacing of atoms in a crystal.

From conservation of energy (nonrelativistic calculation) we have

$$eV = \frac{1}{2} m_0 v^2 = \frac{p^2}{2m_0} = \frac{1}{2m_0} \left(\frac{h}{\lambda} \right)^2$$

$$V = \frac{h^2}{2m_0 e \lambda^2} = \frac{(6.63 \times 10^{-34} \text{ J} \cdot \text{s})^2}{2(9.11 \times 10^{-31} \text{ kg})(1.6 \times 10^{-19} \text{ C})(1 \times 10^{-10} \text{ m})^2} = 151 \text{ V}$$

Note that a kinetic energy of 151 eV is small compared to the rest energy of 0.511 MeV, and this justifies the nonrelativistic calculation.

Accelerating potentials of the order of 150 volts are readily available in the laboratory. Therefore, unlike the macroscopic case of Problem 15.1, conditions are possible for observing de Broglie waves of electrons.

15.3. Calculate the de Broglie wavelength of a 0.05 eV ("thermal") neutron.

Making a nonrelativistic calculation,

$$\lambda = \frac{h}{p} = \frac{h}{\sqrt{2m_0 K}} = \frac{hc}{\sqrt{2(m_0 c^2) K}} = \frac{12.4 \times 10^3 \text{ eV} \cdot \text{Å}}{\sqrt{2(940 \times 10^6 \text{ eV})(0.05 \text{ eV})}} = 1.28 \text{ Å}$$

This convenient wavelength of the order of 1 Å is handy in slow-neutron physics.

15.4. Calculate the energy of a proton of wavelength 0.5 fm (1 fm = 10^{-15} m = 10^{-5} Å = 1 fermi).

From $\lambda = h/p = hc/pc$,

$$0.5 \text{ fm} = \frac{1240 \text{ MeV} \cdot \text{fm}}{pc} \qquad \text{or} \qquad pc = 2480 \text{ MeV}$$

Then, from the relativistic energy-momentum relation,

$$E^2 = (pc)^2 + E_0^2 = (2480 \text{ MeV})^2 + (938 \text{ MeV})^2$$

yielding $E = 2650$ MeV and

$$K = E - E_0 = 2650 \text{ MeV} - 938 \text{ MeV} = 1712 \text{ MeV}$$

In this case, $K \approx E_0$ and a relativistic calculation was indeed necessary.

15.5. If we wish to observe an object which is 2.5 Å in size, what is the minimum-energy photon which can be used?

In order for scattering to occur, the wavelength of the waves must be of the same order of magnitude or smaller than the size of the object being observed (imagine a pea scattering water waves).

Hence the largest possible wavelength we can use in the present problem is $\lambda_{max} = 2.5$ Å. The corresponding minimum energy is then

$$E_{min} = h\nu_{min} = \frac{hc}{\lambda_{max}} = \frac{12.40 \times 10^3 \text{ eV} \cdot \text{Å}}{2.5 \text{ Å}} = 4.96 \times 10^3 \text{ eV}$$

15.6. Rework Problem 15.5 for electrons instead of photons.

As in Problem 15.5, the maximum electron wavelength is $\lambda_{max} = 2.5$ Å. The kinetic energy and momentum are related nonrelativistically by $p = \sqrt{2m_0 K}$. Hence,

$$\lambda = \frac{h}{p} = \frac{h}{\sqrt{2m_0 K}}$$

and

$$K_{min} = \frac{h^2}{2m_0\lambda_{max}^2} = \frac{(hc)^2}{2(m_0 c^2)\lambda_{max}^2} = \frac{(12.4 \times 10^3 \text{ eV} \cdot \text{Å})^2}{2(0.511 \times 10^6 \text{ eV})(2.5 \text{ Å})^2} = 24.1 \text{ eV}$$

Comparison with Problem 15.5 shows that, for a given energy, electrons will have a much higher resolving power than photons. This is why electron microscopes can achieve magnifications much greater than those of optical microscopes.

15.7. At what energy will the nonrelativistic calculation of the de Broglie wavelength of an electron be in error by 5%?

For the nonrelativistic case, the de Broglie wavelength is

$$\lambda_{nr} = \frac{hc}{pc} = \frac{hc}{\sqrt{2m_0 c^2 K}}$$

For the relativistic case,

$$(K + m_0 c^2)^2 = (pc)^2 + (m_0 c^2)^2 \qquad \text{or} \qquad pc = \left[2m_0 c^2 K\left(1 + \frac{K}{2m_0 c^2}\right)\right]^{1/2}$$

and the de Broglie wavelength is

$$\lambda_r = \frac{hc}{pc} = \frac{hc}{\left[2m_0 c^2 K\left(1 + \frac{K}{2m_0 c^2}\right)\right]^{1/2}}$$

For our case, $\lambda_{nr} - \lambda_r = 0.05\lambda_r$; $\lambda_{nr}/\lambda_r = 1.05$.

$$\frac{\lambda_{nr}}{\lambda_r} = \sqrt{1 + \frac{K}{2m_0 c^2}}$$

$$1.05 = \sqrt{1 + \frac{K}{2(0.511 \text{ MeV})}}$$

Solving, $K = 0.105$ MeV.

15.8. Show that the de Broglie wavelength of a particle is approximately the same as that of a photon with the same energy, when the energy of the particle is much greater than its rest energy.

$$E^2 = p^2 c^2 + E_0^2 \qquad \text{or} \qquad p = \frac{E}{c}\sqrt{1 - \left(\frac{E_0}{E}\right)^2} \approx \frac{E}{c}$$

if $E \gg E_0$.   So

$$\lambda = \frac{h}{p} \approx \frac{hc}{E}$$

For a photon, $E = h\nu = hc/\lambda_\gamma$, so

$$\lambda_\gamma = \frac{hc}{E} \approx \lambda$$

15.9.   Determine the phase velocity of the wave corresponding to a de Broglie wavelength of $\lambda = h/p = h/mv$.

The de Broglie frequency is found from

$$E = mc^2 = h\nu \qquad \text{or} \qquad \nu = mc^2/h$$

The phase velocity, u_p, is found from

$$u_p = \nu\lambda = \left(\frac{mc^2}{h} \right)\left(\frac{h}{mv} \right) = \frac{c^2}{v}$$

Note that since $v < c$, $u_p > c$.

15.10.   Determine the group velocity of the wave corresponding to a de Broglie wavelength of $\lambda = h/p$.

The group velocity, u_g, is given by $u_g = d\nu/d(\lambda^{-1})$.   Using the expression for ν found in Problem 15.9, we have

$$u_g = \frac{d(mc^2/h)}{d(p/h)} = \frac{c^2\, dm}{dp}$$

By differentiation of $m^2c^4 = p^2c^2 + m_0^2c^4$, we obtain $c^2 m\, dm = p\, dp$.   Therefore,

$$u_g = \frac{(p/m)dp}{dp} = \frac{p}{m} = v$$

In the theoretical structure of quantum mechanics, a particle is described by associating with it a wave packet formed from the superposition of an infinite number of plane waves.   Each plane wave moves with a phase velocity which may exceed the velocity of light, as shown in Problem 15.9.   The individual phase velocities, however, are not observable.   The quantity that is observable is the velocity of the localized disturbance, or group velocity, which, as just shown, is equal to the velocity one normally associates with a particle and is less than the speed of light.

Supplementary Problems

15.11.   Calculate the de Broglie wavelength of a 2 kg mass whose velocity is 25 m/s.   *Ans.* 1.33×10^{-25} Å

15.12.   Calculate the de Broglie wavelength of a 0.08 eV neutron.   *Ans.*   1.01 Å

15.13.   Calculate the kinetic energy of a neutron whose de Broglie wavelength is 0.7 Å.   *Ans.*   0.167 eV

15.14.   What is the minimum-energy electron needed to observe a 5 Å object?   *Ans.*   6.02 eV

15.15.   For the object of Problem 15.14, what is the minimum-energy proton that can be used?
Ans. 3.28×10^{-3} eV

15.16. A proton is accelerated from rest through a potential of 1 kV.   Find its de Broglie wavelength.
Ans. 9.05×10^{-3} Å

15.17. Find the de Broglie wavelength of a 1 keV α-particle ($m_0 = 3728$ MeV).   *Ans.* 4.54×10^{-3} Å

15.18. At what kinetic energy will the nonrelativistic calculation of the de Broglie wavelength of a proton be in error by 5%?   *Ans.*   192 MeV

15.19. What is the ratio of a particle's Compton and de Broglie wavelengths?   *Ans.* $\dfrac{\lambda_c}{\lambda_d} = \sqrt{\left(\dfrac{E}{E_0}\right)^2 - 1}$

Chapter 16

Experimental Verification of De Broglie's Hypothesis

16.1 THE BRAGG LAW OF DIFFRACTION

Max von Laue, in 1912, suggested that because of their regular arrangement of atoms, crystals might be used as diffraction gratings for X-rays. X-rays are electromagnetic radiation of about 1 Å in wavelength, the same order of size as the interatomic spacing in a typical crystal.

The theory of X-ray diffraction was developed by Sir William H. Bragg in 1913. Bragg showed that a plane of atoms in a crystal, called a *Bragg plane*, would reflect radiation in exactly the same manner that light is reflected from a plane mirror, as shown in Fig. 16-1.

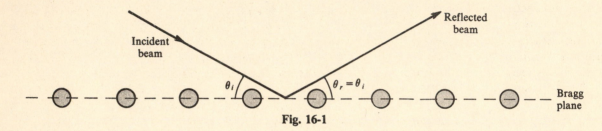

Fig. 16-1

If one considers radiation that is reflected from successive parallel Bragg planes spaced a distance d apart, it is seen from Fig. 16-2 that it is possibie for the beams reflected from each plane to interfere constructively to produce an enhanced overall reflected beam. The condition for constructive interference is that the path difference between the two rays, $2d \sin \theta$, be equal to an integral number of wavelengths, thereby giving *Bragg's law* as

$$n\lambda = 2d \sin \theta$$

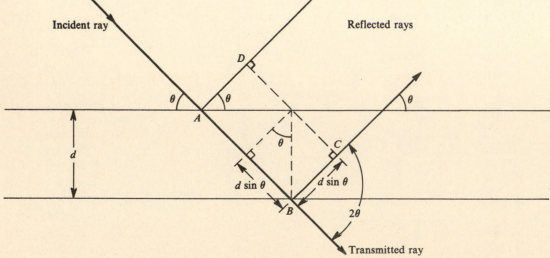

Fig. 16-2

If n and d are known, the wavelength of the incident beam can be determined by measuring the scattering angle 2θ between the transmitted and diffracted beams.

In any crystal many different families of Bragg planes, each with its own spacing, can be formed by taking slices through the crystal in various manners. Each of these families can give rise to diffraction. Hence, if an X-ray beam is passed through crystals that are randomly oriented, as in a powder sample or in a thin foil, a diffraction pattern of concentric circles will be observed on a film placed behind the sample. A given circle will correspond to diffraction of a particular order by a particular family of planes. In the Solved Problems that follow, we shall, except in Problem 16.8, consider only diffraction from the *principal* Bragg planes, whose spacing is the interatomic spacing.

16.2 ELECTRON DIFFRACTION EXPERIMENTS

The first experiments to observe electron diffraction were performed by C. J. Davisson and L. H. Germer at Bell Telephone Laboratories. They directed a beam of 54 eV electrons at a single crystal of nickel, whose interatomic spacing was known from X-ray diffraction measurements to be 2.15 Å, and measured the intensity of the scattered electrons as a function of the scattering angle. If there were no diffraction effects one would expect that the intensity of the scattered electrons would decrease monotonically with the scattering angle, with no large fraction of the electrons coming out at any single angle. Instead, it was found that there was a pronounced peak in the electron intensity at a scattering angle of 50°. With a small correction (see Problems 16.8 and 16.9), the computed wavelength agreed with the de Broglie wavelength, thereby verifying de Broglie's hypothesis.

Shortly after the experiments of Davisson and Germer, G. P. Thomson, in 1927, studied the transmission of electrons through thin metal foils. If the electrons behaved like particles, a blurred image would have resulted in the transmitted beam. Instead, Thomson found circular diffraction patterns, which can be explained only in terms of a wave picture, further confirming de Broglie's hypothesis.

Subsequently, thermal (low-energy) neutron diffraction experiments were performed that further upheld the de Broglie hypothesis.

Solved Problems

16.1. A 0.083 eV neutron beam scatters from an unknown sample and a Bragg reflection peak is observed centered at 22°. What is the Bragg plane spacing?

The wavelength of the neutron beam is found from

$$\lambda = \frac{h}{p} = \frac{h}{\sqrt{2m_0 K}} = \frac{hc}{\sqrt{2(m_0 c^2)K}} = \frac{12.40 \times 10^3 \text{ eV} \cdot \text{Å}}{\sqrt{2(940 \times 10^6 \text{ eV})(0.083 \text{ eV})}} = 0.993 \text{ Å}$$

Assuming the peak corresponds to first-order diffraction ($n = 1$), we have

$$d = \frac{\lambda}{2 \sin \theta} = \frac{0.993 \text{ Å}}{2 \sin 22°} = 1.33 \text{ Å}$$

16.2. Thermal neutrons incident on a sodium chloride crystal (interatomic spacing 2.81 Å) undergo first-order diffraction from the principal Bragg planes at an angle of 20°. What is the energy of the thermal neutrons?

For a first-order Bragg reflection,

$$\lambda = 2d \sin \theta = 2(2.81 \text{ Å}) \sin 20° = 1.922 \text{ Å}$$

From the de Broglie relationship, $\lambda = h/p = hc/\sqrt{2(m_0 c^2)K}$, so

$$1.922 \text{ Å} = \frac{12.40 \times 10^3 \text{ eV} \cdot \text{Å}}{\sqrt{2(940 \times 10^6 \text{ eV})K}} \qquad \text{or} \qquad K = 0.0221 \text{ eV}$$

16.3. A narrow beam of 60 keV electrons passes through a thin silver polycrystalline foil. The interatomic spacing of silver crystals is 4.08 Å. Calculate the radius of the first-order diffraction pattern from the principal Bragg planes on a screen placed 40 cm behind the foil.

The de Broglie wavelength for the electron beam is

$$\lambda = \frac{hc}{pc} = \frac{hc}{\sqrt{E^2 - E_0^2}} = \frac{hc}{\sqrt{(K + E_0)^2 - E_0^2}} = \frac{12.4 \times 10^3 \text{ eV} \cdot \text{Å}}{\sqrt{(60 \times 10^3 \text{ eV} + 511 \times 10^3 \text{ eV})^2 - (511 \times 10^3 \text{ eV})^2}}$$

$$= 0.0487 \text{ Å}$$

For first-order Bragg reflections,

$$\sin \theta = \frac{\lambda}{2d} = \frac{0.0487 \text{ Å}}{2(4.08 \text{ Å})}$$

from which $\theta = 0.342°$. From Fig. 16-3, the radius of the first-order diffraction pattern is given by

$$R = D \tan 2\theta = (40 \text{ cm}) \tan 0.684° = 0.478 \text{ cm}$$

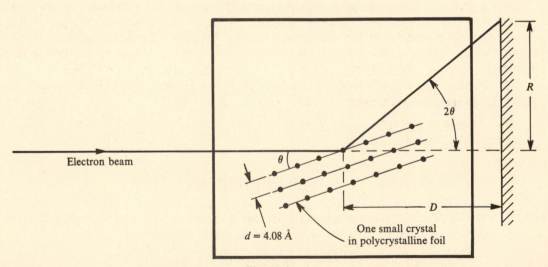

Fig. 16-3

16.4. A crystalline material has a set of Bragg planes separated by 1.1 Å. For 2 eV neutrons, what is the highest-order Bragg reflection?

The wavelength of the neutrons is

$$\lambda = \frac{h}{m_0 v} = \frac{h}{\sqrt{2m_0 K}} = \frac{hc}{\sqrt{2(m_0 c^2)K}} = \frac{12.40 \times 10^3 \text{ eV} \cdot \text{Å}}{\sqrt{2(940 \times 10^6 \text{ eV})(2 \text{ eV})}} = 0.202 \text{ Å}$$

The maximum angle that can be attained is 90°. Then, from Bragg's law,

$$2(1.1 \text{ Å}) \sin 90° = n(0.202 \text{ Å}) \qquad \text{or} \qquad n = 10.89$$

Since n must be an integer, the highest order is $n = 10$.

16.5.   A large crystal is used to extract single-energy neutrons from a beam of neutrons emerging from a reactor.   The spacing of the Bragg planes is 1.1 Å.   If the Bragg angle is set to be 30°, what is the energy of neutrons seen at this angle for a first-order reflection?

$$\lambda = 2d \sin \theta = 2(1.1 \text{ Å})\sin 30° = 1.1 \text{ Å}$$

The wavelength of the neutrons is related to their kinetic energy by

$$\lambda = \frac{h}{m_0 v} = \frac{hc}{\sqrt{2(m_0 c^2)K}} \quad \text{or} \quad K = \frac{(hc)^2}{2(m_0 c^2)\lambda^2} = \frac{(12.40 \times 10^3 \text{ eV} \cdot \text{Å})^2}{2(940 \times 10^6 \text{ eV})(1.1 \text{ Å})^2} = 0.0676 \text{ eV}$$

16.6.   If the crystal of Problem 16.5 is not perfect, so that its Bragg plane spacing varies by ± 0.01 Å, calculate the spread in energy in the diffracted beam.

The differential of $n\lambda = 2d \sin \theta$ is $n \, \delta\lambda = 2 \, \delta d \sin \theta$, so that

$$\delta\lambda = \frac{2 \, \delta d \sin \theta}{n} = \frac{2(\pm 0.01 \text{ Å})\sin 30°}{1} = \pm 0.01 \text{ Å}$$

As in Problem 16.5, $K \propto \lambda^{-2}$, so that

$$\delta K = \frac{-2K}{\lambda} \delta\lambda = \frac{-2(0.0676 \text{ eV})}{1.1 \text{ Å}} (\pm 0.01 \text{ Å}) = \mp 1.23 \times 10^{-3} \text{ eV}$$

16.7.   Determine the interatomic spacing of a NaCl crystal if the density of NaCl is 2.16×10^3 kg/m^3 and the atomic weights of sodium and chlorine are 23.00 and 35.46, respectively.

The molecular weight of NaCl is $23.00 + 35.46 = 58.46$.   The number of molecules per 58.46 kg of NaCl is

$$\frac{1 \text{ kmol}}{58.46 \text{ kg}} \times 6.025 \times 10^{26} \frac{\text{molecules}}{\text{kmol}} = \frac{6.025 \times 10^{26} \text{ molecules}}{58.46 \text{ kg}}$$

Since there are two atoms per molecule, we have

$$\frac{\text{number of atoms}}{\text{volume}} = \frac{\text{number of atoms}}{\text{mass}} \times \frac{\text{mass}}{\text{volume}} = \frac{2 \times 6.025 \times 10^{26} \text{ atoms}}{58.46 \text{ kg}} \times 2.16 \times 10^3 \frac{\text{kg}}{\text{m}^3}$$

$$= 4.45 \times 10^{28} \frac{\text{atoms}}{\text{m}^3}$$

To relate this to the interatomic spacing d, consider the NaCl unit cell shown in Fig. 16-4 (differences between the Na$^+$ and Cl$^-$ ions are ignored).   The volume of the cube is $(2d)^3$.   As to the number of ions to be assigned to the cube, there are: 8 corner ions, each shared by 8 of the cubes; 12 edge ions, each shared by 4 of the cubes; 6 face ions, each shared by 2 of the cubes; and 1 unshared center ion.   Thus

$$\text{number of ions} = 8\left(\frac{1}{8}\right) + 12\left(\frac{1}{4}\right) + 6\left(\frac{1}{2}\right) + 1 = 8$$

and

$$\frac{\text{number of ions}}{\text{volume}} = \frac{8}{(2d)^3} = \frac{1}{d^3}$$

Equating this to the above result, we have

$$\frac{1}{d^3} = 4.45 \times 10^{28} \text{ m}^{-3}$$

or

$$d = 2.82 \times 10^{-10} \text{ m} = 2.82 \text{ Å}$$

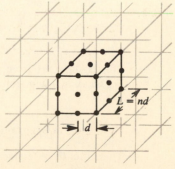

Fig. 16-4

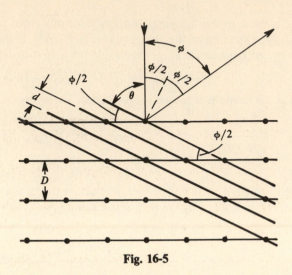

Fig. 16-5

16.8. In one of their experiments Davisson and Germer used electrons incident normally on a nickel crystal surface cut parallel to the principal Bragg planes. They observed constructive interference at an angle of 50.0° to the normal to the surface. Find the wavelength associated with the electron beam. (The interatomic spacing of nickel is 2.15 Å.)

We first find the relation between the scattering angle ϕ to the normal and the interatomic spacing D. From Fig. 16-5 it is seen that $\theta + \phi/2 = 90°$, so that

$$\sin \theta = \cos \frac{\phi}{2}$$

Also from the figure we see the spacing d between the Bragg planes is

$$d = D \sin \frac{\phi}{2}$$

Substituting these results in the Bragg relation, $2d \sin \theta = n\lambda$, and using the half-angle formula

$$2 \sin \frac{\phi}{2} \cos \frac{\phi}{2} = \sin \phi$$

we obtain

$$D \sin \phi = n\lambda$$

Taking $n = 1$, we then have

$$(2.15 \text{ Å}) \sin 50.0° = (1)\lambda \qquad \text{or} \qquad \lambda = 1.65 \text{ Å}$$

16.9. In the experiment described in Problem 16.8 Davisson and Germer used 54.0 eV electrons. Determine the effective accelerating potential of the nickel crystal.

The de Broglie wavelength of 54 eV electrons is

$$\lambda_d = \frac{h}{m_0 v} = \frac{h}{\sqrt{2m_0 K}} = \frac{hc}{\sqrt{2(m_0 c^2)K}} = \frac{12.4 \times 10^3 \text{ eV} \cdot \text{Å}}{\sqrt{2(0.511 \times 10^6 \text{ eV})(54 \text{ eV})}} = 1.67 \text{ Å}$$

This is different from the observed wavelength of 1.65 Å. The kinetic energy corresponding to λ = 1.65 Å is found from

$$\lambda = \frac{hc}{\sqrt{2(m_0 c^2)K'}} \qquad \text{or} \qquad K' = \frac{(hc)^2}{2(m_0 c^2)\lambda^2} = \frac{(12.4 \times 10^3 \text{ eV} \cdot \text{Å})^2}{2(0.511 \times 10^6 \text{ eV})(1.65 \text{ Å})^2} = 55.3 \text{ eV}$$

Therefore, the effective accelerating potential of the nickel crystal is

$$V_e = 55.3 \text{ V} - 54.0 \text{ V} = 1.3 \text{ V}$$

Supplementary Problems

16.10. The spacing between the nuclei in a certain crystal is 1.2 Å. At what angle will first-order Bragg reflection occur for neutrons with kinetic energy of 0.020 eV?   *Ans.*   57.4°

16.11. A 0.1 eV neutron beam scatters from an unknown sample. If a first-order Bragg reflection is found at 28°, what is the Bragg plane spacing?   *Ans.*   0.963 Å

16.12. Thermal neutrons are incident on a crystal whose interatomic spacing is 1.8 Å. If a first-order Bragg reflection from the principal Bragg planes is found at 22°, what is the kinetic energy of the thermal neutrons?   *Ans.* 4.50×10^{-2} eV

16.13. For the crystal of Problem 16.2, what would be the energy of thermal neutrons observed at 30° if this were a second-order Bragg reflection?   *Ans.* 4.14×10^{-2} eV

16.14. Refer to Problem 16.3. Determine the radius of the second-order diffraction pattern from the principal Bragg planes.   *Ans.*   9.6 mm

16.15. A beam of neutrons with kinetic energy 0.020 eV is incident on KCl powder. The lattice spacing of KCl is 3.14 Å. What is the radius of the circle on a flat photographic plate placed 5 cm behind the target from first-order reflections from the Bragg planes that are 3.14 Å apart?   *Ans.*   3.85 cm

16.16. Refer to Problem 16.15. What is the radius of the circle due to second-order reflections from the same Bragg planes?   *Ans.*   28.9 cm

Chapter 17

The Probability Interpretation of De Broglie Waves

We here address ourselves to the question of what it is that does the waving when a massive object such as an electron exhibits wave properties. The probabilistic interpretation that we shall be led to is perhaps disturbing at first; in fact, it is still being debated today. However, with such an interpretation, many otherwise unexplainable experimental results can be resolved.

17.1 A PROBABILITY INTERPRETATION FOR ELECTROMAGNETIC RADIATION

Consider the interference pattern obtained in a double-slit experiment. According to the wave picture, the intensity I (energy per unit area per unit time) at a point on the screen is given by

$$I = \epsilon_0 c \mathcal{E}^2$$

where $\mathcal{E}$ is the value of the electric field at the particular point, ϵ_0 is the permittivity of free space, and c is the velocity of light. In terms of the photon picture, on the other hand, the intensity at a point on the screen is given by

$$I = h\nu N$$

where $h\nu$ is the energy per photon and N is the photon flux (the number of photons per unit area per unit time) striking the particular point on the screen.

There is no way of predicting in advance where any individual photon will strike the screen, producing a single flash. However, since the final pattern consists of alternating bright and dark bands, any single photon has a very high *probability* of arriving at a bright band and zero *probability* of arriving at a dark band. The photon flux N at a point on the screen is therefore a measure of the *probability* of finding a photon near that point.

Since $I = \epsilon_0 c \mathcal{E}^2 = h\nu N$, it follows that $N \propto \mathcal{E}^2$. Hence, in terms of the quantum interpretation of electromagnetic radiation, the quantity that is undergoing the oscillations, namely the electric field $\mathcal{E}$, is that function whose square gives the probability of finding a photon at a given place.

17.2 A PROBABILITY INTERPRETATION OF MATTER

The interference pattern discussed above could have been produced with matter waves instead of light waves. For this case the probability interpretation based upon the wave-particle duality of light is carried over directly to explain the wave-particle duality of matter. Thus, with electron waves, the quantity oscillating with the de Broglie wavelength $\lambda = h/mv$ is that *wave function* whose square gives the probability of finding an electron at a given place. In order to reconcile the wave and particle pictures of matter, we must give up the idea that the location of a single material particle can be specified exactly. Instead, we can talk only of the probability of finding a particle at a particular location at a particular time, as is illustrated in Problem 17.2.

The wave function is commonly denoted ψ. For a photon, the de Broglie wave represented by ψ is an electromagnetic wave; but for an electron or other material body, ψ is a nonelectromagnetic de Broglie wave.

Solved Problems

17.1. Determine the photon flux associated with a beam of monochromatic light of wavelength 3000 Å and intensity 3×10^{-14} W/m².

$$E = h\nu = \frac{hc}{\lambda} = \frac{(6.63 \times 10^{-34} \text{ J} \cdot \text{s})(3 \times 10^8 \text{ m/s})}{3 \times 10^{-7} \text{ m}} = 6.63 \times 10^{-19} \text{ J/photon}$$

$$N = \frac{I}{h\nu} = \frac{3 \times 10^{-14} \text{ J/s} \cdot \text{m}^2}{6.63 \times 10^{-19} \text{ J/photon}} = 4.5 \times 10^4 \frac{\text{photons}}{\text{s} \cdot \text{m}^2} = 4.5 \frac{\text{photons}}{\text{s} \cdot \text{cm}^2}$$

On the average, 4.5 photons will strike a 1 cm² area (of photographic film, say) during a period of 1 s. Of course, only integral numbers of photons can be observed. Thus, for a given 1 cm² area, we might observe 3 photons or 5 photons in a one-second interval, but never 4.5 photons. Only if an average is taken over many intervals will the average number approach 4.5 photons. Also, for a given one-second interval, the arriving photons may cluster within a fixed 1 cm² area. Only after a long period of time will the photon positions approach a uniform distribution.

17.2. Suppose $h = 6.625 \times 10^{-3}$ J·s instead of 6.625×10^{-34} J·s. Balls of mass 66.25 grams are thrown with a speed of 5 m/s into a house through two tall, narrow, parallel windows spaced 0.6 m apart, the choice of window as target being random at each toss. Determine the spacing between the fringes that would be formed on a wall 12 m behind the windows.

The de Broglie wavelength of the balls is

$$\lambda = \frac{h}{mv} = \frac{6.625 \times 10^{-3} \text{ J} \cdot \text{s}}{(6.625 \times 10^{-2} \text{ kg})(5 \text{ m/s})} = 0.02 \text{ m}$$

From interference theory, the angles θ_n to the lines of zero intensity in a double-slit interference pattern are given by

$$d \sin \theta_n = \frac{2n + 1}{2} \lambda \qquad n = 0, 1, 2, \ldots$$

The corresponding y-distance is given from Fig. 17-1 as

$$y_n = L \tan \theta_n \approx L \sin \theta_n = L \frac{2n + 1}{2} \frac{\lambda}{d}$$

The distance between adjacent fringes is then given by

$$\Delta y = y_{n+1} - y_n = \frac{L\lambda}{2d} \left\{ [2(n + 1) + 1] - (2n + 1) \right\}$$

$$= \frac{L\lambda}{d}$$

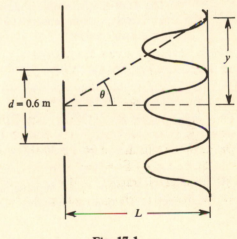

Fig. 17-1

Substituting the values for our problem, we have

$$\Delta y = \frac{(12 \text{ m})(0.02 \text{ m})}{0.6 \text{ m}} = 0.4 \text{ m}$$

This problem illustrates the probabilistic interpretation of de Broglie waves. Any single ball will strike the wall at a specific, although undeterminable, position. Although it cannot be predicted in advance where any ball will strike, each ball has a high probability of arriving at a maximum, and zero probability of arriving at a minimum, of the interference pattern.

The actual interference pattern is experimentally determined by counting the number of balls that strike each part of the wall. In the beginning of the experiment the balls will hit the wall in a more or less sporadic fashion. Only after a large number of balls have been thrown through the windows will the interference pattern become discernible, since the number of hits at the eventual maxima will increase, while the number of hits at the minima will remain zero.

17.3.   A particle of mass m is confined to a one-dimensional line of length L.   From arguments based on the wave interpretation of matter, show that the energy of the particle can have only discrete values and determine these values.

If the particle is confined to a line segment, say from $x = 0$ to $x = L$, the probability of finding the particle outside this region must be zero.   Therefore, the wave function ψ must be zero for $x \leqslant 0$ or $x \geqslant L$, since the square of ψ gives the probability for finding the particle at a certain location.   Inside the limited region the wavelength of ψ must be such that ψ vanishes at the boundaries $x = 0$ and $x = L$, so that it can vary continuously to the outside region.   Hence only those wavelengths will be possible for which an integral number of half wavelengths fit between $x = 0$ and $x = L$, i.e. $L = n\lambda/2$, where n is an integer, called the *quantum number*, with values $n = 1, 2, 3, \ldots$.   From the de Broglie relationship $\lambda = h/p$ we then find that the particle's momentum can have only discrete values given by

$$p = \frac{h}{\lambda} = \frac{nh}{2L}$$

Since the particle is not acted upon by any forces inside the region, its potential energy will be a constant which we set equal to zero.   Therefore the energy of the body is entirely kinetic and will have the discrete values obtained from

$$E = K = \frac{1}{2} mv^2 = \frac{p^2}{2m} = \frac{(nh/2L)^2}{2m}$$

i.e.

$$E_n = n^2 \frac{h^2}{8mL^2} \qquad n = 1, 2, 3, \ldots$$

This very simple problem illustrates one of the basic features of the probability interpretation of matter; namely, that the energy of a bound system can take on only discrete values, with zero energy not being a possible value.

Supplementary Problems

17.4.   Do Problem 17.1 for $\lambda = 4000$ Å and an intensity of 5×10^{-15} W/m².   *Ans.* 1×10^4 photons/s·m²

17.5.   Refer to Problem 17.3.   Suppose that the particle is an electron confined on a line of length $L = 5$ Å (which is of atomic dimensions).   Determine the lowest energy.   *Ans.*   1.5 eV

17.6.   Refer to Problem 17.3.   Suppose that the particle is a small but macroscopic body of mass 0.1 milligram confined to a length $L = 0.1$ mm.   Determine its lowest energy.   *Ans.* 3.43×10^{-34} eV

Chapter 18

The Heisenberg Uncertainty Principle

18.1 MEASUREMENTS AND UNCERTAINTIES

Suppose it is desired to determine the location of a material body such as an electron. In order to measure the body's position an experiment of some type must be performed. We can, for example, place a slit in the suspected path of a body moving parallel to the y-axis with known energy, as shown in Fig. 18-1. If a mark is made by the particle on a screen placed behind the slit, we will then know that the body passed through the slit. Thus, to within the width d of the slit, we will have determined the particle's x-location. Saying this another way, we will have measured the x-position of the particle, upon (and before) entering the slit, up to an *uncertainty* Δx given by $\Delta x = d$. The smaller we make the slit width, the smaller is the uncertainty in the x-position of the body, and hence the more precisely is its location known.

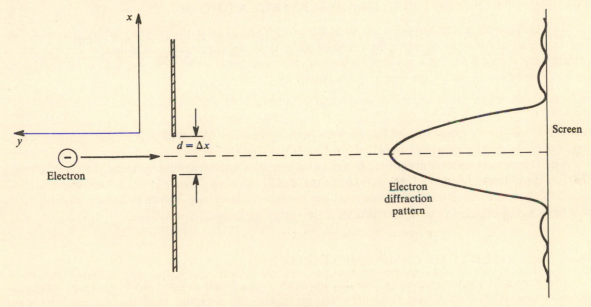

Fig. 18-1

Because of the wave nature of matter we know that the particle will be diffracted as it passes through the slit. However, even though we will not be able to predict where on the screen it will strike, as long as the body strikes the screen *somewhere* we will have ascertained that it has gone through the slit.

The diffraction process, however, has an effect on the momentum of the particle. Before the particle passed through the slit its position was completely unknown, but its momentum was known both in magnitude (since it had a fixed energy) and direction (perpendicular to the slit). When the particle passes through the slit, thereby determining its position, the x-component, p_x, of its momentum may no longer be zero, because the particle will be moving toward some arbitrary point on the diffraction pattern. Because it is not known just where the particle will strike the screen, there is a corresponding uncertainty Δp_x in the x-component of its momentum when at the slit.

An analysis (Problem 18.11) shows that the uncertainty Δp_x can be made as small as desired by *increasing* the slit width d. If the slit width is increased, however, the uncertainty in the particle's position will also increase!

It is therefore seen that with a single experiment the uncertainties in a particle's x-position and x-momentum cannot both be made arbitrarily small; accuracy in one of these quantities can be obtained only at the expense of accuracy in the other.

18.2   THE UNCERTAINTY RELATION FOR POSITION AND MOMENTUM

The above example illustrates the *Heisenberg uncertainty principle*, first set forth in 1927 by W. Heisenberg. A quantum mechanical analysis shows that for all types of experiments the uncertainties Δx and Δp_x will always be related by

$$\Delta p_x \, \Delta x \geqslant \frac{h}{4\pi}$$

It should be noted that this relationship holds both theoretically and experimentally.

18.3   THE UNCERTAINTY RELATION FOR ENERGY AND TIME

The Heisenberg uncertainty relation can also be formulated in terms of other conjugate variables. For example, in order to measure the energy E of a body an experiment must be performed over a certain time interval Δt. An analysis shows that the uncertainty in the energy, ΔE, is related to the time interval Δt over which the energy is measured by

$$\Delta E \, \Delta t \geqslant \frac{h}{4\pi}$$

Thus the energy of a body can be known with perfect precision ($\Delta E = 0$) only if the measurement is made over an infinite period of time ($\Delta t = \infty$).

The Heisenberg uncertainty principle has an important consequence for systems like excited atoms that, on the average, live for a finite period of time, called the *mean lifetime τ*. Since the mean lifetime limits the length of time one has to measure the energy of the system before it decays, these systems will have a natural minimum uncertainty in their energy given by $\Delta E = h/(4\pi\tau)$.

18.4   THE PRINCIPLE OF COMPLEMENTARITY

The uncertainty principle shows that it is impossible in a single experiment to measure conjugate variables (e.g. p_x and x, E and t) to arbitrary precision. As a result, both the particle and wave aspects of matter cannot be measured in the same experiment. Suppose, for example, that an experiment is designed to measure the particle properties of a body. Then necessarily in this experiment Δx and Δt must be zero, since a particle, by definition, can be located with infinite precision at any particular time. The momentum and energy, and hence the wave aspects ($\lambda = h/p$, $\nu = E/h$), will then, according to the uncertainty principle, be completely unknown. Thus, when the particle aspect of matter is displayed, the wave nature is necessarily suppressed. Likewise, if the wave aspects are measured exactly, so that $\Delta \lambda$ and $\Delta \nu$, and therefore Δp and ΔE, are zero, then the particle aspects will not be observed.

The inability to observe the wave and particle aspects of matter at the same time illustrates the *principle of complementarity*, enunciated in 1928 by N. Bohr. The wave and particle aspects of matter complement each other since both pictures are necessary to understand completely the properties of matter, but both aspects cannot simultaneously be observed.

Solved Problems

18.1. Suppose that the momentum of a certain particle can be measured to an accuracy of one part in a thousand. Determine the minimum uncertainty in the position of the particle if the particle is (a) a 5×10^{-3} kg mass moving with a speed of 2 m/s, (b) an electron moving with a speed of 1.8×10^8 m/s.

$$\frac{\Delta p}{p} = 10^{-3} \qquad \text{or} \qquad \Delta p = 10^{-3}p = 10^{-3}mv$$

Then, from $\Delta x \, \Delta p \geqslant h/4\pi$,

$$\Delta x \geqslant \frac{h}{4\pi \, \Delta p} = \frac{h}{4\pi 10^{-3}mv} \tag{1}$$

(a) $$\Delta x \geqslant \frac{6.63 \times 10^{-34} \text{ J} \cdot \text{s}}{4\pi 10^{-3}(5 \times 10^{-3} \text{ kg})(2 \text{ m/s})} = 5.28 \times 10^{-30} \text{ m} = 5.28 \times 10^{-20} \text{ Å}$$

The minimum uncertainty is 5.28×10^{-20} Å, a value that is clearly unmeasurable.

(b) The relativistic mass of the electron, $m = m_0/\sqrt{1 - (v^2/c^2)}$, must be used in (1).

$$\Delta x \geqslant \frac{h\sqrt{1 - (v^2/c^2)}}{4\pi 10^{-3}m_0v} = \frac{(6.63 \times 10^{-34} \text{ J} \cdot \text{s})\sqrt{1 - (0.6)^2}}{4\pi 10^{-3}(9.11 \times 10^{-31} \text{ kg})(1.8 \times 10^8 \text{ m/s})} = 2.57 \times 10^{-10} \text{ m} = 2.57 \text{ Å}$$

The minimum uncertainty is 2.57 Å.

18.2. What is the uncertainty in the location of a photon of wavelength 3000 Å if this wavelength is known to an accuracy of one part in a million?

The momentum of the photon is given by

$$p = \frac{hc}{\lambda c} = \frac{12.40 \times 10^3 \text{ eV} \cdot \text{Å}}{(3 \times 10^3 \text{ Å})c} = 4.13 \frac{\text{eV}}{c}$$

The uncertainty in the photon momentum is (working with magnitudes only):

$$\Delta p = \left| -\frac{h}{\lambda^2} \right| \Delta\lambda = p\frac{\Delta\lambda}{\lambda} = p \times 10^{-6} = 4.13 \times 10^{-6} \frac{\text{eV}}{c}$$

from which

$$\Delta x \geqslant \frac{h}{4\pi \, \Delta p} = \frac{hc}{4\pi c \, \Delta p} = \frac{12.4 \times 10^3 \text{ eV} \cdot \text{Å}}{4\pi c(4.13 \times 10^{-6} \text{ eV}/c)} = 239 \times 10^6 \text{ Å} = 23.9 \text{ mm}$$

18.3. What is the minimum uncertainty in the energy state of an atom if an electron remains in this state for 10^{-8} s?

The time available for measuring the energy is 10^{-8} s. Therefore, from $\Delta E \, \Delta t \geqslant h/4\pi$,

$$\Delta E \geqslant \frac{h}{4\pi \, \Delta t} = \frac{hc}{4\pi c \, \Delta t} = \frac{12.4 \times 10^3 \text{ eV} \cdot \text{Å}}{4\pi(3 \times 10^8 \text{ m/s})(10^{-8} \text{ s})(10^{10} \text{ Å/m})} = 0.329 \times 10^{-7} \text{ eV}$$

The minimum uncertainty in the energy of a state, $\Gamma = h/(4\pi\tau)$, where τ is the mean lifetime of the excited state, is called the *natural width* of the state. For this problem the mean lifetime is 10^{-8} s and the natural width is 0.329×10^{-7} eV.

18.4. The width of a spectral line of wavelength 4000 Å is measured as 10^{-4} Å. What is the average time that the atomic system remains in the corresponding energy state?

From Problem 18.3, $\tau = h/(4\pi\Gamma)$, where $\Gamma = \Delta E$ is the energy spread corresponding to $\Delta\lambda = 10^{-4}$ Å. From $E = hc/\lambda$,

$$|\Delta E| = \frac{hc}{\lambda^2}\,\Delta\lambda$$

and

$$\tau = \frac{h}{4\pi\left(\dfrac{hc}{\lambda^2}\,\Delta\lambda\right)} = \frac{\lambda^2}{4\pi c\,\Delta\lambda} = \frac{(4\times10^{-7}\,\text{m})^2}{4\pi(3\times10^8\,\text{m/s})(10^{-14}\,\text{m})} = 4.24\times10^{-9}\,\text{s}$$

Note that Planck's constant does not enter into the final expression.

18.5. Suppose the uncertainty in the momentum of a particle is equal to the particle's momentum. How is the minimum uncertainty in the particle's location related to its de Broglie wavelength?

We are given that $\Delta p = p$, so that

$$\Delta x \geqslant \frac{h}{4\pi\,\Delta p} = \frac{h}{4\pi\,p} = \frac{\lambda}{4\pi}$$

since the de Broglie wavelength of a particle is $\lambda = h/p$. Thus the minimum uncertainty in the position is $\lambda/4\pi$.

18.6. From the relation $\Delta p\,\Delta x \geqslant h/4\pi$, show that for a particle moving in a circle, $\Delta L\,\Delta\theta \geqslant h/4\pi$. The quantity ΔL is the uncertainty in the angular momentum and $\Delta\theta$ is the uncertainty in the angle.

Since the particle moves in a circle, the uncertainty principle will apply to directions tangent to the circle. Thus,

$$\Delta p_s\,\Delta s \geqslant \frac{h}{4\pi}$$

where s is measured along the circumference of the circle. The angular momentum is related to the linear momentum by

$$L = mvR = p_s R$$

therefore $\Delta p_s = \Delta L/R$. The angular displacement is related to the arc length by $\theta = s/R$; therefore $\Delta s = R\,\Delta\theta$. Hence

$$\Delta p_s\,\Delta s = (\Delta L/R)(R\,\Delta\theta) = \Delta L\,\Delta\theta \geqslant h/4\pi$$

For a state of fixed angular momentum (e.g. an electron in a Bohr orbit, which will be discussed in Chapter 19) the uncertainty in the angular momentum, ΔL, is zero. Therefore the uncertainty in the angular position, $\Delta\theta$, is infinite, so that the position of the particle in the orbit is undeterminable.

18.7. If we assume that $E = \frac{1}{2}mv^2$ for a particle moving in a straight line, show that $\Delta E\,\Delta t \geqslant h/4\pi$, where $\Delta t = \Delta x/v$.

$$E = \frac{1}{2}mv^2 = \frac{(mv)^2}{2m} = \frac{p^2}{2m}$$

Taking differentials of both sides of this expression, we obtain

$$\Delta E = \frac{p\,\Delta p}{m} = \frac{mv\,\Delta p}{m} = v\,\Delta p$$

Then, from $\Delta p\,\Delta x \geqslant h/4\pi$,

$$\frac{\Delta E}{v}\,\Delta x \geqslant \frac{h}{4\pi} \qquad \text{or} \qquad \Delta E\,\Delta t \geqslant \frac{h}{4\pi}$$

18.8. A particle of mass m is confined to a one-dimensional line of length L. From arguments based upon the uncertainty principle, estimate the value of the smallest energy that the body can have.

Since the particle must be somewhere in the given segment, the uncertainty in its position, Δx, cannot be greater than L. If Δx is set equal to L, the uncertainty relation $\Delta x \, \Delta p_x \geqslant h/4\pi$ in turn implies that the momentum must be uncertain by the amount $\Delta p_x \geqslant h/4\pi L$. We are looking for the smallest possible value of the energy and hence, since $K = p_x^2/2m$, the smallest possible $|p_x|$. We identify the uncertainty in $|p_x|$ with that in p_x, and assume that the uncertainty interval is symmetrical about $|p_x|$. Then (see Fig. 18-2)

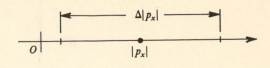

Fig. 18-2

$$|p_x| - \tfrac{1}{2}\Delta|p_x| \geqslant 0 \qquad \text{or} \qquad |p_x| \geqslant \tfrac{1}{2}\Delta|p_x| \geqslant \frac{1}{2}\left(\frac{h}{4\pi L}\right) = \frac{h}{8\pi L}$$

Thus, the minimum magnitude of p_x is $h/8\pi L$, and

$$K_{\min} = \frac{1}{2m}\left(\frac{h}{8\pi L}\right)^2 = \frac{h^2}{128\pi^2 m L^2}$$

This value compares reasonably well, considering the crudeness of our argument, with the value

$$E_1 = \frac{h^2}{8mL^2}$$

from Problem 17.3. The result further illustrates that, under the uncertainty principle, bound systems cannot have zero energy.

18.9. Calculate the minimum kinetic energy of a neutron in a nucleus of diameter 10^{-14} m.

The situation is that of Problem 18.8, with L equal to the nuclear diameter. Thus,

$$K_{\min} = \frac{1}{2m}\left(\frac{h}{8\pi L}\right)^2 = \frac{1}{2(mc^2)}\left(\frac{hc}{8\pi L}\right)^2 = \frac{1}{2(940 \text{ MeV})}\left[\frac{12.4 \times 10^{-3} \text{ MeV} \cdot \text{Å}}{8\pi(10^{-4} \text{ Å})}\right]^2 = 0.013 \text{ MeV}$$

18.10. If an electron were in the nucleus of Problem 18.9, what would be its minimum kinetic energy?

For an electron, a relativistic calculation is necessary. As in Problem 18.8, the minimum magnitude of the momentum is

$$|p|_{\min} = \frac{h}{8\pi L} = \frac{h}{8\pi(10^{-4} \text{ Å})}$$

Then:

$$(K_{\min} + E_0)^2 = (|p|_{\min}c)^2 + E_0^2$$

$$(K_{\min} + 0.511 \text{ MeV})^2 = \left[\frac{12.4 \times 10^{-3} \text{ MeV} \cdot \text{Å}}{8\pi(10^{-4} \text{ Å})}\right]^2 + (0.511 \text{ MeV})^2$$

Solving, $K_{\min} = 4.45$ MeV.

When the emission of electrons (β-rays) from nuclei was first observed, it was believed that the electrons must reside inside the nucleus. The energies of the emitted electrons, however, were often a few hundred keV and not the minimum 4 MeV predicted by the foregoing calculation. It was therefore concluded that electrons are not nuclear building blocks. (See also Problem 26.1.)

18.11. The position of a particle is measured by passing it through a slit of width d. Find the corresponding uncertainty induced in the particle's momentum.

When monochromatic waves of wavelength λ pass through a slit of width d, a diffraction pattern will be produced on a screen as shown in Fig. 18-3. The location of the first point of zero intensity is found from diffraction theory to be $\sin \alpha = \lambda/d$.

Because of its associated de Broglie wave, whose wavelength is $\lambda = h/p$, the particle will be diffracted as it passes through the slit and hence will acquire some unknown momentum in the

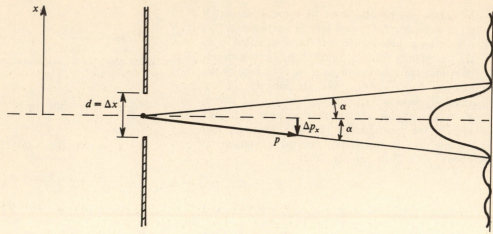

Fig. 18-3

x-direction. Although we do not know exactly where the particle will strike the screen, the most probable place for it to hit will be somewhere within the central region of the diffraction pattern. Therefore we can be reasonably certain that the x-component of the particle's momentum has a magnitude between 0 and $p \sin \alpha$; i.e.

$$\Delta p_x = p \sin \alpha = \frac{h}{\lambda} \frac{\lambda}{d} = \frac{h}{d}$$

This uncertainty can be made as small as desired by increasing d. However, since $d = \Delta x$, the uncertainty in the particle's x-position, we see that

$$\Delta p_x \, \Delta x = h$$

in agreement with Heisenberg's uncertainty principle.

18.12. It is desired to measure the position and momentum of an electron by observing it through a microscope. Analyze the observation process in detail to show that results consistent with the uncertainty principle are obtained.

When light is scattered from the electron during the observation process, the momentum of the electron, which we are trying to measure, will be affected because the incident light itself carries momentum. Hence, we will consider the experiment to be performed with the smallest amount of light possible, namely with a single photon.

When light reflecting from a particle passes through the objective lens of a microscope, a diffraction pattern is produced at the location of the eye (or photographic film). Thus, a "fuzzy" pattern rather than a precise sharp point will be observed with normally intense light consisting of many photons. Diffraction theory of light shows that the diameter of the central disc of the diffraction pattern is given approximately by

$$d = \frac{\lambda}{\sin \alpha}$$

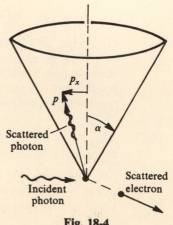

Fig. 18-4

where λ is the wavelength of the light and 2α is the angle subtended at the particle by the microscope objective, as shown in Fig. 18-4. When we observe the single photon in our experiment, we can be reasonably certain only that it will have arrived somewhere in the central disc of the diffraction pattern. Hence, the uncertainty in the electron's position can be taken as

$$\Delta x = d = \frac{\lambda}{\sin \alpha}$$

The uncertainty in the position can be made as small as desired by using a sufficiently small wavelength.

In the scattering process some of the photon's momentum will be transferred to the electron. If the momentum of the scattered photon were known exactly, it would be a relatively easy matter to work backwards to determine how the original momentum of the electron was affected. However, since all we know is that the scattered photon entered the objective lens somewhere, its x-component of momentum could have a magnitude anywhere between 0 and $p \sin \alpha$, where $p = h/\lambda$ is the photon's momentum. Hence, when we finally measure the momentum of the electron, the value of its x-momentum will be uncertain by

$$\Delta p_x = \frac{h}{\lambda} \sin \alpha$$

We can make Δp_x as small as desired by making λ sufficiently large, but then Δx becomes correspondingly larger. Taking the product of the two uncertainties, we obtain

$$\Delta x \, \Delta p_x = h$$

in agreement with Heisenberg's uncertainty principle.

Supplementary Problems

18.13. Suppose that the x-component of the velocity of a 2×10^{-4} kg mass is measured to an accuracy of $\pm 10^{-6}$ m/s. What then is the limit of the accuracy with which we can locate the particle along the x-axis?   *Ans.* 1.32×10^{-25} m

18.14. Repeat Problem 18.13 for an electron.   *Ans.*   29.0 m

18.15. Repeat Problem 18.2 for a gamma-ray photon of wavelength 10^{-5} Å.   *Ans.*   0.796 Å

18.16. What is the minimum uncertainty in the energy of an excited state of a system if, on the average, it remains in that state for 10^{-11} s?   *Ans.* 3.29×10^{-5} eV

18.17. Refer to Problem 18.3. If the transition from the energy state in question to the ground state corresponds to 3.39 eV, determine the minimum uncertainty in the wavelength of the emitted photon.   *Ans.* 3.55×10^{-5} Å

18.18. If the energy width of an excited state of a system is 1.1 eV, what is the average lifetime of that state?   *Ans.* 2.99×10^{-16} s

18.19. If the state of Problem 18.18 is located at an excitation energy of 1.6 keV, what is the minimum uncertainty in the wavelength of the photon emitted when the excited state decays?   *Ans.* 5.33×10^{-3} Å

18.20. If the uncertainty in the energy of a nuclear state is 33 keV, what is its average lifetime?   *Ans.* 9.97×10^{-21} s

18.21.   If the uncertainty in a photon's wavelength is one part in a million, find the minimum value of the uncertainty in its position if the photon's wavelength is (*a*) 3000 Å, (*b*) 0.5 Å, and (*c*) 2×10^{-4} Å.
Ans.   (*a*) 2.39 cm; (*b*) 3.98×10^4 Å; (*c*) 15.9 Å

18.22.   For an object of size 0.5 Å, what is the longest-wavelength photon with which it can be observed?
Ans.   0.5 Å

18.23.   For the object of Problem 18.22, what is the smallest-energy electron which can be used to make the measurement?   *Ans.*   602 eV

18.24.   For the object of Problem 18.22, what is the smallest-energy proton which can be used to make the measurement?   *Ans.*   0.328 eV

18.25.   If a photon was in the nucleus of Problem 18.9, what would be its minimum energy?
Ans. $\dfrac{hc}{8\pi d} = 4.9$ MeV

Chapter 19

The Bohr Atom

19.1 THE HYDROGEN SPECTRUM

By the end of the last century much experimental work had been done on the analysis of the discrete spectrum of the radiation emitted when electrical discharges were produced in gases. The lightest and simplest of all atoms is hydrogen, being composed of a nucleus and one electron. It was perhaps not surprising, then, that very precise spectroscopic measurements showed that hydrogen had the simplest spectrum of all the elements. It was found that the various lines in the optical and nonoptical regions were systematically spaced in various series. Amazingly it turned out that *all* the wavelengths of atomic hydrogen were given by a single empirical relation, the *Rydberg formula*:

$$\frac{1}{\lambda} = R\left(\frac{1}{n_l^2} - \frac{1}{n_u^2}\right) \qquad R = 1.0967758 \times 10^{-3} \text{ Å}^{-1}$$

where $n_l = 1$ and $n_u = 2, 3, 4 \ldots$ gives the *Lyman series* (ultraviolet region)
$n_l = 2$ and $n_u = 3, 4, 5 \ldots$ gives the *Balmer series* (optical region)
$n_l = 3$ and $n_u = 4, 5, 6 \ldots$ gives the *Paschen series* (infrared region)
$n_l = 4$ and $n_u = 5, 6, 7 \ldots$ gives the *Brackett series* (far infrared region)

and so on for other series lying in the farther infrared.

19.2 THE BOHR THEORY OF THE HYDROGEN ATOM

In 1913 Niels Bohr developed a physical theory of atomic hydrogen from which the Rydberg formula could be derived. Bohr's model for atomic hydrogen is based upon a planetary picture where a light negatively charged electron revolves around a heavy positively charged nucleus. The force maintaining the electron in its orbit is the attractive Coulomb force

$$F = k\frac{Ze^2}{r^2} \qquad k = 9.0 \times 10^9 \text{ N} \cdot \text{m}^2/\text{C}^2$$

where for hydrogen $Z = 1$. A straightforward classical calculation (Problem 19.14) then shows that the orbital velocity of the electron is related to the radius of its orbit, assumed by Bohr to be circular, by

$$v^2 = \frac{kZe^2}{mr} \tag{19.1}$$

where m is the mass of the electron, and the total energy of the electron (kinetic energy + potential energy) is given by

$$E = -\frac{kZe^2}{2r} \tag{19.2}$$

Now we come to the point where Bohr's model differs radically from a classical picture. (The presentation that follows, in terms of de Broglie waves, differs from the approach actually employed by Bohr. It was the ability of the de Broglie hypothesis to get the Bohr orbits in a natural way, rather than by Bohr's original arbitrary route, that led to de Broglie's theory being taken seriously.) As the electron moves in its orbit with linear momentum mv, it will have a de Broglie wavelength associated with it, given by $\lambda = h/mv$. Now, a wave can be associated with a given circular orbit only if the circumference of the orbit is an integral number of wavelengths. Thus Bohr effectively postulated that only those orbits are allowed that satisfy the relation

$$n\lambda = \frac{nh}{mv} = 2\pi r \quad \text{or} \quad mvr = n\frac{h}{2\pi} \qquad (19.3)$$

where $n = 1, 2, 3, \ldots$. The quantity $L = mvr$ is the angular momentum of the electron moving in its circular orbit, so it is seen that in the Bohr theory the *electron's angular momentum is quantized*. The integer n is called the *principal quantum number*.

Solving (19.1), (19.2), and (19.3) for the three unknowns r, E, and v, we find the following quantized quantities:

$$r_n = \frac{n^2 r_1^\circ}{Z} \qquad r_1^\circ = \frac{h^2}{4\pi^2 k m e^2} \qquad (19.4)$$

$$E_n = -\frac{Z^2 E_1^\circ}{n^2} \qquad E_1^\circ = \frac{2\pi^2 k^2 e^4 m}{h^2} \qquad (19.5)$$

$$v_n = \frac{Z v_1^\circ}{n} \qquad v_1^\circ = \frac{2\pi k e^2}{h} \qquad (19.6)$$

In the stable states of the atom specified by (19.4), (19.5), and (19.6) the electron is assumed not to radiate; the radiation process is discussed in Section 19.3. The minimum-energy state ($n = 1$) is called the *ground state*.

It is seen that the quantities r_1°, E_1°, and v_1° depend only on the fundamental constants of nature m, e, k, and h. When the numerical values of these constants are used, one obtains (Problems 19.33 through 19.35)

$$r_1^\circ = 0.529 \text{ Å} \qquad E_1^\circ = 13.58 \text{ eV} \qquad v_1^\circ = \frac{c}{137.0}$$

Note that for hydrogen ($Z = 1$), $r_1^\circ = r_1$, $E_1^\circ = -E_1$, and $v_1^\circ = v_1$. The values 0.529 Å and 13.58 eV are in good agreement with experimental determinations of the radius and ionization energy of the hydrogen atom.

19.3 EMISSION OF RADIATION IN BOHR'S THEORY

Classical electrodynamical theory predicted that an orbiting (hence, accelerating) charge should emit radiation whose frequency would be equal to the frequency of revolution. We have already seen in the photoelectric effect, however, that classical electrodynamics must be modified on an atomic scale when *absorption* of electromagnetic radiation is considered. In an analogous fashion Bohr chose to modify classical electrodynamics on an atomic scale when *emission* of electromagnetic radiation occurs.

Bohr postulated that an atom will emit radiation only when the electron, initially in one of the stable allowed orbits where $E = E_u$, changes to another allowed orbit with a smaller energy given by $E = E_l$. The energy of the emitted photon will then be equal to the difference between the electron energies in the two allowed orbits. Thus the wavelength of the emitted photon will be found from

$$E_\gamma = h\nu = h\frac{c}{\lambda} = E_u - E_l \quad \text{or} \quad \frac{1}{\lambda} = \frac{1}{hc}(E_u - E_l) \qquad (19.7)$$

Substituting the values of the orbital energies given by (19.5), we then find

$$\frac{1}{\lambda} = \frac{2\pi^2 k^2 e^4 m Z^2}{h^3 c}\left(\frac{1}{n_l^2} - \frac{1}{n_u^2}\right) = R_\infty Z^2\left(\frac{1}{n_l^2} - \frac{1}{n_u^2}\right)$$

with

$$R_\infty = \frac{2\pi^2 (ke^2)^2 (mc^2)}{(hc)^3} = 1.09737 \times 10^{-3} \text{ Å}^{-1}$$

In this analysis it was assumed that the positively charged nucleus is so massive compared to the electron that it could be considered infinitely heavy. If the finite mass of the nucleus is taken into consideration, the motion of the combined electron (m) and nucleus (M) system, separated by distance r, about its center of mass ($m\rho = MP$, $r = \rho + P$) is equivalent to a particle of *reduced mass*

$$\mu = \frac{m}{1 + \dfrac{m}{M}} = \frac{M}{1 + \dfrac{M}{m}}$$

orbiting the center of mass at a radius r. For hydrogen $m/M = 1/1836$, and using this to modify the Rydberg constant we obtain

$$R_H = \frac{R_\infty}{1 + (m/M)} = \frac{1.09737 \times 10^{-3}\ \text{Å}^{-1}}{1 + (1/1836)} = 1.0968 \times 10^{-3}\ \text{Å}^{-1}$$

in agreement with the experimental value $R = 1.0967758 \times 10^{-3}\ \text{Å}^{-1}$.

19.4 ENERGY LEVEL DIAGRAMS

A convenient way of describing the transitions between allowed states is in terms of *energy level diagrams*. In these, the allowed energy levels, given by (*19.5*), are plotted, as shown in Fig. 19-1 for $Z = 1$. The transitions are then indicated by arrows running from the initial energy state, designated by n_u, to the final energy state, designated by n_l. Thus, for example, the transitions that give rise to the Balmer series are indicated in Fig. 19-1 by the arrows that end on $n_l = 2$. The lines in the Balmer series are called H_α, H_β, H_γ, etc., as indicated in Fig. 19-1.

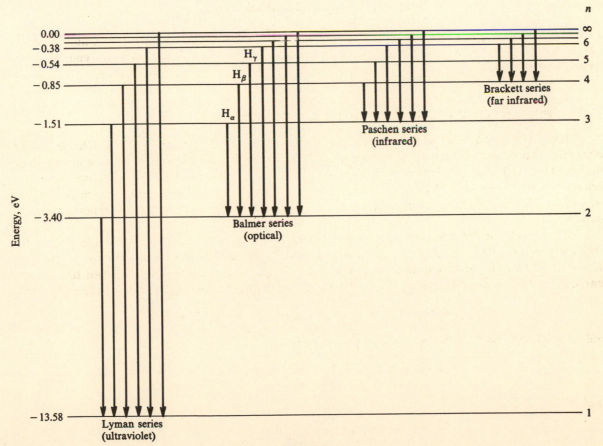

Fig. 19-1

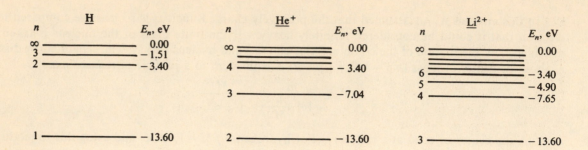

Fig. 19-2

19.5 HYDROGENIC ATOMS

A *hydrogenic atom* is an atom that is stripped of all but one of its electrons. Thus, hydrogenic atoms are singly ionized helium (He^+, $Z = 2$), doubly ionized lithium (Li^{2+}, $Z = 3$), triply ionized beryllium (Be^{3+}, $Z = 4$) and so forth. These atoms behave in all respects like hydrogen except that the nucleus has a positive charge of Ze, where Z is the atomic number of the atom. Equations (*19.1*) through (*19.7*) hold for hydrogenic atoms, provided the appropriate value of Z is used. Figure 19-2 shows the energy levels for H, He^+, and Li^{2+}.

19.6 μ-MESIC AND π-MESIC ATOMS

In Yukawa's explanation of nuclear binding forces (i.e. strong interactions), the existence of a particle called a *meson*, of rest mass 264 times the rest mass of an electron, was predicted. Two years after this prediction, in 1937, a particle with a rest mass of 207 electron masses was discovered. However, in 1946, it was shown that this *μ-meson* was not the predicted particle, and a short time after, the Yukawa particle, called a *π-meson*, was found.

Both π- and μ-mesons can be found with negative charge, and can therefore form hydrogenic atoms. The Bohr orbits of these particles, due to their large masses, are much smaller than the electron's orbits. In fact, for certain nuclei the orbits actually lie within the nuclear charge distribution (see Problems 19.21 through 19.25).

Solved Problems

19.1. Determine, in angstroms, the shortest and longest wavelengths of the Lyman series of hydrogen.

Wavelengths in the Lyman series are given by $n_l = 1$:

$$\frac{1}{\lambda} = (1.097 \times 10^{-3}\,\text{Å}^{-1})\left(\frac{1}{1^2} - \frac{1}{n_u^2}\right) \qquad n_u = 2, 3, 4, \ldots$$

The longest wavelength corresponds to $n_u = 2$:

$$\frac{1}{\lambda_{max}} = (1.097 \times 10^{-3}\,\text{Å}^{-1})\left(1 - \frac{1}{2^2}\right) \qquad \text{or} \qquad \lambda_{max} = 1215\,\text{Å}$$

The shortest wavelength corresponds to $n_u = \infty$:

$$\frac{1}{\lambda_{min}} = 1.097 \times 10^{-3}\,\text{Å}^{-1}\left(1 - \frac{1}{\infty^2}\right) \qquad \text{or} \qquad \lambda_{min} = 912\,\text{Å}$$

19.2. Determine the wavelength of the second line of the Paschen series for hydrogen.

$$\frac{1}{\lambda} = (1.097 \times 10^{-3}\,\text{Å}^{-1})\left(\frac{1}{n_l^2} - \frac{1}{n_u^2}\right)$$

The Paschen series is defined by $n_l = 3$, and the second line corresponds to $n_u = 5$. Hence,

$$\frac{1}{\lambda} = (1.097 \times 10^{-3}\,\text{Å}^{-1})\left(\frac{1}{3^2} - \frac{1}{5^2}\right) \qquad \text{or} \qquad \lambda = 12{,}820\,\text{Å}$$

19.3. The longest wavelength in the Lyman series for hydrogen is 1215 Å. What is the Rydberg constant?

$$\frac{1}{\lambda} = R\left(\frac{1}{n_l^2} - \frac{1}{n_u^2}\right)$$

For the Lyman series, $n_l = 1$; the longest wavelength will correspond to the value $n_u = 2$.

$$\frac{1}{1215 \text{ Å}} = R\left(\frac{1}{1^2} - \frac{1}{2^2}\right) \quad \text{or} \quad R = 1.097 \times 10^{-3} \text{ Å}^{-1}$$

19.4. Determine the wavelengths of hydrogen that lie in the optical spectrum (3800 Å to 7700 Å).

The wavelengths for hydrogen are given by

$$\frac{1}{\lambda} = (1.097 \times 10^{-3} \text{ Å}^{-1})\left(\frac{1}{n_l^2} - \frac{1}{n_u^2}\right)$$

In Problem 19.1 it was found that when $n_l = 1$ the wavelengths range from 912 Å to 1215 Å, so that none of these lie in the optical region. For $n_l = 2$ the longest wavelength corresponds to $n_u = 3$, giving

$$\frac{1}{\lambda} = (1.097 \times 10^{-3} \text{ Å}^{-1})\left(\frac{1}{2^2} - \frac{1}{3^2}\right) \quad \text{or} \quad \lambda = 6563 \text{ Å}$$

and the shortest wavelength corresponds to $n_u = \infty$, giving

$$\frac{1}{\lambda} = (1.097 \times 10^{-3} \text{ Å}^{-1})\left(\frac{1}{2^2} - \frac{1}{\infty^2}\right) \quad \text{or} \quad \lambda = 3646 \text{ Å}$$

Hence, some of the wavelengths in the Balmer series ($n_l = 2$) lie in the optical region. To determine these wavelengths set $\lambda = 3800$ Å and solve for n_u.

$$\frac{1}{3.8 \times 10^3 \text{ Å}} = (1.097 \times 10^{-3} \text{ Å}^{-1})\left(\frac{1}{4} - \frac{1}{n_u^2}\right) \quad \text{or} \quad n_u = 9.9$$

Therefore, the lines in the optical region are given by

$$\frac{1}{\lambda} = (1.097 \times 10^{-3} \text{ Å}^{-1})\left(\frac{1}{4} - \frac{1}{n_u^2}\right) \quad n_u = 3, 4, 5, \ldots, 9$$

Since the shortest wavelength of the Paschen series ($n_l = 3$) is

$$\frac{1}{\lambda} = (1.097 \times 10^{-3} \text{ Å}^{-1})\left(\frac{1}{3^2} - \frac{1}{\infty^2}\right) \quad \text{or} \quad \lambda = 8200 \text{ Å}$$

all other series will give rise to lines lying outside the optical region.

19.5. Evaluate the ionization potential of hydrogen, E_1°, in units of eV.

$$E_1^\circ = \frac{2\pi^2 k^2 e^4 m}{h^2} = \frac{2\pi^2 (ke^2)^2 (mc^2)}{(hc)^2} = \frac{2\pi^2 (14.40 \text{ eV} \cdot \text{Å})^2 (0.511 \times 10^6 \text{ eV})}{(12.40 \times 10^3 \text{ eV} \cdot \text{A})^2} = 13.6 \text{ eV}$$

19.6. Find the wavelength of the photon that is emitted when a hydrogen atom undergoes a transition from $n_u = 5$ to $n_l = 2$.

From the Bohr model the energy levels are $E_n = (-13.6 \text{ eV})/n^2$. Hence,

$$E_2 = -\frac{13.6 \text{ eV}}{2^2} = -3.40 \text{ eV} \qquad E_5 = -\frac{13.6 \text{ eV}}{5^2} = -0.544 \text{ eV}$$

From the Bohr postulates the energy of the emitted photon is

$$E_\gamma = -0.544 \text{ eV} - (-3.40 \text{ eV}) = 2.86 \text{ eV}$$

The wavelength of this photon is given by

$$\lambda = \frac{hc}{E_\gamma} = \frac{12.4 \times 10^3 \text{ eV} \cdot \text{Å}}{2.86 \text{ eV}} = 4340 \text{ Å}$$

This problem can also be solved using Rydberg's formula,

$$\frac{1}{\lambda} = R\left(\frac{1}{n_l^2} - \frac{1}{n_u^2}\right) = 1.097 \times 10^{-3}\left(\frac{1}{2^2} - \frac{1}{5^2}\right)$$

Solving, $\lambda = 4340$ Å.

19.7. Determine the ionization energy of hydrogen if the shortest wavelength in the Balmer series is found to be 3650 Å.

The Balmer series is given by $n_l = 2$. The shortest wavelength will correspond to $n_u = \infty$. Consequently, from $E_n = -E_1^\circ/n^2$, where E_1° is the ionization energy, we have

$$\frac{hc}{\lambda} = E_u - E_l = 0 - \left(-\frac{E_1^\circ}{4}\right) \quad \text{or} \quad E_1^\circ = \frac{4hc}{\lambda} = \frac{4(12.4 \times 10^3 \text{ eV} \cdot \text{Å})}{3650 \text{ Å}} = 13.6 \text{ eV}$$

19.8. How many different photons can be emitted by hydrogen atoms that undergo transitions to the ground state from the $n = 5$ state?

We can consider the problem for arbitrary n. If $n_u > n_l$ is any pair of unequal integers in the range 1 to n, it is clear that there is at least one route from state n down to the ground state that includes the transition $n_u \to n_l$. Thus, the number of photons is equal to the number of such pairs, which is

$$\binom{n}{2} = \frac{n(n-1)}{2}$$

For $n = 5$, there are $5(4)/2 = 10$ photons.

The above reasoning fails if there is "degeneracy," i.e. if two different pairs of quantum numbers correspond to the same energy difference. In that case the number of distinct photons is smaller than $n(n-1)/2$.

19.9. In a transition to a state of excitation energy 10.19 eV a hydrogen atom emits a 4890 Å photon. Determine the binding energy of the initial state.

The energy of the emitted photon is

$$h\nu = \frac{hc}{\lambda} = \frac{12.40 \times 10^3 \text{ eV} \cdot \text{Å}}{4.89 \times 10^3 \text{ Å}} = 2.54 \text{ eV}$$

The excitation energy (E_x) is the energy to excite the atom to a level above the ground state. Therefore the energy of the level is

$$E_n = E_1 + E_x = -13.6 \text{ eV} + 10.19 \text{ eV} = -3.41 \text{ eV}$$

The photon arises from the transition between energy states such that $E_u - E_l = h\nu$; hence

$$E_u - (-3.41 \text{ eV}) = 2.54 \text{ eV} \quad \text{or} \quad E_u = -0.87 \text{ eV}$$

Therefore the binding energy of an electron in the state is 0.87 eV.

Note that the transition corresponds to

$$n_u = \sqrt{\frac{E_1}{E_u}} = \sqrt{\frac{13.6 \text{ eV}}{0.87 \text{ eV}}} = 4 \quad \text{and} \quad n_l = \sqrt{\frac{E_1}{E_l}} = \sqrt{\frac{13.6 \text{ eV}}{3.41 \text{ eV}}} = 2$$

19.10. Electrons of energy 12.2 eV are fired at hydrogen atoms in a gas discharge tube. Determine the wavelengths of the lines that can be emitted by the hydrogen.

The maximum energy that can be absorbed by a hydrogen atom is equal to the electron energy, 12.2 eV. Absorption of this energy would excite the atom into an energy state E_u given by (assuming the atom was initially in the ground state)

$$E_u = E_1 + 12.2 \text{ eV} = -13.6 \text{ eV} + 12.2 \text{ eV} = -1.4 \text{ eV}$$

The value of n corresponding to this state is found from $E_n = -E_1^\circ/n^2$; thus

$$-1.4 \text{ eV} = -\frac{13.6 \text{ eV}}{n^2} \quad \text{or} \quad n = 3.12$$

Since n must be an integer, the highest state that can be reached corresponds to $n = 3$. Hence (Problem 19.8) there are three possible wavelengths that will be emitted as the atom returns to the ground state, corresponding to the transitions $3 \to 2$, $2 \to 1$ and $3 \to 1$. These wavelengths are found from

$$\frac{1}{\lambda} = (1.097 \times 10^{-3} \text{ Å}^{-1})\left(\frac{1}{2^2} - \frac{1}{3^2}\right) \qquad \text{or} \qquad \lambda = 6563 \text{ Å}$$

$$\frac{1}{\lambda} = (1.097 \times 10^{-3} \text{ Å}^{-1})\left(\frac{1}{1^2} - \frac{1}{2^2}\right) \qquad \text{or} \qquad \lambda = 1215 \text{ Å}$$

$$\frac{1}{\lambda} = (1.097 \times 10^{-3} \text{ Å}^{-1})\left(\frac{1}{1^2} - \frac{1}{3^2}\right) \qquad \text{or} \qquad \lambda = 1026 \text{ Å}$$

19.11. According to the Bohr theory, how many revolutions will an electron make in the first excited state of hydrogen if the lifetime in that state is 10^{-8} s?

By (*19.4*) and (*19.6*), the radius and orbital velocity for the state $n = 2$ are given by

$$r_2 = 4r_1^\circ = 4(0.529 \text{ Å}) = 2.12 \text{ Å} = 2.12 \times 10^{-10} \text{ m}$$

$$v_2 = \frac{v_1^\circ}{2} = \frac{c}{2(137)} = \frac{3 \times 10^8 \text{ m/s}}{2(137)} = 1.10 \times 10^6 \text{ m/s}$$

The angular velocity is then

$$\omega = \frac{v_2}{r_2} = \frac{1.10 \times 10^6 \text{ m/s}}{2.12 \times 10^{-10} \text{ m}} = 0.52 \times 10^{16} \text{ rad/s}$$

and the total number of revolutions is

$$N = \frac{\omega t}{2\pi} = \frac{(0.52 \times 10^{16} \text{ rad/s})(10^{-8} \text{ s})}{6.28 \text{ rad/rev}} = 8.3 \times 10^6 \text{ rev}$$

19.12. Determine the correction to the wavelength of an emitted photon when the recoil kinetic energy of the hydrogen nucleus is taken into account.

Assuming that the atom is initially at rest, conservation of energy gives

$$E_u = E_l + E_\gamma + K \qquad \text{or} \qquad \frac{E_u - E_l}{hc} - \frac{E_\gamma}{hc} = \frac{K}{hc}$$

where K is the nuclear kinetic energy. The first term on the left is $1/\lambda_0$, and the second is $1/\lambda$, where λ_0 and λ are the uncorrected and actual wavelengths. Thus

$$\frac{1}{\lambda_0} - \frac{1}{\lambda} = \frac{K}{hc} \qquad \text{or} \qquad \frac{\lambda - \lambda_0}{\lambda_0} = \frac{\lambda K}{hc}$$

The recoil momentum of the nucleus is $p = \sqrt{2MK}$. Then, by conservation of momentum,

$$0 = -\sqrt{2MK} + \frac{h}{\lambda} \qquad \text{or} \qquad K = \frac{h^2}{2M\lambda^2}$$

from which

$$\frac{\lambda - \lambda_0}{\lambda_0} = \frac{\lambda(h^2/2M\lambda^2)}{hc} = \frac{hc}{2(Mc^2)\lambda} = \frac{12.40 \times 10^3 \text{ eV} \cdot \text{Å}}{2(939 \times 10^6 \text{ eV})\lambda} = \frac{6.60 \times 10^{-6} \text{ Å}}{\lambda}$$

Since the wavelengths are of the order $\lambda \sim 10^3$ Å, the fractional change is of the order 10^{-9} and is therefore negligible.

19.13. For hydrogen, show that when $n \gg 1$ the frequency of the emitted photon in a transition from n to $n - 1$ equals the rotational frequency.

The rotational frequency in state n is

$$\frac{\omega}{2\pi} = \frac{v_n}{2\pi r_n} = \frac{2\pi k e^2/nh}{2\pi n^2 h^2/4\pi^2 k m e^2} = \frac{4\pi^2 k^2 m e^4}{n^3 h^3}$$

The frequency of the emitted photon is

$$\nu = c\frac{1}{\lambda} = cR_\infty \left[\frac{1}{(n-1)^2} - \frac{1}{n^2}\right] = cR_\infty \frac{2n-1}{n^2(n-1)^2}$$

For $n \gg 1$,

$$\nu \approx cR_\infty \frac{2n}{n^2 n^2} = c\frac{2\pi^2 k^2 e^4 m}{h^3 c}\frac{2}{n^3} = \frac{4\pi^2 k^2 me^4}{n^3 h^3}$$

which is the same as the rotational frequency given above.

This problem illustrates Bohr's *correspondence principle*, which states that for large n a quantum equation should go over into the corresponding classical equation. According to classical theory, radiation emitted from a rotating charge will have a frequency equal to the rotational frequency.

19.14. An electron rotates in a circle around a nucleus with positive charge Ze. How is the electron's velocity related to the radius of its orbit?

Equating the Coulomb force to the (electron mass) × (centripetal acceleration),

$$\frac{k(e)(Ze)}{r^2} = \frac{mv^2}{r} \quad \text{or} \quad v^2 = \frac{kZe^2}{mr}$$

19.15. How is the total energy of the electron in Problem 19.14 related to the radius of its orbit?

The electrical potential energy of the electron is

$$U = qV = (-e)V = -e\frac{k(Ze)}{r} = -\frac{kZe^2}{r}$$

The kinetic energy of the electron is found by using the result of Problem 19.14:

$$K = \frac{1}{2}mv^2 = \frac{1}{2}m\frac{kZe^2}{mr} = \frac{kZe^2}{2r}$$

The total energy is then

$$E = K + U = \frac{kZe^2}{2r} - \frac{kZe^2}{r} = -\frac{kZe^2}{2r} = \frac{1}{2}U$$

19.16. Assuming that all transitions are possible, will the optical spectrum (3800 Å to 7700 Å) of hydrogen have more or fewer lines than the optical spectrum of doubly ionized lithium?

For hydrogen: $E_{nH} = -\dfrac{E_1^\circ}{n^2}$

For Li^{2+}: $E_{nLi} = -\dfrac{Z^2 E_1^\circ}{n^2} = -\dfrac{E_1^\circ}{(n/3)^2}$

Hence the energy level diagram for Li^{2+} contains all the energy levels of hydrogen, plus two extra levels for each hydrogen level. Since there are more levels available, there will be more lines in the optical spectrum of Li^{2+} than in the optical spectrum of hydrogen.

19.17. Determine the mass ratio of deuterium and hydrogen if, respectively, their H_α lines have wavelengths of 6561.01 Å and 6562.80 Å. (It was through measurements of this type that deuterium was discovered.)

In terms of the reduced mass of the atom, the Rydberg formula is

$$\frac{1}{\lambda} = \frac{R_\infty Z^2}{1 + \dfrac{m}{M}}\left(\frac{1}{n_l^2} - \frac{1}{n_u^2}\right)$$

where m and M are the electronic and nuclear masses. For a fixed transition and fixed Z, this implies that λ is proportional to $1 + (m/M)$, so that

$$\frac{\lambda_D}{\lambda_H} = \frac{1 + \dfrac{m}{M_D}}{1 + \dfrac{m}{M_H}}$$

or

$$\frac{\lambda_D - \lambda_H}{\lambda_H} = \frac{\dfrac{m}{M_D} - \dfrac{m}{M_H}}{1 + \dfrac{m}{M_H}} = \frac{m}{M_H} \frac{M_H - M_D}{M_D\left(1 + \dfrac{m}{M_H}\right)} \approx \frac{m}{M_H} \frac{M_H - M_D}{M_D}$$

Substituting the data and $m/M_H = 1/1836$,

$$\frac{-1.79 \text{ Å}}{6562.80 \text{ Å}} \approx \frac{1}{1836}\left(\frac{M_H}{M_D} - 1\right)$$

Solving,

$$\frac{M_H}{M_D} \approx 0.5 \qquad \text{or} \qquad \frac{M_D}{M_H} \approx 2.0$$

19.18. Find the difference in wavelength between the line from the $3 \rightarrow 2$ transition in hydrogen ($R_H = 1.09678 \times 10^{-3} \text{ Å}^{-1}$) and the line from the $6 \rightarrow 4$ transition in singly ionized helium ($R_{He} = 1.09722 \times 10^{-3} \text{ Å}^{-1}$).

We have

$$\frac{1}{\lambda} = R\left[\left(\frac{Z}{n_l}\right)^2 - \left(\frac{Z}{n_u}\right)^2\right] \qquad \text{where} \qquad R = \frac{R_\infty}{1 + (m/M)}$$

Since $Z_H = 1$, $Z_{He} = 2$, the expression in brackets has the same value, 5/36, for the two transitions, and the difference in wavelength arises solely from the difference in Rydberg constants. Taking differentials,

$$-\frac{1}{\lambda^2} d\lambda = dR \frac{5}{36} \qquad \text{or} \qquad -d\lambda = \lambda^2 dR \frac{5}{36} = \frac{dR}{\frac{5}{36} R^2}$$

Therefore, approximately,

$$\lambda_H - \lambda_{He} = \frac{R_{He} - R_H}{\frac{5}{36} R_H^2} = \frac{0.00044 \times 10^{-3} \text{ Å}^{-1}}{\frac{5}{36}(1.09678 \times 10^{-3} \text{ Å}^{-1})^2} = 2.63 \text{ Å}$$

19.19. Determine the Rydberg constant for positronium (a bound system composed of a positron and an electron).

The mass of a positron is the same as the mass of an electron, so that

$$R_P = \frac{R_\infty}{1 + (m/M)} = \frac{R_\infty}{1 + (m/m)} = \frac{R_\infty}{2} = 0.5485 \times 10^{-3} \text{ Å}^{-1}$$

19.20. Refer to Problem 19.19. Find the ionization potential of positronium.

$$\frac{1}{\lambda} = R_P\left(\frac{1}{n_l^2} - \frac{1}{n_u^2}\right) \qquad \text{or} \qquad \frac{hc}{\lambda} = h\nu = hcR_P\left(\frac{1}{n_l^2} - \frac{1}{n_u^2}\right)$$

The ionization energy is the energy required to excite positronium from its ground state ($n_l = 1$) to the state $n_u = \infty$. Thus

$$E_{ion} = hcR_P = (12.40 \times 10^3 \text{ eV} \cdot \text{Å})(0.5485 \times 10^{-3} \text{ Å}^{-1}) = 6.8 \text{ eV}$$

19.21. Determine the ionization energy of a μ-mesic atom that is formed when a μ-meson is captured by a proton. A μ-meson is an elementary particle with a charge of $-e$ and a rest mass equal to 207 times the rest mass of an electron.

The analysis is identical in all respects to that of a hydrogen atom, with the mass, m, of the electron replaced by $207m$.

$$E_{\text{ion}} = 207(13.6 \text{ eV}) = 2.82 \text{ keV}$$

19.22. Refer to Problem 19.21. Calculate the radius of the first Bohr orbit in ^{208}Pb ($Z = 82$) for a μ-mesic atom.

By (19.4), r_1 varies inversely with Zm. Hence,

$$r_{1\mu} = \frac{1}{(82)(207)} (0.529 \text{ Å}) = 3.12 \times 10^{-5} \text{ Å} = 3.12 \text{ fm}$$

19.23. For Problem 19.22, calculate the energy of the first Bohr orbit.

By (19.5), E_1 varies directly with $Z^2 m$. Hence,

$$E_{1\mu} = (82)^2 (207)(-13.58 \text{ eV}) = -19.0 \text{ MeV}$$

19.24. Refer to Problems 19.21 through 19.23. For a ^{208}Pb μ-mesic atom, what is the energy of the photon given off in the first Lyman transition ($n_u = 2$ to $n_l = 1$)?

$$E_\gamma = \Delta E = E_u - E_l = -Z^2 E_1^\circ \left(\frac{1}{n_u^2} - \frac{1}{n_l^2} \right) = E_1 \left(\frac{1}{n_u^2} - \frac{1}{n_l^2} \right) = (-19.0 \text{ MeV}) \left(\frac{1}{2^2} - \frac{1}{1^2} \right) = 14.25 \text{ MeV}$$

19.25. The transition in Problem 19.24 is measured experimentally to be 6.0 MeV. The ^{208}Pb nucleus is found to have a radius of 7.1 fm. From Problem 19.22 it is then seen that the first Bohr orbit lies *inside* the nucleus. Assuming the charge of the nucleus to be uniformly distributed, determine the new first Bohr orbit that will be consistent with the above data.

The potential of a uniformly charged sphere of radius R and total charge $Q = Ze$ is

$$V = \begin{cases} \dfrac{kQ}{R} \left[\dfrac{3}{2} - \dfrac{r^2}{2R^2} \right] & r < R \\[3mm] \dfrac{kQ}{r} & r > R \end{cases}$$

For a μ-meson moving inside the uniform charge distribution, we have for the energy,

$$E = K + U = \frac{1}{2} mv^2 + (-e) \frac{kZe}{R} \left[\frac{3}{2} - \frac{r^2}{2R^2} \right]$$

If the μ-meson moves in a circular orbit of radius r, we have

$$F = ma \qquad \text{or} \qquad e\mathcal{E} = \frac{mv^2}{r}$$

where $\mathcal{E}$, the radial electric field inside the charge distribution, is given by

$$\mathcal{E} = -\frac{dV}{dr} = \frac{kQr}{R^3} = \frac{kZer}{R^3}$$

Hence

$$\frac{mv^2}{2} = \frac{re\mathcal{E}}{2} = \frac{kZe^2 r^2}{2R^3}$$

and for the energy we get

$$E = \frac{kZe^2 r^2}{2R^3} - \frac{kZe^2}{R} \left[\frac{3}{2} - \frac{r^2}{2R^2} \right] = Z \frac{ke^2}{R} \left[\frac{r^2}{R^2} - \frac{3}{2} \right]$$

The second Bohr orbit lies outside the nucleus; therefore $E_{2\mu}$ can be calculated *as if* $E_{1\mu} = -19.0$ MeV (Problem 19.23), i.e.

$$E_{2\mu} = \frac{-19.0 \text{ MeV}}{2^2} = -4.75 \text{ MeV}$$

Using the experimental data, we then have

$$E_{1\mu} = E_{2\mu} - \Delta E = -4.75 \text{ MeV} - 6.0 \text{ MeV} = -10.75 \text{ MeV}$$

and so

$$-10.75 \text{ MeV} = 82 \, \frac{1.44 \text{ MeV} \cdot \text{fm}}{7.1 \text{ fm}} \left[\frac{r_{1\mu}^2}{(7.1 \text{ fm})^2} - \frac{3}{2} \right]$$

Solving, $r_{1\mu} = 6.56$ fm, an increase of 3.44 fm (see Problem 19.22) due to the orbit's being inside the nuclear charge distribution.

19.26. An electron moves in a uniform spherical charge distribution of total charge Ze and radius R. Using the Bohr postulates, calculate the allowed energy levels assuming that the Bohr orbits lie inside the charge.

From Problem 19.25, the energy of the electron is given by

$$E = \frac{kZe^2}{R} \left[\frac{r^2}{R^2} - \frac{3}{2} \right] \tag{1}$$

Bohr's first postulate gives ($\hbar = h/2\pi$)

$$mvr = n\hbar \qquad \text{or} \qquad v^2 = \frac{n^2\hbar^2}{m^2 r^2} \tag{2}$$

From Newton's second law and assuming the electron to move in a circular orbit, we have

$$\frac{kZe^2 r}{R^3} = m \frac{v^2}{r} \tag{3}$$

Eliminating v^2 between (2) and (3), we obtain

$$r^2 = \frac{n\hbar}{e} \sqrt{\frac{R^3}{mkZ}} \tag{4}$$

Putting (4) into (1), we find for each integer value of n

$$E_n = \frac{kZe^2}{R} \left[\frac{n\hbar}{e} \sqrt{\frac{1}{mkZR}} - \frac{3}{2} \right]$$

Supplementary Problems

19.27. Repeat Problem 19.1 for the Balmer series.   *Ans.*   3646 Å; 6563 Å

19.28. Determine in angstroms the wavelength of the photon emitted in the transition $n_u = 6$ to $n_l = 3$. (This is the third transition in the Paschen series.)   *Ans.* 1.094×10^4 Å

19.29. Calculate the shortest-wavelength photon in the series of transitions with $n_l = 4$ (the Brackett series). *Ans.* 1.459×10^4 Å

19.30. The shortest wavelength in the Balmer series for hydrogen is 3646 Å. Determine the Rydberg constant from this value.   *Ans.* 1.097×10^{-3} Å^{-1}

19.31. Find the value of n_u in the series that gives rise to the line in the hydrogen spectrum at 1026 Å.   (Note: this is in the Lyman series.)   *Ans.*   3

19.32. Repeat Problem 19.31 for the hydrogen spectral line at 4861 Å.   (Note: this is in the Balmer series.) *Ans.*   4

19.33. Evaluate ke^2 in units of eV · Å.   *Ans.*   14.40 eV · Å

19.34. Show that $v_1^\circ/c = 2\pi\, ke^2/hc \equiv \alpha$ is 1/137.   The dimensionless quantity α is called the *fine structure constant*.

19.35. From (*19.4*) evaluate the radius of the first Bohr orbit of hydrogen in angstroms.   *Ans.*   0.529 Å

19.36. Determine the ratio of the Compton wavelength of an electron (Chapter 12) to the radius of the first Bohr orbit of hydrogen.   *Ans.*   21.8

19.37. What is the minimum accelerating potential that will enable an electron to excite a hydrogen atom out of its ground state?   *Ans.*   10.2 V

19.38. Determine the minimum energy that must be given to a hydrogen atom so that it can emit the H_β line. (The H_β line corresponds to a $4 \to 2$ transition.)   *Ans.*   2.55 eV

19.39. Determine the binding energy of an electron in the third excited state of hydrogen.   *Ans.*   0.85 eV

19.40. What accelerating potential will enable an electron to ionize a hydrogen atom?   *Ans.*   13.6 V

19.41. What is the highest state that unexcited hydrogen atoms can reach when they are bombarded with 12.6 eV electrons?   *Ans.* $n = 3$

19.42. Find the recoil energy of a hydrogen atom when a photon is emitted in a transition from $n_u = 10$ to $n_l = 1$.   *Ans.* 9.6×10^{-8} eV

19.43. Calculate the fractional change in the wavelength of a spectral line that arises from a small change in the reduced mass of the atom.   *Ans.* $\Delta\lambda/\lambda = -\Delta\mu/\mu$

19.44. Determine the radius of the second Bohr orbit for doubly ionized lithium.   *Ans.*   0.705 Å

19.45. For triply ionized beryllium ($Z = 4$) determine the first Bohr orbit radius.   *Ans.*   0.132 Å

19.46. Determine the wavelength of the H_β line of deuterium if the H_β line of hydrogen is 4862.6 Å ($R_D = 1.09707 \times 10^{-3}$ Å^{-1}).   *Ans.*   4861.3 Å

19.47. Calculate the first and second Bohr radii for positronium.   *Ans.*   1.06 Å; 4.23 Å

19.48. (*a*) Calculate the first three energy levels for positronium.   (*b*) Find the wavelength of the H_α line ($3 \to 2$ transition) of positronium.   *Ans.*   (*a*) -6.8 eV, -1.7 eV, -0.76 eV; (*b*) 1313 Å

19.49. For the π-mesic atom of ^{208}Pb ($m_\pi = 273m_e$), calculate (*a*) the first two Bohr radii, (*b*) the energies of the first two Bohr orbits, (*c*) the energy of the photon released when the π-meson makes a transition from the second to the first Bohr orbit.
Ans.   (*a*) 5.39 fm (inside the nucleus), 9.45 fm; (*b*) -15.35 MeV, -6.25 MeV; (*c*) 9.1 MeV

Chapter 20

Electron Orbital Motion and the Zeeman Effect

20.1 ORBITAL ANGULAR MOMENTUM FROM A CLASSICAL VIEWPOINT

Consider a particle of mass m moving in an elliptical orbit under the influence of a central force, as shown in Fig. 20-1. The vector angular momentum about the force center, $\mathbf{L}$, has magnitude mvd, where d is the perpendicular distance between the velocity direction and the force center, and v is the particle's speed. The direction of $\mathbf{L}$ is given by the usual right-hand rule, as shown in the figure. From Newton's second law, the net torque τ on the particle will be equal to the rate at which its angular momentum changes: $\tau = d\mathbf{L}/dt$. However, since the force on the particle is a central force, the torque exerted will be zero. Hence the particle's angular momentum $\mathbf{L}$ will have a constant magnitude and direction at every point along its elliptical trajectory.

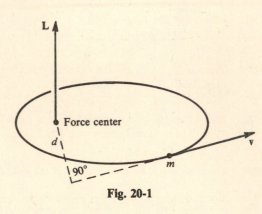

Fig. 20-1

Figure 20-2 shows various possible elliptical motions, ranging from a circle to nearly a straight line, all with the same major axis, $2a$. It can be shown that the total energy E (kinetic and potential) depends only on the value of the major axis and hence will have the *same value* (e.g. $E = -ke^2/2a$ for the Coulomb force; see Problem 19.15) for all these ellipses. The orbital angular momentum, however, will vary from ellipse to ellipse, ranging continuously from a maximum value of $a\sqrt{-2mE}$ for the circle to nearly zero for the approximately straight line. (A straight-line ellipse has zero angular momentum because $d = 0$ everywhere along the trajectory.)

Since $\mathbf{L}$ is constant, the component $L_z = L \cos \theta$ in any direction in space will also be a constant during the elliptical motion. In the classical picture there is no restriction on θ; it can take on any value from 0° to 180°.

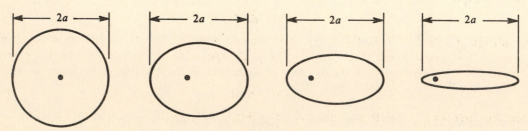

Fig. 20-2

20.2 CLASSICAL MAGNETIC DIPOLE MOMENT

An electron moving in a circular path will produce a current given by

$I = $ (charge on electron) $\times$ (number of times per second electron passes a given point) $= ef$

where f is the frequency of rotation of the electron.   The circular current loop in turn will produce a magnetic field very similar to the field produced by a small bar magnet, as shown in Fig. 20-3.   Just as with the bar magnet, there will be a *magnetic dipole moment* μ associated with the orbiting electron, whose magnitude is given by

$$|\mu| = IA = (ef)(\pi r^2) \qquad (20.1)$$

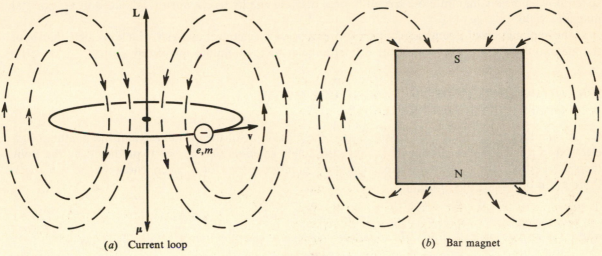

(*a*)   Current loop   (*b*)   Bar magnet

Fig. 20-3

and whose direction is opposite to that of **L** (because the electron has a negative charge).   Since

$$|\mathbf{L}| = mvr = m(2\pi rf)r = 2mf\pi r^2 = \frac{2m}{e}|\mu|$$

we have

$$\mu = -\frac{e}{2m}\mathbf{L} \qquad (20.2)$$

20.3   CLASSICAL ENERGY OF A MAGNETIC DIPOLE MOMENT IN AN EXTERNAL MAGNETIC FIELD

Suppose that either the current loop or the small bar magnet is placed in an *external* magnetic field **B**.   The loop experiences a torque,

$$\tau = \mu \times \mathbf{B} \qquad (20.3)$$

tending to align μ with **B**.   Thus the system has potential energy, E_B, the change in which gives the work done by the torque when the orientation of μ changes.   By integration of the torque we can show that

$$E_B = -\mu \cdot \mathbf{B} \qquad (20.4)$$

Choosing the direction of **B** as the z-direction and using (*20.2*), we have for an orbiting electron:

$$E_B = \frac{e}{2m}\mathbf{L} \cdot \mathbf{B} = \frac{e}{2m}L_z B \qquad (20.5)$$

20.4   THE ZEEMAN EXPERIMENT

An experiment measuring the effects of the interaction between an atom's internal magnetic moment and an external magnetic field was performed before the advent of quantum mechanics by the Dutch physicist Pieter Zeeman in 1896.   In a Zeeman experiment an atom is placed in an external magnetic field and its excitation spectrum is measured and compared with the spectrum when there is no magnetic field present. This could be accomplished, for example, by measuring the wavelengths of the radiation emitted from a discharge tube when it is placed in a magnetic field.

When the experiment is performed it is found that in the presence of the external field each spectral line is *split* into a number of discrete lines. Further, the observed change in frequency of the lines is directly proportional to the magnitude of the applied magnetic field. This observation of extra spectral lines means that an atom has additional discrete energy levels when it is placed in an external magnetic field.

The explanation of Zeeman splitting requires a wave mechanical analysis which predicts that both the magnitude and the direction of the orbital angular momentum are quantized.

20.5 QUANTIZATION OF THE MAGNITUDE OF THE ORBITAL ANGULAR MOMENTUM

A quantum mechanical analysis shows that the magnitude of the orbital angular momentum of an electron in a one-electron atom will *not* have the single value $n\hbar$ ($\hbar = h/2\pi$) as predicted by the Bohr theory. Instead, for a given principal quantum number n (i.e. for a given energy $E_n = -E_1^\circ/n^2$), there are n possible values of

$$|\mathbf{L}| = \sqrt{l(l+1)}\,\hbar \qquad (20.6)$$

where l is an integer, called the *orbital angular momentum quantum number*, with the range

$$l = 0, 1, 2, \ldots, n-1 \qquad (20.7)$$

In particular, for the lowest value of energy, corresponding to $n = 1$, the value for l is zero, and therefore the orbital angular momentum is also zero.

20.6 QUANTIZATION OF THE DIRECTION OF THE ORBITAL ANGULAR MOMENTUM

Suppose that a one-electron atom is placed in an external magnetic field, whose direction we take as the z-direction. A wave mechanical analysis shows that the *direction* of the orbital angular momentum vector $\mathbf{L}$ cannot be arbitrary. Instead, $\mathbf{L}$ will be oriented such that the component L_z along the z-direction will be quantized, with discrete values

$$L_z = m_l \hbar \qquad (20.8)$$

where m_l is an integer, called the *magnetic quantum number*, with the range

$$m_l = l, l-1, l-2, \ldots, 0, \ldots, -(l-1), -l \qquad (20.9)$$

Note that for a given l, the maximum value of $L_z (= l\hbar)$ is *less than* the magnitude of $\mathbf{L}$ $(= \sqrt{l(l+1)}\,\hbar)$.

20.7 EXPLANATION OF THE ZEEMAN EFFECT

Quantum mechanics states that the energy of a single-electron atom placed in an external magnetic field is changed by the potential energy (20.5). Now, however, L_z is quantized according to (20.8), so that the total energy is

$$E = E_0 + E_B = E_0 + m_l \frac{e\hbar}{2m} B \qquad (20.10)$$

where E_0 is the quantized energy before the field B is turned on. Thus, in the presence of a magnetic field, each energy level E_0 will be split into $2l + 1$ equal-spaced sublevels, with the spacing proportional to B. The factor $e\hbar/2m$ is called the *Bohr magneton*; its value is

$$\frac{e\hbar}{2m} = 5.79 \times 10^{-5} \text{ eV/T} = 9.27 \times 10^{-24} \text{ J/T}$$

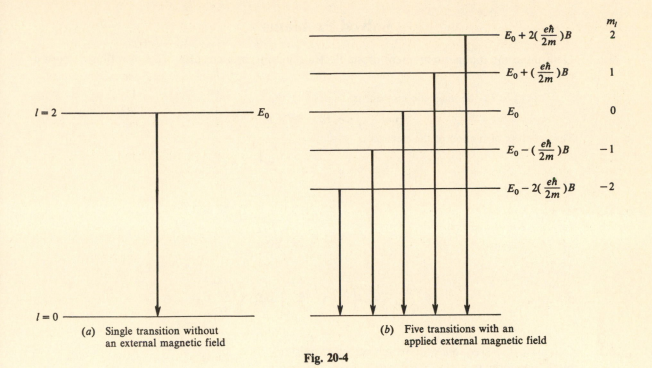

Fig. 20-4

(a) Single transition without an external magnetic field

(b) Five transitions with an applied external magnetic field

Since there are more discrete energy levels available after a magnetic field is turned on, there will be additional discrete lines seen in the excitation spectrum of an atom when it is placed in an external magnetic field, as illustrated in Fig. 20-4.

It is found that the more intense transitions in atoms obey the following selection rules:

$$\Delta l = \pm 1 \qquad \Delta m_l = \pm 1 \text{ or } 0 \qquad (20.11)$$

For these *electric dipole transitions*, (20.10) gives $\Delta E = \Delta E_0$ (the zero-field spectral line) and

$$\Delta E = \Delta E_0 \pm \frac{e\hbar}{2m} B$$

i.e. two new lines shifted in energy from the zero-field line by the absolute amount

$$\Delta E_{Zee} = \frac{e\hbar}{2m} B \qquad (20.12)$$

Other transitions may occur, but they result in much weaker spectral lines. In any case, the energy or frequency differences among the new lines will be proportional to the magnitude of the applied field.

The above predictions correspond exactly to what is observed in the "normal" Zeeman effect. The discrete splittings are clear experimental evidence of the phenomenon of orbital angular momentum quantization. If the orientation of the angular momentum were not quantized, then L_z could take on all possible values, as in the Bohr theory, and the lines would be broadened into a continuous band instead of the discrete values that are experimentally observed. The above analysis, however, does not explain *all* the lines observed in Zeeman experiments. There are additional transitions that fall into the category of the *anomalous Zeeman effect*, which involves the concept of electron spin. The anomalous Zeeman effect will be treated in Problems 24.17 through 24.21; it reduces to the normal Zeeman effect for sufficiently strong magnetic fields (Problem 24.22).

Solved Problems

20.1. Determine the magnetic moment of an electron moving in a circular orbit of radius r about a proton.

From (20.1), the magnetic moment is

$$\mu = IA = (ef)(\pi r^2)$$

The equation of motion of the electron is

$$F_{\text{rad}} = ma_{\text{rad}} \quad \text{or} \quad \frac{ke^2}{r^2} = \frac{mv^2}{r}$$

from which

$$f = \frac{v}{2\pi r} = \frac{1}{2\pi}\sqrt{\frac{ke}{mr}}$$

Therefore,

$$\mu = \pi e r^2 \left(\frac{1}{2\pi}\sqrt{\frac{ke}{mr}} \right) = \frac{e^2}{2}\sqrt{\frac{kr}{m}}$$

20.2. Calculate the frequency at which an electron's orbital magnetic moment μ precesses in a magnetic field **B**.

A magnetic moment in a magnetic field will experience a torque τ, given by (20.3) as

$$\tau = \mu \times \mathbf{B} = -\frac{e}{2m}\mathbf{L} \times \mathbf{B}$$

This torque will cause a change in the angular momentum given by

$$\tau = \frac{d\mathbf{L}}{dt} = -\frac{e}{2m}\mathbf{L} \times \mathbf{B}$$

The change in **L**, $d\mathbf{L}$, is perpendicular to both **L** and **B**, as shown in Fig. 20-5, resulting in a precession of **L** about the direction of **B**. From the figure it is seen that

$$d\phi = \frac{|d\mathbf{L}|}{L \sin \theta}$$

from which

$$\omega_p = \frac{d\phi}{dt} = \frac{\left| \dfrac{d\mathbf{L}}{dt} \right|}{L \sin \theta} = \frac{\dfrac{e}{2m} LB \sin \theta}{L \sin \theta} = \frac{e}{2m} B$$

This is known as the *Larmor precession*, and ω_p is equal to the frequency difference observed in the normal Zeeman effect.

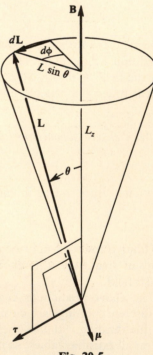

Fig. 20-5

20.3. Using the results of quantum mechanics, calculate the magnetic moments that are possible for an $n = 3$ level.

For $n = 3$ the possible values of l are 2, 1, 0; and $L = \sqrt{l(l+1)}\,\hbar$.
For $l = 2$:

$$\mu = \frac{e}{2m} L = \frac{e\hbar}{2m}\sqrt{l(l+1)} = \left(0.927 \times 10^{-23}\ \frac{\text{J}}{\text{T}} \right)\sqrt{2(2+1)} = 2.27 \times 10^{-23}\ \frac{\text{J}}{\text{T}}$$

For $l = 1$:

$$\mu = \frac{e\hbar}{2m} \sqrt{l(l+1)} = \left(0.927 \times 10^{-23} \ \frac{J}{T}\right)(\sqrt{2}) = 1.31 \times 10^{-23} \ \frac{J}{T}$$

For $l = 0$, $\mu = 0$.

Notice that none of these results agrees with what the Bohr theory predicts.   From the Bohr theory $L = n\hbar$, so

$$\mu_B = \frac{e}{2m} L = \frac{e}{2m} (3\hbar) = 3\left(\frac{e\hbar}{2m}\right)$$

$$= 3\left(0.927 \times 10^{-23} \ \frac{J}{T}\right) = 2.78 \times 10^{-23} \ \frac{J}{T}$$

20.4.   Show the possible orientations of the orbital angular momentum vector **L** for $l = 0, 1, 2, 3$ and 4.

The possible values of L_z are $m_l\hbar$, with m_l taking on all integer values between $+l$ and $-l$.   The corresponding possible orientations for the orbital angular momentum vector are shown in Fig. 20-6.

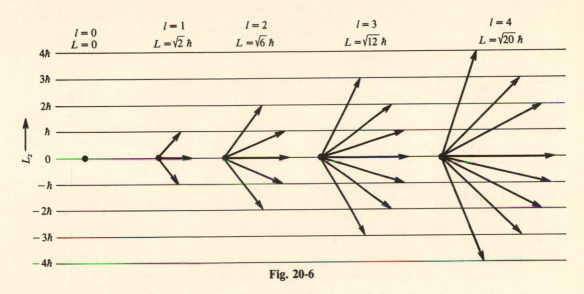

Fig. 20-6

20.5.   Determine the normal Zeeman splitting of the cadmium red line of 6438 Å when the atoms are placed in a magnetic field of 0.009 T.

The change in wavelength is found by taking the differential of $E = hc/\lambda$:

$$dE = -hc \frac{d\lambda}{\lambda^2} \qquad \text{or} \qquad |d\lambda| = \frac{\lambda^2 |dE|}{hc}$$

The energy shift is found from (20.12):

$$|dE| = \Delta E_{Zee} = \frac{e\hbar}{2m} B = \left(5.79 \times 10^{-5} \ \frac{eV}{T}\right)(0.009 \ T) = 5.21 \times 10^{-7} \ eV$$

giving

$$|d\lambda| = \frac{\lambda^2 |dE|}{hc} = \frac{(6438 \ \text{Å})^2 (5.21 \times 10^{-7} \ eV)}{12.4 \times 10^3 \ eV \cdot \text{Å}} = 1.74 \times 10^{-3} \ \text{Å}$$

20.6.   What magnetic flux density B is required to observe the normal Zeeman effect if a spectrometer can resolve spectral lines separated by 0.5 Å at 5000 Å?

From Problem 20.5,

$$\frac{|d\lambda|}{\lambda} = \frac{|dE|}{hc/\lambda} = \frac{(e\hbar/2m)B}{hc/\lambda}$$

giving

$$B = \frac{|d\lambda|}{\lambda}\left(\frac{hc}{\lambda}\right)\left(\frac{2m}{e\hbar}\right) = \left(\frac{0.5\ \text{Å}}{5000\ \text{Å}}\right)\left(\frac{12.4 \times 10^3\ \text{eV}\cdot\text{Å}}{5000\ \text{Å}}\right)\left(\frac{1}{5.79 \times 10^{-5}\ \text{eV/T}}\right) = 4.28\ \text{T}$$

20.7. In a normal Zeeman experiment the calcium 4226 Å line splits into 3 lines separated by 0.25 Å in a magnetic field of 3 T. Determine e/m for the electron from these data.

From Problem 20.6 ($\hbar = h/2\pi$):

$$\frac{|d\lambda|}{\lambda} = \frac{(e/4\pi m)B}{c/\lambda}$$

Solving for e/m, one obtains

$$\frac{e}{m} = \frac{4\pi}{B}\left(\frac{c}{\lambda^2}\right)|d\lambda| = \frac{4\pi}{3\ \text{T}}\left[\frac{3 \times 10^8\ \text{m/s}}{(4226 \times 10^{-10}\ \text{m})^2}\right](0.25 \times 10^{-10}\ \text{m}) = 1.76 \times 10^{11}\ \text{C/kg}$$

where we have used the conversion

$$1\ \text{T} \equiv 1\ \frac{\text{kg}}{\text{A}\cdot\text{s}^2} = 1\ \frac{\text{kg}}{\text{C}\cdot\text{s}}$$

20.8. Transitions occur in an atom between $l = 2$ and $l = 1$ states in a magnetic field of 0.6 T. If the wavelength before the field was turned on was 5000 Å, determine the wavelengths that are observed.

The energy separation between adjacent levels is given by

$$\Delta E_{Zee} = \frac{e\hbar}{2m}\,B = \left(5.79 \times 10^{-5}\ \frac{\text{eV}}{\text{T}}\right)(0.6\ \text{T}) = 3.47 \times 10^{-5}\ \text{eV}$$

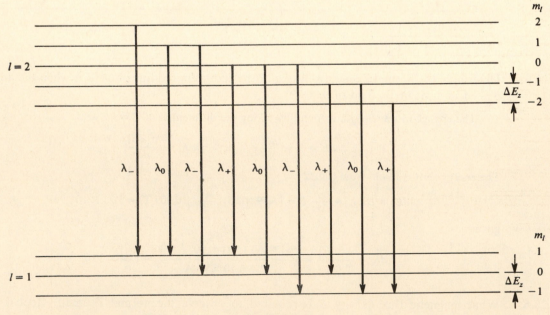

Fig. 20-7

Therefore, as in Problem 20.5,

$$|d\lambda| = \frac{\lambda^2 \, \Delta E_{Zee}}{hc} = \frac{(5000 \text{ Å})^2 (3.47 \times 10^{-5} \text{ eV})}{12.4 \times 10^3 \text{ eV} \cdot \text{Å}} = 0.07 \text{ Å}$$

The transitions must obey the selection rule $\Delta m_l = +1, 0, -1$; they are shown in Fig. 20-7. Only three different wavelengths are observed from the nine possible transitions:

$$\lambda_+ = 5000.07 \text{ Å} \qquad \lambda_0 = 5000 \text{ Å} \qquad \lambda_- = 4999.93 \text{ Å}$$

Supplementary Problems

20.9. An electron in He$^+$ is in an $n = 2$ orbit. What is its magnetic moment due to its orbital motion according to the Bohr theory?   *Ans.* 1.85×10^{-23} J/T

20.10. Do Problem 20.9 using quantum mechanical theory.   *Ans.* 1.31×10^{-23} J/T or 0

20.11. An electron in a circular orbit has an angular momentum of $\sqrt{2} \, \hbar$. In a field of 0.5 T, what is its Larmor frequency?   *Ans.* 6.99×10^9 Hz

20.12. For $l = 3$ calculate the possible angles **L** makes with the z-axis.   *Ans.*   16.8°; 35.3°; 60°

20.13. Determine the normal Zeeman splitting in the mercury 4916 Å line when in a magnetic field of 0.3 T.   *Ans.* 3.38×10^{-2} Å

20.14. What will the separation be between adjacent normal Zeeman components for emitted radiation of 4500 Å in a magnetic field of 0.4 T?   *Ans.* 3.78×10^{-2} Å

20.15. A 5000 Å line exhibits a normal Zeeman splitting of 1.1×10^{-3} Å. Find the magnetic field.   *Ans.* 9.42×10^{-3} T

20.16. What is the frequency difference in the photons emitted in a normal Zeeman effect corresponding to transitions from adjacent magnetic sublevels to the same final state in a magnetic field of 1.2 T?   *Ans.* 1.68×10^{10} Hz

20.17. Transitions occur in an atom between an $l = 3$ and an $l = 2$ state in a field of 0.2 T. If the wavelength before the field was turned on was 4000 Å, determine the final wavelengths observed.   *Ans.*   4000.0149 Å; 4000 Å; 3999.9851 Å

Chapter 21

The Stern–Gerlach Experiment and Electron Spin

21.1 THE STERN–GERLACH EXPERIMENT

In the *Stern–Gerlach experiment*, performed in 1921, a beam of silver atoms having zero total orbital angular momentum passes through an *inhomogeneous* magnetic field and strikes a photographic plate, as shown in Fig. 21-1. Any deflection of the beam when the magnetic field is turned on is measured on the photographic plate.

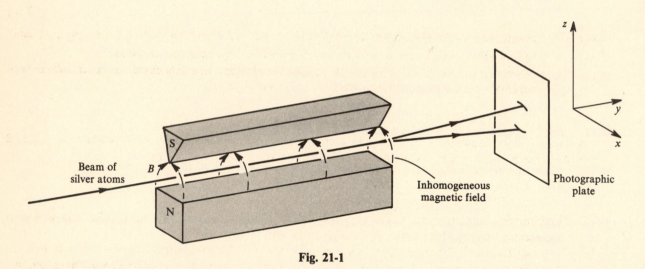

Fig. 21-1

The purpose of the *inhomogeneous* magnetic field is to produce a deflecting force on any magnetic moments that are present in the beam. If a homogeneous magnetic field were used, each magnetic moment would experience only a torque and no deflecting force. In an inhomogeneous magnetic field, however, a net deflecting force will be exerted on each magnetic moment μ_s. For the situation of Fig. 21-1,

$$F_z = \mu_s \cos \theta \, \frac{dB}{dz} \qquad (21.1)$$

where θ is the angle between μ_s and **B**, and dB/dz is the gradient of the inhomogeneous field (see Problem 21.1).

In the experiment it is found that when the beam strikes the photographic plate it has split into two distinct parts, with equal numbers of atoms deflected above and below the point where the beam strikes when there is no magnetic field. Because the atoms have zero total orbital angular momentum and therefore zero magnetic moment due to orbital electron motion, the magnetic interaction that produced the deflections must come from another type of magnetic moment.

21.2 ELECTRON SPIN

In 1925, S. A. Goudsmit and G. E. Uhlenbeck suggested that an electron possesses an *intrinsic* angular momentum called its *spin*. The extra magnetic moment μ_s associated with the intrinsic spin angular momentum **S** of the electron accounts for the deflection of the beam observed in the Stern–Gerlach experiment.

Similar to the orbital angular momentum, the electron's intrinsic angular momentum and associated magnetic moment are quantized both in magnitude and direction. The two equally spaced lines observed in the Stern–Gerlach experiment show that the intrinsic angular momentum can assume only two orientations with respect to the direction of the impressed magnetic field. In Section 20.6 it was shown that for orbital motion specified by the quantum number l, the component of the orbital magnetic moment along the magnetic field can have $2l + 1$ discrete values. Similarly, if the quantum number for the spin angular momentum is specified by s, we have, since there are only two orientations possible, $2 = 2s + 1$, giving the unique value $s = 1/2$. The magnitude of the spin angular momentum **S** is then

$$|\mathbf{S}| = \sqrt{s(s+1)}\, \hbar = \sqrt{\tfrac{1}{2}\left(\tfrac{1}{2}+1\right)}\, \hbar = \frac{\sqrt{3}}{2}\, \hbar \qquad (21.2)$$

The component S_z along the z-direction is

$$S_z = m_s \hbar \qquad m_s = s, s - 1 = \tfrac{1}{2}, -\tfrac{1}{2} \qquad (21.3)$$

The two orientations of **S** are commonly referred to as "spin up" ($m_s = +\tfrac{1}{2}$) and "spin down" ($m_s = -\tfrac{1}{2}$) (although the spin can never point in the positive or negative z-direction).

It is also found that the electron's intrinsic magnetic moment μ_s and intrinsic angular momentum **S** are proportional to each other; their relationship can be written as

$$\mu_s = -g_s \frac{e}{2m} \mathbf{S} \qquad (21.4)$$

The dimensionless quantity g_s is called the *gyromagnetic ratio*; for the electron, it has the value 2.002 (we shall use $g_s = 2.0$ in the problems). A comparison of (21.4) with (20.2) gives

$$g_s = \frac{|\mu_s|/|\mathbf{S}|}{|\mu|/|\mathbf{L}|}$$

Thus the ratio of magnetic moment to angular momentum is about twice as great for electron spin as it is for the electron's orbital motion.

The unique value $1/2$ for the spin quantum number is a characteristic as basic to the electron as its unique charge and mass. The properties of electron spin were first explained by Dirac around 1928, by combining the principles of wave mechanics with the theory of relativity. It should be noted that particles other than electrons, e.g. protons and neutrons, also possess an intrinsic angular momentum.

Solved Problems

21.1. Derive (21.1).

The potential energy of an electron in a magnetic field is [cf. (20.4)]

$$E_B = -\mu_s \cdot \mathbf{B} = -\mu_{sx} B_x - \mu_{sy} B_y - \mu_{sz} B_z$$

For the field of Fig. 21-1, $B_y \equiv 0$, and B_x and B_z depend only on x and z. Therefore,

$$F_x = -\frac{\partial E_B}{\partial x} = \mu_{sx} \frac{\partial B_x}{\partial x} + \mu_{sz} \frac{\partial B_z}{\partial x}$$

$$F_y = -\frac{\partial E_B}{\partial y} = 0$$

$$F_z = -\frac{\partial E_B}{\partial z} = \mu_{sx} \frac{\partial B_x}{\partial z} + \mu_{sz} \frac{\partial B_z}{\partial z}$$

But, along the beam axis, $\partial B_z / \partial x = 0$ (by symmetry) and $B_x = \partial B_x / \partial z = 0$ (by antisymmetry); also, $\partial B_x / \partial x$ will be very small.   Consequently,

$$F_x \approx 0 \qquad F_y = 0 \qquad F_z = \mu_{sz} \frac{dB}{dz} = \mu_s \cos \theta \frac{dB}{dz}$$

21.2.   Determine the maximum separation of a beam of hydrogen atoms that moves a distance of 20 cm with a speed of 2×10^5 m/s perpendicular to a magnetic field whose gradient is 2×10^2 T/m.   Neglect the magnetic moment of the proton (see Problem 26.9).

In the ground state, hydrogen atoms have zero orbital angular momentum.   From Problem 21.1, the force on a hydrogen atom is

$$F_z = \mu_{sz} \frac{dB}{dz}$$

By (21.3) and (21.4), with $g_s = 2$, $\mu_{sz} = -(e/m)m_s \hbar$, so that

$$|F_z| = \frac{e\hbar}{m} |m_s| \frac{dB}{dz} = \frac{e\hbar}{2m} \frac{dB}{dz} = \left(9.27 \times 10^{-24} \frac{\text{J}}{\text{T}}\right)\left(2 \times 10^2 \frac{\text{T}}{\text{m}}\right) = 1.85 \times 10^{-21} \text{ N}$$

Using the constant-acceleration formulas $\Delta z = \frac{1}{2} a_z t^2$ and $\Delta y = vt$ (see Fig. 21-1 for the coordinates), we obtain

$$\Delta z = \frac{1}{2} a_z t^2 = \frac{1}{2} \left(\frac{F_z}{m_{\text{H}}} \right)\left(\frac{\Delta y}{v} \right)^2$$

The mass of hydrogen is 1.67×10^{-27} kg, so

$$\Delta z = \frac{1}{2} \left(\frac{1.85 \times 10^{-21} \text{ N}}{1.67 \times 10^{-27} \text{ kg}} \right)\left(\frac{0.20 \text{ m}}{2 \times 10^5 \text{ m/s}} \right)^2 = 5.54 \times 10^{-7} \text{ m}$$

Since this is the displacement up or down, the total separation is $2 \Delta z = 1.11 \times 10^{-6}$ m.

21.3.   Determine the energy difference between the electrons that are "aligned" and "anti-aligned" with a uniform magnetic field of 0.8 T when a beam of free electrons moves perpendicular to the field.

From Problem 21.1 with $B_x = B_y = 0$

$$E_B = - B\mu_{sz} = - B\left(- \frac{e\hbar}{m} \right)m_s$$

Hence,

$$\Delta E_B = B \frac{e\hbar}{m} \Delta m_s = (0.8 \text{ T})\left(2 \times 5.79 \times 10^{-5} \frac{\text{eV}}{\text{T}}\right)\left[\frac{1}{2} - \left(- \frac{1}{2} \right)\right] = 9.26 \times 10^{-5} \text{ eV}$$

21.4.   The 21 cm line is used in radioastronomy to map the galaxy.   The line arises from the emission of a photon when the electron in a galactic hydrogen atom "flips" its spin from being aligned to being anti-aligned with the spin of the proton in the hydrogen atom.   What is the magnetic field the electron experiences?

$$\Delta E = \frac{hc}{\lambda} = \frac{12.4 \times 10^3 \text{ eV} \cdot \text{Å}}{21 \times 10^8 \text{ Å}} = 5.9 \times 10^{-6} \text{ eV}$$

From Problem 21.3 we have

$$\Delta E_B = B \frac{e\hbar}{m} \Delta m_s$$

$$5.9 \times 10^{-6} \text{ eV} = B\left[2 \times 5.79 \times 10^{-5} \frac{\text{eV}}{\text{T}} \right]\left[\frac{1}{2} - \left(- \frac{1}{2} \right)\right]$$

$$B = 0.0510 \text{ T}$$

Supplementary Problems

21.5. In a Stern–Gerlach experiment silver atoms traverse a distance of 0.1 m through an inhomogeneous magnetic field with a gradient of 60 T/m. If the separation observed on the collector plate is 0.15 mm, determine the velocity of the silver atoms.   The mass of a silver atom is 1.79×10^{-25} kg.
Ans.   455 m/s

21.6. What is the energy difference between the two electron spin orientations when the electrons are in a magnetic field of 0.5 T?   *Ans.* 5.79×10^{-5} eV

21.7. Refer to Problem 21.6.   Determine the wavelength of the radiation that can cause the electrons to "flip" their spins.   *Ans.*   2.14 cm

21.8. The wavelength needed to cause an electron-spin "flip" is 1.5 cm.   Calculate the magnetic field the electron is in.   *Ans.*   0.714 T

Chapter 22

Electron Spin and Fine Structure

22.1 SPIN-ORBIT COUPLING

In Section 20.3 it was shown that a magnetic moment μ placed in a magnetic field **B** has a potential energy E_B given by

$$E_B = -\mu \cdot \mathbf{B} \qquad (22.1)$$

When this expression was developed, only *external* magnetic fields were considered, but the result holds in general.

In a semiclassical Bohr picture, the electron revolves around the nucleus with an orbital angular momentum **L**. From the point of view of the electron, however, the positively charged nucleus revolves around the electron with the same angular velocity. The revolving nucleus will therefore produce a magnetic field **B** at the location of the electron that will be parallel to the electron's orbital angular momentum **L**. This internal magnetic field in turn will interact with the electron's intrinsic magnetic moment μ_s described in Section 21.2. Since μ_s is proportional to the electron's intrinsic spin **S**, and since **B** and **L** are proportional for a given orbit, it is seen that there will be a potential energy, E_s, of the form

$$E_s = K\mathbf{L} \cdot \mathbf{S} \qquad (22.2)$$

where the precise value of the quantity K need not be considered here.

Effectively, the spin-orbit interaction behaves like an internal Zeeman effect, splitting each energy level for which $\mathbf{L} \neq 0$ into two sublevels, corresponding to the two values of S_z allowed by (21.3).

22.2 FINE STRUCTURE

Since there are more energy levels available than were previously considered, there should be additional lines seen in the spectrum of hydrogen, as indicated in Fig. 22-1. Such additional lines, or *fine structure*, are readily observed with spectrometers of moderately high resolution. With such an instrument it is found that many spectral lines that were previously seen as single are actually composed of two or more distinct lines, separated from each other by a few angstroms in wavelength.

It was the observation of the fine structure of spectral lines that originally motivated Uhlenbeck and Goudsmit to introduce the concept of electron spin.

22.3 TOTAL ANGULAR MOMENTUM (THE VECTOR MODEL)

In classical mechanics the *total* angular momentum (orbital plus spin) is an important quantity because its rate of change is equal to the net torque applied to the system. Similarly, in wave mechanics, the *total* angular momentum **J**, found from the vector addition

$$\mathbf{J} = \mathbf{L} + \mathbf{S}$$

plays an important role. Because the vector model applies to many-electron, as well as one-electron, atoms, we here introduce the following notation: *Quantum numbers specifying states of individual electrons will be denoted, as previously, by small letters; quantum numbers representing atomic states will be denoted by capital letters. In the special case of a one-electron atom the electronic state is the atomic state, and capital letters will be used.*

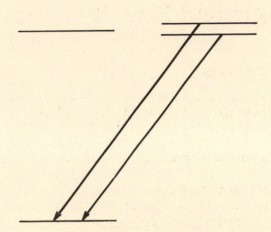

Energy levels with
L·S term neglected

Energy levels with L·S term included

Fig. 22-1

In this notation, then, the magnitude of **J** is quantized according to

$$|\mathbf{J}| = \sqrt{J(J+1)}\, \hbar \qquad (22.3)$$

The quantum number J has the possible values

$$J = L + S, L + S - 1, \ldots, |L - S| \qquad (22.4)$$

where L and S are the orbital and spin quantum numbers. As is the case for the orbital and spin angular momenta, the component of **J** in a physically defined z-direction is separately quantized. We have

$$J_z = M_J \hbar \qquad M_J = J, J - 1, J - 2, \ldots, -J \qquad (22.5)$$

For a hydrogenlike atom, $S = 1/2$ and (22.4) becomes

$$J = \begin{cases} L + S, L - S & L > 0 \\ S & L = 0 \end{cases} \qquad (22.6)$$

Solved Problems

22.1. Express **L·S** in terms of J, L, and S.

Evaluating $\mathbf{J} \cdot \mathbf{J} \equiv |\mathbf{J}|^2$, where $\mathbf{J} = \mathbf{L} + \mathbf{S}$, we have

$$|\mathbf{J}|^2 = (\mathbf{L} + \mathbf{S}) \cdot (\mathbf{L} + \mathbf{S}) = \mathbf{L} \cdot \mathbf{L} + 2\mathbf{L} \cdot \mathbf{S} + \mathbf{S} \cdot \mathbf{S} = |\mathbf{L}|^2 + 2\mathbf{L} \cdot \mathbf{S} + |\mathbf{S}|^2$$

or

$$\mathbf{L} \cdot \mathbf{S} = \tfrac{1}{2}\left(|\mathbf{J}|^2 - |\mathbf{L}|^2 - |\mathbf{S}|^2\right)$$

Substituting $|\mathbf{J}|^2 = J(J+1)\hbar^2$, $|\mathbf{L}|^2 = L(L+1)\hbar^2$, $|\mathbf{S}|^2 = S(S+1)\hbar^2$, one finds

$$\mathbf{L} \cdot \mathbf{S} = \tfrac{1}{2}\left[J(J+1) - L(L+1) - S(S+1)\right]\hbar^2$$

22.2. Calculate the possible values of $\mathbf{L} \cdot \mathbf{S}$ for $L = 1$ and $S = 1/2$.

From (22.4), the possible values of J are

$$J = L + S = 1 + \frac{1}{2} = \frac{3}{2}, \qquad J = L + S - 1 = |L - S| = 1 - \frac{1}{2} = \frac{1}{2}$$

From Problem 22.1,

$$\mathbf{L} \cdot \mathbf{S} = \frac{1}{2} [J(J+1) - L(L+1) - S(S+1)]\hbar^2$$

For $J = 3/2$:

$$\mathbf{L} \cdot \mathbf{S} = \frac{1}{2} \left[\frac{3}{2}\left(\frac{3}{2}+1\right) - 1(1+1) - \frac{1}{2}\left(\frac{1}{2}+1\right) \right]\hbar^2 = \frac{1}{2}\hbar^2$$

For $J = 1/2$:

$$\mathbf{L} \cdot \mathbf{S} = \frac{1}{2} \left[\frac{1}{2}\left(\frac{1}{2}+1\right) - 1(1+1) - \frac{1}{2}\left(\frac{1}{2}+1\right) \right]\hbar^2 = -\hbar^2$$

Figure 22-2 illustrates the relative orientations of the three vectors.

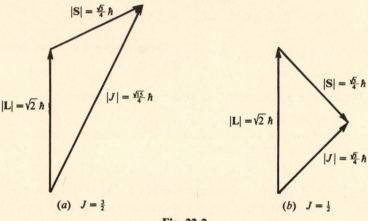

$(a)\quad J = \frac{3}{2}$ $\qquad\qquad\qquad\qquad\qquad (b)\quad J = \frac{1}{2}$

Fig. 22-2

22.3. Estimate the strength of the magnetic field produced by the electron's orbital motion which results in the two sodium D lines (5889.95 Å, 5895.92 Å).

This transition occurs between an $L = 1$ state and an $L = 0$ state; only the $L = 1$ state is split.   The difference in energy between the $L = 1$ substates can be obtained from the wavelength difference by

$$E = \frac{hc}{\lambda}$$

$$|dE| = \frac{hc|d\lambda|}{\lambda^2} = \frac{(12.4 \times 10^3 \text{ eV} \cdot \text{Å})(5.97 \text{ Å})}{(5890 \text{ Å})^2} = 2.13 \times 10^{-3} \text{ eV}$$

The energy difference can also be found from (22.1).   We have, using (21.4) with $g_s = 2$,

$$E_B = -\boldsymbol{\mu}_s \cdot \mathbf{B} = -\mu_{sz}B = \frac{e}{m} m_s \hbar B$$

so that

$$\Delta E_B = \frac{e\hbar}{m} B \, \Delta m_s = \frac{e\hbar}{m} B \left[\frac{1}{2} - \left(-\frac{1}{2}\right) \right] = 2\left(\frac{e\hbar}{2m}\right)B$$

Hence,

$$2.13 \times 10^{-3} \text{ eV} = 2\left(5.79 \times 10^{-5} \frac{\text{eV}}{\text{T}}\right)B$$

$$B = 18.4 \text{ T}$$

22.4. Calculate the value of K in (22.2) using the information given in Problem 22.3 and $S = 1/2$.

For $S = 1/2$, the two $L = 1$ states are $J = 3/2$ and $J = 1/2$. From Problem 22.2 we have

$$J = \frac{3}{2} : \quad \mathbf{L} \cdot \mathbf{S} = \frac{1}{2}\hbar^2$$

$$J = \frac{1}{2} : \quad \mathbf{L} \cdot \mathbf{S} = -\hbar^2$$

From (22.2) we have $\Delta E_s = K(\mathbf{L} \cdot \mathbf{S})_{upper} - K(\mathbf{L} \cdot \mathbf{S})_{lower}$, or

$$2.13 \times 10^{-3} \text{ eV} = K\left(\frac{1}{2}\hbar^2\right) - K(-\hbar^2) = K\left(\frac{3}{2}\hbar^2\right) = K\frac{3}{2}(0.658 \times 10^{-15} \text{ eV} \cdot \text{s})^2$$

Solving,

$$K = 3.28 \times 10^{27} \frac{1}{\text{eV} \cdot \text{s}^2}$$

Supplementary Problems

22.5. Calculate the possible values of J for $L = 3$ and $S = 1/2$.   *Ans.* $\frac{7}{2}; \frac{5}{2}$

22.6. For Problem 22.5 calculate $\mathbf{L} \cdot \mathbf{S}$.   *Ans.* $\frac{3}{2}\hbar^2; -2\hbar^2$

22.7. Estimate the strength of the magnetic field produced by the electron's orbital motion which results in the 7664.1 Å and 7699.0 Å lines observed in the $L = 1$ to $L = 0$ transition in potassium.   *Ans.* 63.3 T

22.8. Repeat Problem 22.4 with the data of Problem 22.7 if $S = 1/2$.   *Ans.* $1.13 \times 10^{28} \frac{1}{\text{eV} \cdot \text{s}^2}$

PART V: *Many-Electron Atoms*

Chapter 23

The Pauli Exclusion Principle

23.1 INTRODUCTION

To this point we have considered quantum mechanical systems possessing many energy levels but only one effective electron, namely hydrogenlike atoms. We have found that in the absence of strong spin-orbit coupling the behavior of the electron is described by specifying the values of its four quantum numbers (n, l, m_l, m_s) that are respectively associated with its energy, orbital angular momentum, the z-component of the orbital angular momentum, and the z-component of its spin. When these four quantum numbers are given it is said that the *state* of the (one-electron) system has been specified.

In this and the following chapters we shall describe quantum mechanical systems that have many energy levels and *more than one electron*.

23.2 THE PAULI EXCLUSION PRINCIPLE

By analyzing spectroscopic data from atoms with more than one electron, Wolfgang Pauli in 1924 came to the conclusion that in a quantum mechanical system *no two electrons can occupy the same state*. This result is called the *Pauli exclusion principle*; put another way, it states that no two electrons can have the same set of quantum numbers (n, l, m_l, m_s).

The Pauli exclusion principle correlates many important experimental facts of atomic structure and provides the explanation of the periodic table of the elements, the subject of Chapter 24. To illustrate the Pauli exclusion principle, however, we will here discuss the simple problem of one or more particles of mass m moving in a straight line and confined between the points 0 and a, i.e. particles in a one-dimensional "box" of length a.

23.3 A SINGLE PARTICLE IN A ONE-DIMENSIONAL BOX

The problem of a single particle in a one-dimensional box was solved previously in Problem 17.3. It was found that the energy of the particle could not vary continuously, but could have only the discrete values given by

$$E_n = n^2 \frac{h^2}{8ma^2} \qquad n = 1, 2, 3, \dots$$

Figure 23-1(a) shows these energy levels. We now take the particle in the box to be an electron with intrinsic spin. The state of the system is then specified by the pair of quantum numbers (n, m_s). In Fig. 23-1(b) the electron is in the $n = 1$ state with "spin up" $(m_s = +\frac{1}{2})$; in Fig. 23-1(c) it is in the $n = 3$ state with "spin down" $(m_s = -\frac{1}{2})$.

23.4 MANY PARTICLES IN A ONE-DIMENSIONAL BOX

The Pauli exclusion principle will have an important effect on the situation when more than one particle is in the one-dimensional box. In the following we assume that the energy levels are not altered when more than one particle is present.

With two electrons, the ground (lowest-energy) state of the system will have both electrons in the $n = 1$ energy level, one with spin up $(1, +\frac{1}{2})$ and one with spin down $(1, -\frac{1}{2})$, as shown in Fig. 23-2(a). Note that the two electrons do *not* have the same set of quantum numbers (n, m_s).

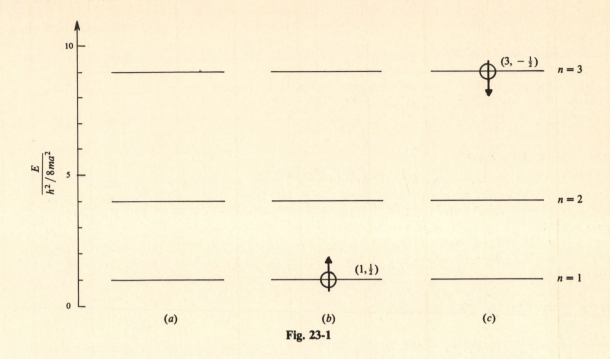

Fig. 23-1

Now consider what happens when a third electron is added to the system.   The Pauli exclusion principle prohibits this electron from occupying the $n = 1$ energy level; for if it were in the $n = 1$ level, two of the three electrons would have the same set of quantum numbers (n, m_s).   The third electron must therefore go to a *different* energy level, the $n = 2$ level if the system is in its ground state, as shown in Fig. 23-2(*b*) for a spin up configuration.

A similar line of reasoning shows that a fourth electron can be put into the $n = 2$ level, but when a fifth is added it must go into the $n = 3$ level, as shown in Fig. 23-2(*c*) for a spin down configuration. Thus it is seen that the Pauli exclusion principle has the effect of increasing the total energy of the

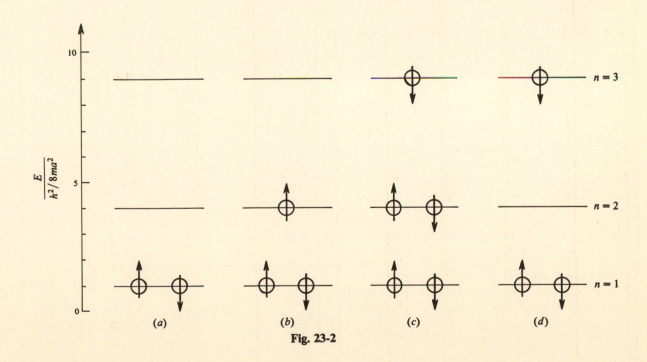

Fig. 23-2

ground state of the system to a much higher value than it would have if all the electrons occupied the $n = 1$ energy level.

Excited states of the above systems occur when the electrons do not occupy all the lowest available energy levels, as shown for a three-electron system in Fig. 23-2(d). As with one-electron systems, it is possible for energy in the form of photons to be emitted when excited electrons seek their ground state configurations.

Solved Problems

23.1. Calculate the first three energy levels for noninteracting electrons in an infinite square well of length 6 Å.

The energy levels are given by

$$E_n = \frac{n^2 h^2}{8ma^2} = \frac{n^2(hc)^2}{8(mc^2)a^2} = \frac{n^2(12.4 \times 10^3 \text{ eV} \cdot \text{Å})^2}{8(0.511 \times 10^6 \text{ eV})(6 \text{ Å})^2} = 1.04n^2 \text{ eV}$$

Therefore, $E_1 = 1.04$ eV, $E_2 = 4.16$ eV, $E_3 = 9.36$ eV.

23.2. What are the energies of the photons that would be emitted when the four-electron system in Fig. 23-3(a) returns to its ground state?

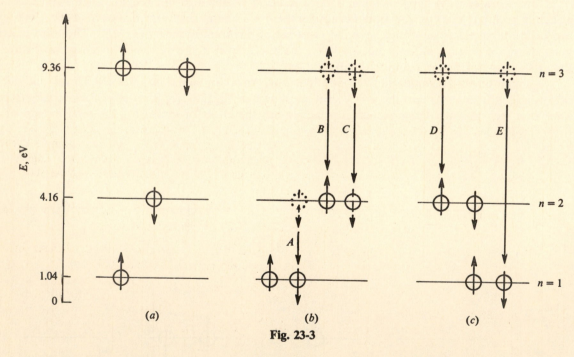

Fig. 23-3

The possible transitions that will return the system to its ground state are shown in Fig. 23-3(b) and (c). The energy of the emitted photon will be equal to the energy difference between the initial and final levels. Transitions B, C, and D have the same energy difference and therefore give rise to the same-energy photon.

$$E_A = E_2 - E_1 = 4.16 \text{ eV} - 1.04 \text{ eV} = 3.12 \text{ eV}$$
$$E_B = E_C = E_D = E_3 - E_2 = 9.36 \text{ eV} - 4.16 \text{ eV} = 5.20 \text{ eV}$$
$$E_E = E_3 - E_1 = 9.36 \text{ eV} - 1.04 \text{ eV} = 8.32 \text{ eV}$$

23.3. Consider three noninteracting particles in their ground states in an infinite square well [Fig. 23-4(a)]. What happens when a magnetic field is turned on which interacts with the spins of the particles?

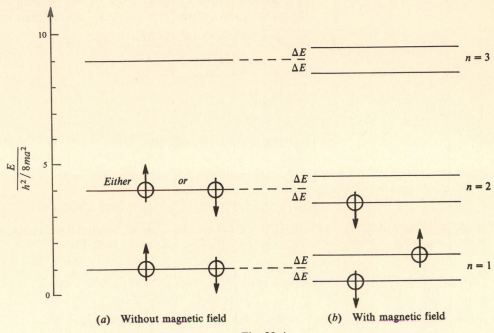

(a) Without magnetic field   (b) With magnetic field

Fig. 23-4

After the external magnetic field is applied, the new value (E_i) of each particle's energy equals the original value (E_n) plus the interaction energy:

$$E_i = E_n - \boldsymbol{\mu}_s \cdot \mathbf{B} = E_n - \mu_{sz} B = E_n + m_s \frac{e\hbar}{m} B$$

(see Problem 22.3). Since $m_s = \pm \frac{1}{2}$, the new levels will be displaced from the old by an amount $\Delta E = \pm e\hbar B/2m$, with a spin $-\frac{1}{2}$ particle occupying the lower sublevel and a spin $+\frac{1}{2}$ particle occupying the upper sublevel [Fig. 23-4(b)]. In contrast to the situation without the magnetic field present, the particle in the $n = 2$ level will have its spin $-\frac{1}{2}$ if the system is in the ground state.

23.4. In an infinite square well of length a there are 5×10^9 electrons per meter. If all the lowest energy levels are filled, determine the energy of the most energetic electron.

Since there are two electrons in each energy level, the total number of electrons up to and including the last or nth level is $N = 2n$. The number of electrons per unit length is therefore

$$\frac{N}{a} = \frac{2n}{a} = 5 \times 10^9 \text{ m}^{-1}$$

so that

$$\frac{n}{a} = (2.5 \times 10^9 \text{ m}^{-1})\left(10^{-10} \frac{\text{m}}{\text{Å}}\right) = 0.25 \text{ Å}^{-1}$$

The energy of the nth energy level is thus

$$E_n = \frac{n^2 h^2}{8ma^2} = \left(\frac{n}{a}\right)^2 \frac{(hc)^2}{8(mc^2)} = (0.25 \text{ Å}^{-1})^2 \frac{(12.4 \times 10^3 \text{ eV} \cdot \text{Å})^2}{8(0.511 \times 10^6 \text{ eV})} = 2.35 \text{ eV}$$

23.5. If a nucleus is approximated by a square well, there will be about 1 neutron per 10^{-15} m. In this approximation, determine the energy of the most energetic neutron. The neutron rest mass is 938 MeV.

Refer to the Problem 23.4.

$$\frac{N}{a} = \frac{2n}{a} = 10^{15} \text{ m}^{-1}$$

$$\frac{n}{a} = (0.5 \times 10^{15} \text{ m}^{-1})\left(10^{-10} \frac{\text{m}}{\text{Å}}\right) = 0.5 \times 10^5 \text{ Å}^{-1}$$

$$E_n = \left(\frac{n}{a}\right)^2 \frac{(hc)^2}{8(mc^2)} = (0.5 \times 10^5 \text{ Å}^{-1})^2 \frac{(12.4 \times 10^3 \text{ eV} \cdot \text{Å})^2}{8(938 \times 10^6 \text{ eV})} = 51.2 \times 10^6 \text{ eV} = 51.2 \text{ MeV}$$

Supplementary Problems

23.6.   What is the energy of the photon that would be emitted in a transition from the $n = 3$ to the $n = 2$ level in the infinite square well of Problem 23.1?   *Ans.*   5.20 eV

23.7.   An infinite square well has length 10^{-14} m.   What are the values of the first three energy levels for a neutron ($m_n = 938$ MeV) in the well?   *Ans.*   2.05 MeV; 8.20 MeV; 18.45 MeV

23.8.   Repeat Problem 23.7 for an electron in a 1 Å well.   *Ans.*   37.6 eV; 150 eV; 338 eV

23.9.   Repeat Problem 23.2 for neutrons.   *Ans.*   615 MeV; 1024 MeV; 1634 MeV

23.10.   Determine the energy required to cause the uppermost electron in Problem 23.3 to "flip" its spin if the magnetic field is 2 T.   *Ans.* 2.32×10^{-4} eV

Chapter 24

Many-Electron Atoms and the Periodic Table

24.1 SPECTROSCOPIC NOTATION FOR ELECTRON CONFIGURATIONS IN ATOMS

A great deal of information about the character of many-electron atomic states can be found by assuming in a first approximation that each electron moves independently in the field of the nucleus and the average field produced by the other electrons. The other existing interactions are treated separately, as will be described below. In such an *independent particle model* the quantum numbers n, l, m_l, and m_s are used to describe each electron's state.

For a given n the integer values that the quantum number l can take on are

$$l = 0, 1, 2, \ldots, n-1$$

The value of l will be designated by a lowercase letter according to the following scheme.

value of l:	0	1	2	3	4	5	$\ldots$
letter symbol:	s	p	d	f	g	h	$\ldots$

The convention for specifying the number of electrons in a particular *orbit*, defined by the quantum numbers (n, l), is to give n, followed by the letter symbol for l, with the number of electrons as a post-superscript. Various orbits are sequentially written one after the other, thereby defining an electron *configuration*. As an example, the configuration for the five electrons in the ground state of boron is $1s^2 2s^2 2p^1$.

Electrons with the same value of n are said to occupy the same electron *shell*. The various shells are designated by capital letters according to the following scheme.

value of n:	1	2	3	4	$\ldots$
shell letter:	K	L	M	N	$\ldots$

Within a given shell different values of l are possible; each value of l defines a *subshell* (which is thus equivalent to an orbit). For example, in the boron ground-state configuration, $1s^2 2s^2 2p^1$, there are two electrons in the s subshell and one electron in the p subshell of the L shell.

24.2 THE PERIODIC TABLE AND AN ATOMIC SHELL MODEL

The lowest-energy, or ground-state, electron configuration for many-electron atoms can be explained using the Pauli exclusion principle together with an atomic shell model. The Pauli principle states that no two electrons can have the same set of quantum numbers (n, l, m_l, m_s). Therefore the number of combinations of m_l and m_s for a given subshell (n, l) gives the maximum number of electrons in that subshell.

For each value of l there are $2l + 1$ values of m_l, and for each value of l and m_l there are two values of m_s ($m = \pm \frac{1}{2}$). Thus the maximum number of electrons that can be placed in a given subshell without violating the Pauli exclusion principle is $2(2l + 1)$, as shown in the following table.

value of l:	0	1	2	3
letter symbol of subshell:	s	p	d	f
maximum number of electrons:	2	6	10	14

133

The order in which the various subshells are filled for most atoms is shown by reading upwards in Fig. 24-1. The figure also indicates the relative energies of the electrons in any particular atom. The gaps observed in Fig. 24-1 occur at $Z = 2$, 10, 18, 36, 54, and 86, which are the inert or noble gases that are chemically inactive and very difficult to ionize. With the exception of He ($Z = 2$), the gaps correspond to the complete filling of a p subshell. The other properties of the periodic table of the elements can also be explained by the manner in which the different subshells fill, as shown in the Solved Problems.

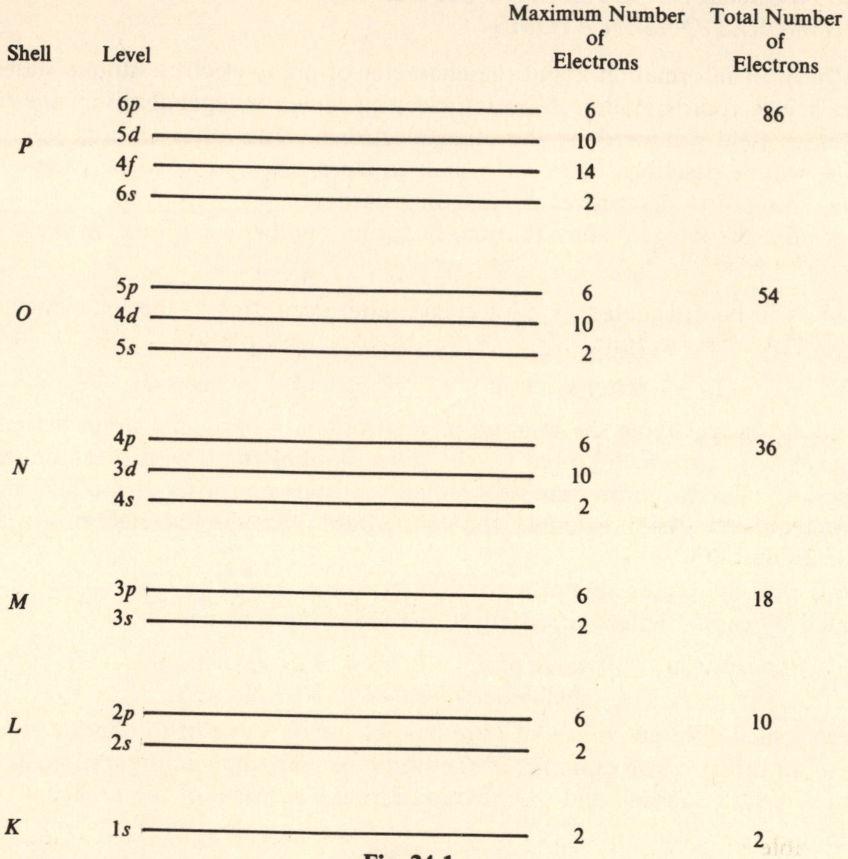

Shell	Level		Maximum Number of Electrons	Total Number of Electrons
P	6p		6	86
	5d		10	
	4f		14	
	6s		2	
O	5p		6	54
	4d		10	
	5s		2	
N	4p		6	36
	3d		10	
	4s		2	
M	3p		6	18
	3s		2	
L	2p		6	10
	2s		2	
K	1s		2	2

Fig. 24-1

24.3 SPECTROSCOPIC NOTATION FOR ATOMIC STATES

Each state of an atom is characterized by giving the set of quantum numbers L, S, J related respectively to the atom's total orbital angular momentum, total spin angular momentum, and total angular momentum (see Section 22.3).

The particular value of L for the atomic state is designated in spectroscopic notation by a capital letter according to the following scheme.

value of L:	0	1	2	3	4	5	...
letter symbol:	S	P	D	F	G	H	...

Atomic states are specified by giving the letter symbol of L with the value $2S + 1$ as a pre-superscript and the value of J as a post-subscript. As an example, in the ground state of boron $L = 1$, $S = \frac{1}{2}$, $J = \frac{1}{2}$ and the spectroscopic notation is $^2P_{1/2}$.

24.4 ATOMIC EXCITED STATES AND *LS* COUPLING

The mathematical analysis of many-electron atomic states is complicated by the fact that besides the Coulomb interaction between the electrons and the nucleus there are also residual Coulomb interactions between the individual electrons, interactions between the electron orbital angular momenta and the electron spins, and interactions between the spins of the different electrons. For light and medium-heavy atoms it is found that a scheme called "*LS* coupling," developed by Russell and Saunders in 1925, provides a method for understanding the observed atomic states. For *LS* coupling, the atom's orbital angular momentum **L** is the vector sum of the orbital angular momenta of the individual electrons,

$$\mathbf{L} = \sum_i \mathbf{L}_i \qquad (24.1)$$

Similarly, the atom's spin angular momentum **S** is the vector sum of the spin angular momenta of the individual electrons,

$$\mathbf{S} = \sum_i \mathbf{S}_i \qquad (24.2)$$

The atom's total angular momentum is then given by

$$\mathbf{J} = \mathbf{L} + \mathbf{S} \qquad (24.3)$$

as in Section 22.3. The magnitudes of the three atomic momentum vectors are quantized according to

$$|\mathbf{L}|^2 = L(L+1)\hbar^2 \qquad |\mathbf{S}|^2 = S(S+1)\hbar^2 \qquad |\mathbf{J}|^2 = J(J+1)\hbar^2 \qquad (24.4)$$

and their *z*-components are quantized according to

$$L_z = M_L \hbar \qquad S_z = M_S \hbar \qquad J_z = M_J \hbar \qquad (24.5)$$

The *z*-component quantum numbers are related to those of the individual electrons by the following addition rules:

$$M_L = \sum_i (m_l)_i \qquad M_S = \sum_i (m_s)_i \qquad M_J = M_L + M_S \qquad (24.6)$$

Knowing M_L, M_S, and M_J, one can infer L, S, and J from the conditions

$$M_L = L, L-1, L-2, \ldots, -L \qquad (24.7)$$

$$M_S = S, S-1, S-2, \ldots, -S \qquad (24.8)$$

$$M_J = J, J-1, J-2, \ldots, -J \qquad (24.9)$$

It is possible to excite an atom into energy levels above the ground state. In returning to its ground state the atom will emit radiation with a corresponding line spectrum. For strong transitions the following selection rules apply:

$$\Delta J = 0, \pm 1 \quad (\text{but } J = 0 \rightarrow J = 0 \text{ is not allowed})$$

$$\Delta L = 0, \pm 1 \qquad \Delta S = 0 \qquad \Delta M_J = 0, \pm 1 \quad (\text{but if } \Delta J = 0, M_J = 0 \rightarrow M_J = 0 \text{ is not allowed})$$

These are electric dipole transitions; other transitions occur but they are found to be much weaker. If the electric dipole transition involves just one electron, $\Delta L \neq 0$.

24.5 THE ANOMALOUS ZEEMAN EFFECT

In a semiclassical picture, the normal Zeeman effect (triple line-splitting) is associated with the precession of the atomic magnetic moment **μ** about an external magnetic field **B** (Problem 20.2). The stronger the field, the faster the precession and the greater the separation between the three spectral lines into which the zero-field line is split. When the **L·S** interaction is strong compared to the interaction of either vector with **B**, then **S** and **L** precess rapidly about **J**, producing a rapid precession of $|\mu$ about **J**; this system then precesses slowly about **B**. In this way arises the anomalous Zeeman effect, whose strength depends on the component of μ along **J**. In Problems 24.17 through 24.21 it is shown that anomalous Zeeman splitting produces more than three spectral lines.

Solved Problems

24.1. Show that under LS coupling the total angular momentum quantum number has the values given by (22.4), i.e.

$$J = L + S, L + S - 1, L + S - 2, \ldots, |L - S|$$

By (24.7) and (24.8),

$$M_L = L, L - 1, L - 2, \ldots, -(L - 2), -(L - 1), -L$$
$$M_S = S, S - 1, S - 2, \ldots, -(S - 2), -(S - 1), -S$$

and M_J is calculated from $M_J = M_L + M_S$.

Table 24-1

	Value of M_J	$L+S$	$L+S-1$	$\cdots$	$L-S+1$	$L-S$	$\cdots$	$-(L-S)$	$-(L-S+1)$	$\cdots$	$-(L+S-1)$	$-(L+S)$
(1) (2)	Multiplicity	1	2	(steps of $+1$)	$2S$	$2S+1$	(constant)	$2S+1$	$2S$	(steps of -1)	2	1
	$J = L+S$	1	1	$\cdots$	1	1	$\cdots$	1	1	(steps of -1)	1	1
	$J = L+S-1$		1	$\cdots$	1	1	$\cdots$	1	1	$\cdots$	1	
(3) Multiplicity Assigned To:	$\vdots$			$\vdots$	$\vdots$	$\vdots$		$\vdots$	$\vdots$			
	$J = L-S+1$				1	1	$\cdots$	1	1			
	$J = L-S$					1	$\cdots$	1	1			
(4)	Remaining Unassigned Multiplicity	0	0	$\cdots$	0	0	$\cdots$	0	0	$\cdots$	0	0

Assume that $L \geqslant S$. Row (1) of Table 24-1 shows the calculated values of M_J, and row (2) shows the number of (M_L, M_S) combinations that give rise to each M_J-value. For instance, the multiplicity of $L + S - 2$ is 3 because

$$L + S - 2 = \begin{cases} (L) + (S - 2) \\ (L - 1) + (S - 1) \\ (L - 2) + (S) \end{cases}$$

Observe that the sum of the multiplicities is

$$2(1 + 2 + 3 + \cdots + 2S) + (2S + 1)[2(L - S) + 1] = 2\frac{2S(2S + 1)}{2} + (2S + 1)(2L - 2S + 1)$$
$$= (2S + 1)(2L + 1)$$

or the total number of (M_L, M_S) combinations.

The desired values of J are such that to each one of them corresponds a range of M_J-values:

$$M_J = J, J - 1, \ldots, -(J - 1), -J$$

Rows (3) and (4) of Table 24-1 show how the values

$$J = L + S, L + S - 1, \ldots, L - S = |L - S|$$

precisely exhaust the (M_L, M_S) combinations.

For the case $S > L$, simply interchange L and S in the above argument.

24.2.   Find the maximum number of electrons that can occupy a d subshell.

For a d subshell, $l = 2$. As shown in Section 24.2, the maximum number of electrons in the subshell is given by

$$2(2l + 1) = 2(2 \times 2 + 1) = 10$$

corresponding to the 10 combinations of m_l and m_s shown in Table 24-2.

Table 24-2

l	2	2	2	2	2	2	2	2	2	2
m_l	2	2	1	1	0	0	-1	-1	-2	-2
m_s	$\frac{1}{2}$	$-\frac{1}{2}$	$\frac{1}{2}$	$-\frac{1}{2}$	$\frac{1}{2}$	$-\frac{1}{2}$	$\frac{1}{2}$	$-\frac{1}{2}$	$\frac{1}{2}$	$-\frac{1}{2}$

24.3.   Show that the maximum number of electrons that can lie in a shell specified by a quantum number n is $2n^2$.

The total number of substates with a given l is $2(2l + 1)$. The values of l are

$$l = 0, 1, 2, \ldots, n - 1$$

so the number of electrons in a filled n shell is

$$N = \sum_{l=0}^{n-1} 2(2l + 1) = 2[1 + 3 + \cdots + (2n - 1)]$$

Let us define

$$\mathcal{S} = 1 + 3 + \cdots + (2n - 1)$$

which written backwards is

$$\mathcal{S} = (2n - 1) + (2n - 3) + \cdots + 1$$

Adding these two expressions term by term gives

$$2\mathcal{S} = 2n + 2n + \cdots + 2n = (2n)n$$

Therefore, $N = 2\mathcal{S} = 2n^2$.

24.4.   Show that atoms composed of filled subshells will have a ${}^1S_0$ ground state.

The z-component of the atom's total orbital angular momentum, $M_L\hbar$, and the z-component of the total spin angular momentum, $M_S\hbar$, are found from

$$M_L = \sum m_l \qquad M_S = \sum m_s$$

the summations being over all electrons. The electrons in a complete l subshell have the following values of m_l and m_s:

$$(m_l, m_s) = \left(l, \pm\tfrac{1}{2}\right), \left(l - 1, \pm\tfrac{1}{2}\right), \ldots, \left(-l, \pm\tfrac{1}{2}\right)$$

Thus, summing over all the electrons in the atom always gives $M_L = 0$ and $M_S = 0$. Since these are the only possible values of M_L and M_S, we can only have $L = 0$ and $S = 0$, which in turn imply that $J = 0$. The state is therefore ${}^1S_0$.

24.5.   For hydrogen ($l = L$) the energy states shown in Fig. 24-2 are found. What are the possible electric dipole transitions for these states?

The transitions must obey the selection rule $\Delta l = \pm 1$. Thus only the transitions shown in Fig. 24-3 are allowed.

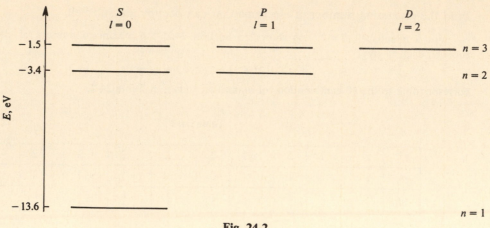

Fig. 24-2

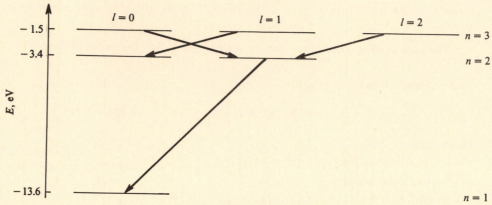

Fig. 24-3

24.6. Calculate $\mathbf{L} \cdot \mathbf{S}$ for a $^3F_2$ state.

For a $^3F_2$ state, $S = 1$, $L = 3$ and $J = 2$. From the result of Problem 22.1 we have

$$\mathbf{L} \cdot \mathbf{S} = \frac{1}{2}[J(J+1) - L(L+1) - S(S+1)]\hbar^2 = \frac{1}{2}[2(2+1) - 3(3+1) - 1(1+1)]\hbar^2 = -4\hbar^2$$

24.7. Determine the transitions occurring from a ^3F state to a ^3D state with an $\mathbf{L} \cdot \mathbf{S}$ interaction present.

The relative sizes of the spin-orbit splittings are determined by evaluating

$$\mathbf{L} \cdot \mathbf{S} = \frac{1}{2}[J(J+1) - L(L+1) - S(S+1)]\hbar^2$$

See Table 24-3, in which J is evaluated from (22.4).

Table 24-3

State	L	S	J	$\mathbf{L} \cdot \mathbf{S}$
^3F	3	1	4, 3, 2	$3\hbar^2,\ -1\hbar^2,\ -4\hbar^2$
^3D	2	1	3, 2, 1	$2\hbar^2,\ -1\hbar^2,\ -3\hbar^2$

In this case the $\mathbf{L} \cdot \mathbf{S}$ interactions split each energy level into three parts, as shown in Fig. 24-4. The transitions satisfying the selection rule $\Delta J = 0, \pm 1$ (but no $0 \to 0$) are also shown. The transition rule $\Delta L = \pm 1$ is automatically satisfied.

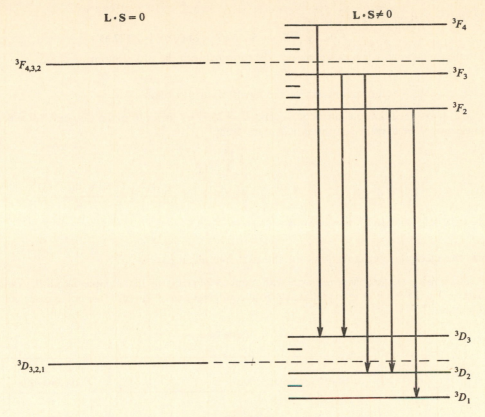

Fig. 24-4

24.8. Assuming a ^{3}Li atom to be hydrogenlike, determine the ionization energy of the $2s$ electron. Explain qualitatively the difference from the experimental value of 5.39 eV.

If the two inner electrons were neglected, the valence electron would be in the $n = 2$ Bohr orbit with $Z = 3$. From (19.5),

$$E_n = - \frac{(13.58)Z^2}{n^2} \text{ eV} \qquad \text{so} \qquad E_2 = - \frac{(13.58)(3)^2}{2^2} \text{ eV} = -30.6 \text{ eV}$$

so the ionization energy is 30.6 eV. On the other hand, if we consider that the two inner electrons completely shield the ^{3}Li nucleus so the outer electron only sees $Z = 1$, then

$$E_2 = - \frac{13.58}{2^2} \text{ eV} = -3.4 \text{ eV}$$

and the ionization energy is 3.4 eV. The actual answer, 5.39 eV, lies between these two values. This shows that the valence electron penetrates the helium core and thus sees some but not all of the nuclear charge. The inner electrons partially shield the valence electron from the nuclear charge.

24.9. The measured ionization energy of He is $E_M = 24.60$ eV. Suppose that the interaction energy between the two electrons of an He atom is taken to be the difference between their common binding energy, assuming each moves independently in a Bohr orbit, and the measured ionization energy. Determine this interaction energy.

According to Bohr's theory the ground-state energy of an electron in the field of a nucleus of charge Ze is [see (19.5)]

$$E_1 = - Z^2(13.58 \text{ eV})$$

For He, $Z = 2$, so that if each electron is treated as being completely independent of the other electron, its binding energy by the Bohr theory would be

$$E_B = - E_1 = (2)^2(13.58 \text{ eV}) = 54.32 \text{ eV}$$

and the interaction energy would be

$$E_i = E_B - E_M = 54.32 \text{ eV} - 24.60 \text{ eV} = 29.72 \text{ eV}$$

The interaction energy is positive because the force between the two electrons is repulsive.

24.10. Calculate the average separation of the electrons in Problem 24.9.

If it is assumed that the interaction energy $E_i = 29.72$ eV arises from the Coulomb force between the two electrons, the average separation d is found from

$$E_i = k \frac{e^2}{d}$$

$$(29.72 \text{ eV})\left(1.60 \times 10^{-19} \frac{\text{J}}{\text{eV}}\right) = \left(8.998 \times 10^9 \frac{\text{N} \cdot \text{m}^2}{\text{C}}\right) \frac{(1.60 \times 10^{-19} \text{ C})^2}{d}$$

Solving, $d = 0.484 \times 10^{-10}$ m $= 0.484$ Å.

24.11. Give the electron configurations for the first five *noble gases*.

The noble gases are those elements with numbers of electrons that completely fill the various shells shown in Fig. 24-1. The ground-state electron configurations of the first five noble gases are shown in Table 24-4.

Table 24-4

Noble Gas	Electron Configuration	Number of Electrons (Z)
He	$1s^2$	2
Ne	$1s^2\,2s^2\,2p^6$	10
Ar	$1s^2\,2s^2\,2p^6\,3s^2\,3p^6$	18
Kr	$1s^2\,2s^2\,2p^6\,3s^2\,3p^6\,4s^2\,3d^{10}\,4p^6$	36
Xe	$1s^2\,2s^2\,2p^6\,3s^2\,3p^6\,4s^2\,3d^{10}\,4p^6\,5s^2\,4d^{10}\,5p^6$	54

24.12. *Alkali metals* have one electron more than a noble gas. Give the electron configurations of the first four alkali metals.

When a particular np subshell (that is, the p subshell of a particular shell n) is completely filled, corresponding to one of the noble gases, the next-higher-Z atom will have an $(n+1)s$ electron added to the noble gas core. Thus we have the ground-state electron configurations given in Table 24-5.

Table 24-5

Alkali Metal	Electron Configuration	Noble Gas Core
Li ($Z=3$)	$1s^2\,2s^1$	He
Na ($Z=11$)	$1s^2\,2s^2\,2p^6\,3s^1$	Ne
K ($Z=19$)	$1s^2\,2s^2\,2p^6\,3s^2\,3p^6\,4s^1$	Ar
Rb ($Z=37$)	$1s^2\,2s^2\,2p^6\,3s^2\,3p^6\,4s^2\,3d^{10}\,4p^6\,5s^1$	Kr

Because alkali metals have one s electron added to a relatively inert core, their spectra are qualitatively similar to that of hydrogen and their ground states are all $^2S_{1/2}$.

24.13. Members of the *halogen family* have one electron less than a noble gas. Give the electron configurations of the first three halogens.

In an atom one Z-unit lower than a noble gas the "missing" electron (except for H) will be an np electron.   The ground-state electron configurations are as shown in Table 24-6.

Table 24-6

Halogen	Electron Configuration
F ($Z = 9$)	$1s^2\, 2s^2\, 2p^5$
Cl ($Z = 17$)	$1s^2\, 2s^2\, 2p^6\, 3s^2\, 3p^5$
Br ($Z = 35$)	$1s^2\, 2s^2\, 2p^6\, 3s^2\, 3p^6\, 4s^2\, 3d^{10}\, 4p^5$

24.14.   Following the filling of the $4s$ subshell, the $3d$ subshell is filled.   The 10 elements thereby formed are called the *transition elements*.   Give the electron configurations for the first three elements of the transition group ($_{21}$Sc, $_{22}$Ti, $_{23}$V).

See Table 24-7.

Table 24-7

Transition Element	Electron Configuration
Sc ($Z = 21$)	$1s^2\, 2s^2\, 2p^6\, 3s^2\, 3p^6\, 4s^2\, 3d^1$
Ti ($Z = 22$)	$1s^2\, 2s^2\, 2p^6\, 3s^2\, 3p^6\, 4s^2\, 3d^2$
V ($Z = 23$)	$1s^2\, 2s^2\, 2p^6\, 3s^2\, 3p^6\, 4s^2\, 3d^3$

24.15.   What are the possible states of an atom if they are determined by two equivalent np electrons?

For each electron, n is fixed; $l = 1$; $m_l = 1, 0, -1$; $m_s = +\frac{1}{2}, -\frac{1}{2}$.   Thus there are $3 \times 2 = 6$ ways of choosing each electron's state (m_l, m_s), and so there are $6 \times 6 = 36$ ways of specifying the state of the two electrons.   Of these 36, 6 put both electrons in the same state, in violation of the Pauli principle. The remaining 30 fall into fifteen pairs, wherein the members of a pair differ only in respect to which electron is labeled "electron 1" and which "electron 2."   Since the two electrons are indistinguishable, we must drop one member of each pair, and are left with 15 possible atomic substates [$(m_l, m_s), (m_l', m_s')$]. In Table 24-8 we calculate the values of the atomic z-component quantum numbers corresponding to these 15 substates.

Table 24-8

Substate	(m_l, m_s)	(m_l', m_s')	$M_L = m_l + m_l'$	$M_S = m_s + m_s'$
#1	$(1, -\frac{1}{2})$	$(1, \frac{1}{2})$	2	0
#2	$(0, \frac{1}{2})$	$(1, \frac{1}{2})$	1	1
#3	$(0, \frac{1}{2})$	$(1, -\frac{1}{2})$	1	0
#4	$(0, -\frac{1}{2})$	$(1, \frac{1}{2})$	1	0
#5	$(0, -\frac{1}{2})$	$(1, -\frac{1}{2})$	1	-1
#6	$(0, -\frac{1}{2})$	$(0, \frac{1}{2})$	0	0
#7	$(-1, \frac{1}{2})$	$(1, \frac{1}{2})$	0	1
#8	$(-1, \frac{1}{2})$	$(1, -\frac{1}{2})$	0	0
#9	$(-1, \frac{1}{2})$	$(0, \frac{1}{2})$	-1	1
#10	$(-1, \frac{1}{2})$	$(0, -\frac{1}{2})$	-1	0
#11	$(-1, -\frac{1}{2})$	$(1, \frac{1}{2})$	0	0
#12	$(-1, -\frac{1}{2})$	$(1, -\frac{1}{2})$	0	-1
#13	$(-1, -\frac{1}{2})$	$(0, \frac{1}{2})$	-1	0
#14	$(-1, -\frac{1}{2})$	$(0, -\frac{1}{2})$	-1	-1
#15	$(-1, -\frac{1}{2})$	$(-1, \frac{1}{2})$	-2	0

We must now find values of L and S that, consistent with (24.7) and (24.8), account for the (M_L, M_S) values of all fifteen substates (and for these only). First, find the largest entry in the last two columns of Table 24-8; this is $M_L = 2$ in substate #1. (If, in the more general case, more than one (M_L, M_S) pair contained the maximal entry, we would further maximize over the other member of the pair.) Then (24.7) shows that the largest value of L that must be considered is $L = 2$. Correspondingly, $S = 0$. (By (24.8), $S \geqslant 0$, but if $S > 0$ then substate #1 would not be maximal.) To $L = 2$, $S = 0$ we assign without repetition the (M_L, M_S) pairs with $+2 \geqslant M_L \geqslant -2$, $+0 \geqslant M_S \geqslant -0$; e.g. substates #1, #3, #6, #10, #15. (We could have chosen #4 instead of #3, #8 or #11 instead of #6, #13 instead of #10.)

Now, repeat the process for the remaining 10 substates. The new maximal substate is #2, giving $L = 1$, $S = 1$; to these we assign substates #2, #4, #5, #7, #8, #9, #12, #13, #14.

Now only substate #11 remains; it corresponds to $L = 0$, $S = 0$.

Table 24-9 displays the final results, wherein J is calculated from L and S via (22.4).

Table 24-9

Substate	L	S	J	Designation of Atomic State
#1, #3, #6, #10, #15	2	0	2	$^1D_2$
#2, #4, #5, #7, #8, #9, #12, #13, #14	1	1	2, 1, 0	$^3P_{2,1,0}$
#11	0	0	0	$^1S_0$

24.16. The ground-state configuration for $_6C$ is $1s^2\,2s^2\,2p^2$. Calculate the possible atomic states from this configuration and compare them to the observed spectrum.

For the given six electrons we wish to calculate the possible values of L, S and J. The two $1s$ electrons will not contribute to the atomic state since they form a closed shell (Problem 24.4). The $2s$ electrons both have $l = 0$, so that, according to the Pauli exclusion principle, their spins must be antiparallel, giving a net spin for these two of zero. Thus the atomic state is determined entirely by the two $2p$ electrons, and Problem 24.15 gives the possibilities as $^1D_2$, $^3P_2$, $^3P_1$, $^3P_0$, and $^1S_0$.

The actual ground state of carbon turns out to be $^3P_0$. The first five states, as inferred from the observed spectrum, are indicated in Fig. 24-5.

24.17. Knowing that

$$\mu_L = -\left(\frac{e}{2m}\right)\mathbf{L} \quad \text{and} \quad \mu_S = -2\left(\frac{e}{2m}\right)\mathbf{S}$$

show in a vector diagram that μ and $\mathbf{J}$ are not parallel.

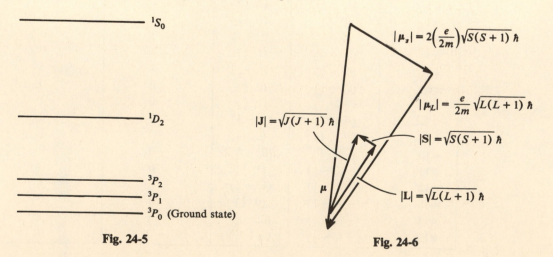

Fig. 24-5   **Fig. 24-6**

The vector relations $\mathbf{J} = \mathbf{L} + \mathbf{S}$ and $\boldsymbol{\mu} = \boldsymbol{\mu}_L + \boldsymbol{\mu}_S$ are shown in Fig. 24-6. Because

$$\frac{|\boldsymbol{\mu}_S|}{|\mathbf{S}|} = 2 \frac{|\boldsymbol{\mu}_L|}{|\mathbf{L}|}$$

the two triangles are not similar, and $\boldsymbol{\mu}$ and $\mathbf{J}$ are not parallel.

24.18. Refer to Problem 24.17. Calculate the projection of the total magnetic moment vector $\boldsymbol{\mu}$ on the vector $\mathbf{J}$.

From Problem 24.17,

$$\boldsymbol{\mu} = \boldsymbol{\mu}_L + \boldsymbol{\mu}_S = -\frac{e}{2m}(\mathbf{L} + 2\mathbf{S}) = -\frac{e}{2m}(\mathbf{J} + \mathbf{S})$$

The projection of $\boldsymbol{\mu}$ on $\mathbf{J}$ is

$$\frac{\boldsymbol{\mu} \cdot \mathbf{J}}{|\mathbf{J}|} = \left(-\frac{e}{2m}\right) \frac{\mathbf{J} \cdot \mathbf{J} + \mathbf{J} \cdot \mathbf{S}}{|\mathbf{J}|}$$

Now,

$$\mathbf{L} \cdot \mathbf{L} = (\mathbf{J} - \mathbf{S}) \cdot (\mathbf{J} - \mathbf{S}) = \mathbf{J} \cdot \mathbf{J} + \mathbf{S} \cdot \mathbf{S} - 2\mathbf{J} \cdot \mathbf{S}$$

or

$$\mathbf{J} \cdot \mathbf{S} = \tfrac{1}{2}(\mathbf{J} \cdot \mathbf{J} + \mathbf{S} \cdot \mathbf{S} - \mathbf{L} \cdot \mathbf{L})$$

Therefore,

$$\frac{\boldsymbol{\mu} \cdot \mathbf{J}}{|\mathbf{J}|} = \left(-\frac{e}{2m}\right) \frac{\mathbf{J} \cdot \mathbf{J} + \tfrac{1}{2}(\mathbf{J} \cdot \mathbf{J} + \mathbf{S} \cdot \mathbf{S} - \mathbf{L} \cdot \mathbf{L})}{|\mathbf{J}|}$$

$$= \left(-\frac{e}{2m}\right) \frac{J(J+1)\hbar^2 + \tfrac{1}{2}\left[J(J+1)\hbar^2 + S(S+1)\hbar^2 - L(L+1)\hbar^2\right]}{\sqrt{J(J+1)}\,\hbar}$$

$$= \left(-\frac{e\hbar}{2m}\right)\sqrt{J(J+1)}\left[1 + \frac{J(J+1) + S(S+1) - L(L+1)}{2J(J+1)}\right]$$

$$\equiv \left(-\frac{e\hbar}{2m}\right)\sqrt{J(J+1)}\,g$$

The quantity

$$g = 1 + \frac{J(J+1) + S(S+1) - L(L+1)}{2J(J+1)}$$

is called the *Landé g-factor*. As will be seen in the following problems, the g-factor is needed to calculate the relative splitting of different energy levels in weak magnetic fields.

24.19. Determine the value of the energy splitting of an atom in a magnetic field B if it is assumed that the splitting depends only on the component of $\boldsymbol{\mu}$ along $\mathbf{J}$.

From Problem 24.18, the component of $\boldsymbol{\mu}$ along $\mathbf{J}$ is

$$\mu_J = \left(-\frac{e\hbar}{2m}\right)\sqrt{J(J+1)}\,g$$

or in vector notation

$$\boldsymbol{\mu}_J = \mu_J \frac{\mathbf{J}}{|\mathbf{J}|} = \left(-\frac{e\hbar}{2m}\right)\sqrt{J(J+1)}\,g\,\frac{\mathbf{J}}{\sqrt{J(J+1)}\,\hbar} = -\frac{e}{2m}g\mathbf{J}$$

The splitting in energy is given by (*22.1*) as

$$\Delta E = -\boldsymbol{\mu}_J \cdot \mathbf{B} = \frac{e}{2m}g\mathbf{J} \cdot \mathbf{B} = \frac{e}{2m}g\,BJ_z = \frac{e}{2m}g\,BM_J\hbar$$

Since $M_J = J, J-1, \ldots, -J+1, -J$, it is seen that for a given field B each energy level will split into $2J + 1$ sublevels, with the amount of splitting being determined by the g-factor associated with that level.

24.20. Assuming the $\mathbf{L} \cdot \mathbf{S}$ interaction to be much stronger than the interaction with an external magnetic field, calculate the anomalous Zeeman splitting of the lowest states ($^2S_{1/2}$, $^2P_{1/2}$, $^2P_{3/2}$) in hydrogen for a field of 0.05 T.

From Problem 24.19 we have

$$\Delta E = \frac{e\hbar}{2m} g B M_J = \left(5.79 \times 10^{-5} \; \frac{\text{eV}}{\text{T}}\right) g (0.05 \; \text{T}) M_J$$

with

$$g = 1 + \frac{J(J+1) + S(S+1) - L(L+1)}{2J(J+1)}$$

Table 24-10 shows the calculations.

Table 24-10

State	L	S	J	g	M_J	ΔE, eV $\times 10^{-5}$
$^2P_{3/2}$	1	1/2	3/2	4/3	$\pm 3/2$	± 0.579
					$\pm 1/2$	± 0.193
$^2P_{1/2}$	1	1/2	1/2	2/3	$\pm 1/2$	± 0.097
$^2S_{1/2}$	0	1/2	1/2	2	$\pm 1/2$	± 0.290

24.21. Refer to Problem 24.20. Determine the lines resulting from the transitions $^2P_{3/2} \rightarrow {}^2S_{1/2}$ and $^2P_{1/2} \rightarrow {}^2S_{1/2}$ in hydrogen for a field of 0.05 T. Without a magnetic field these transitions result in $(1210 - 3.54 \times 10^{-3})$ Å and $(1210 + 1.77 \times 10^{-3})$ Å lines, respectively.

The relationship between the splitting of the spectral lines and the applied field is found from

$$E_u - E_l = \frac{hc}{\lambda} \qquad \text{or} \qquad dE_u - dE_l = -\frac{hc}{\lambda^2} \, d\lambda$$

or

$$d\lambda = -\frac{\lambda^2}{hc}(dE_u - dE_l) = -\frac{(1210 \; \text{Å})^2}{12.4 \times 10^3 \; \text{eV} \cdot \text{Å}}(dE_u - dE_l) = \left(-118 \; \frac{\text{Å}}{\text{eV}}\right)(dE_u - dE_l)$$

The values for dE_u and dE_l are given in Table 24-10. There are 10 transitions that satisfy the rule $\Delta M_J = \pm 1, 0$ (see Fig. 24-7). The deviation of each of these lines from $\lambda_0 = 1210$ Å is calculated in Table 24-11.

This problem illustrates the anomalous Zeeman effect in that 10 lines are observed instead of 3 lines as in the normal Zeeman effect.

Table 24-11

$d\lambda_0$, Å $\times 10^{-3}$	Transition	dE_u, eV $\times 10^{-5}$	dE_l, eV $\times 10^{-5}$	$d\lambda$, Å $\times 10^{-3}$	$d\lambda_T = d\lambda_0 + d\lambda$, Å $\times 10^{-3}$
-3.54	a	$+0.579$	$+0.290$	-0.341	-3.88
-3.54	b	$+0.193$	$+0.290$	$+0.114$	-3.43
-3.54	c	$+0.193$	-0.290	-0.570	-4.11
-3.54	d	-0.193	$+0.290$	$+0.570$	-2.97
-3.54	e	-0.193	-0.290	-0.114	-3.65
-3.54	f	-0.579	-0.290	$+0.341$	-3.20
1.77	g	$+0.097$	$+0.290$	$+0.228$	$+2.00$
1.77	h	$+0.097$	-0.290	-0.457	$+1.31$
1.77	i	-0.097	$+0.290$	$+0.457$	$+2.23$
1.77	j	-0.097	-0.290	-0.228	$+1.54$

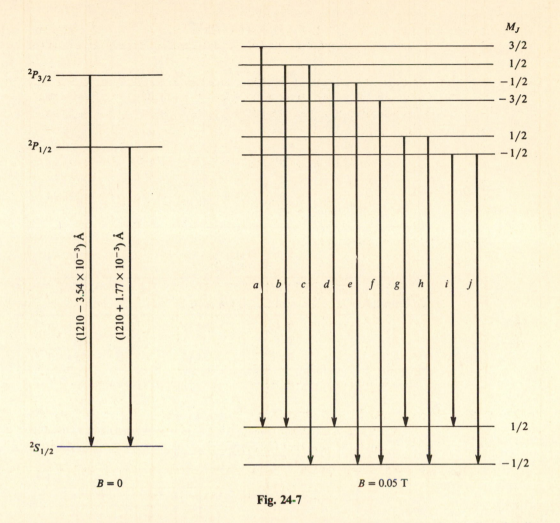

Fig. 24-7

24.22. Neglecting the spin-orbit interaction in a strong external field of 5 T, determine the lines resulting from the $2p \to 1s$ transition ($\lambda_0 = 1210$ Å) in hydrogen.

The total magnetic moment is the vector sum of the orbital and spin magnetic moments:

$$\boldsymbol{\mu} = \boldsymbol{\mu}_L + \boldsymbol{\mu}_S = -\frac{e}{2m}\mathbf{L} - 2\frac{e}{2m}\mathbf{S} = -\frac{e}{2m}(\mathbf{L} + 2\mathbf{S})$$

If the spin-orbit interaction is neglected, the change in the energy of the system will be determined only by the interaction of the total magnetic moment with the external field, giving

$$dE = -\boldsymbol{\mu}\cdot\mathbf{B} = -\mu_z B = \frac{e}{2m}(L_z + 2S_z)B = \frac{e\hbar}{2m}(M_L + 2M_S)B$$

$$= \left(5.79 \times 10^{-5}\ \frac{\text{eV}}{\text{T}}\right)(5\ \text{T})(M_L + 2M_S) = (28.94 \times 10^{-5}\ \text{eV})(M_L + 2M_S)$$

We therefore have the splittings given in Table 24-12.

The shift in the wavelengths is found from Problem 24.21 to be

$$d\lambda = \left(-118\ \frac{\text{Å}}{\text{eV}}\right)(dE_u - dE_l)$$

Since the separation between the upper energy levels is the same as that between the lower levels (28.94×10^{-5} eV), only three lines result from the transitions satisfying the selection rules $\Delta M_L = \pm 1, 0$, $\Delta M_S = 0$ (see Fig. 24-8 and Table 24-13).

It is seen that in strong external magnetic fields where the spin-orbit interaction is negligible, one has a situation similar to the normal Zeeman effect, where only three lines are present.

Table 24-12

State	L	M_L	M_S	$M_L + 2M_S$	$dE,$ eV $\times 10^{-5}$
$2p$	1	$+1$	$+\frac{1}{2}$	$+2$	$+57.88$
$2p$	1	0	$+\frac{1}{2}$	$+1$	$+28.94$
$2p$	1	-1	$+\frac{1}{2}$	0	0.00
$2p$	1	$+1$	$-\frac{1}{2}$	0	0.00
$2p$	1	0	$-\frac{1}{2}$	-1	-28.94
$2p$	1	-1	$-\frac{1}{2}$	-2	-57.88
$1s$	0	0	$+\frac{1}{2}$	$+1$	$+28.94$
$1s$	0	0	$-\frac{1}{2}$	-1	-28.94

Table 24-13

Transitions	$dE_u - dE_l,$ eV $\times 10^{-5}$	$d\lambda,$ Å $\times 10^{-3}$	$\lambda = \lambda_0 + d\lambda,$ Å
a, d	28.94	-34.15	$1210 - 34.15 \times 10^{-3}$
b, e	0	0	1210
c, f	-28.94	$+34.15$	$1210 + 34.15 \times 10^{-3}$

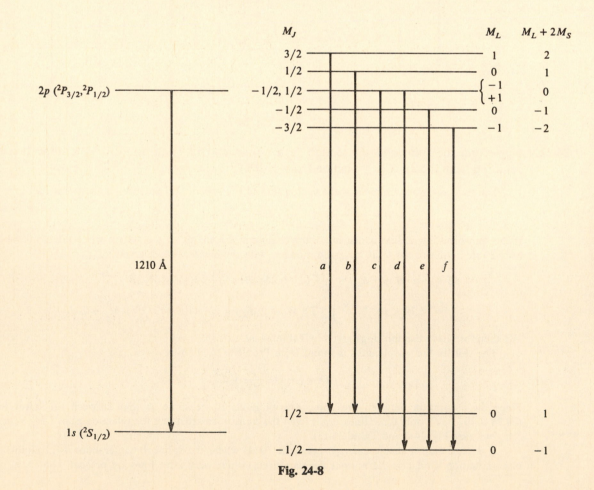

Fig. 24-8

Supplementary Problems

24.23. Suppose that an atom's state is determined by a single electron. What are the possible values of the atom's total angular momentum in an (a) S state, (b) P state, (c) D state?

Ans. (a) $\frac{\sqrt{3}}{2}\hbar$; (b) $\frac{\sqrt{15}}{2}\hbar$, $\frac{\sqrt{3}}{2}\hbar$; (c) $\frac{\sqrt{35}}{2}\hbar$, $\frac{\sqrt{15}}{2}\hbar$

24.24. Find the maximum number of electrons that can occupy a p subshell.   *Ans.*   6 electrons

24.25. Find the maximum number of electrons that can occupy an f subshell and list the values of m_l and m_s for the electrons.   *Ans.*   14 electrons; $m_l = \pm 3, \pm 2, \pm 1, 0$; $m_s = \pm\frac{1}{2}$

24.26. Calculate $\mathbf{L}\cdot\mathbf{S}$ for a $^2D_{3/2}$ state.   *Ans.* $-\frac{3}{2}\hbar^2$

24.27. Give the following states in spectroscopic notation: (a) $L = 0, S = 0, J = 0$; (b) $L = 2, S = 0, J = 2$; (c) $L = 3, S = 1/2, J = 5/2$; (d) $L = 4, S = 1, J = 5$.   *Ans.*   (a) $^1S_0$; (b) $^1D_2$; (c) $^2F_{5/2}$; (d) $^3G_5$

24.28. Give the strong transitions from a ^2D state to a ^2P state with an $\mathbf{L}\cdot\mathbf{S}$ interaction present.
Ans. $^2D_{5/2}\rightarrow{}^2P_{3/2}$; $^2D_{3/2}\rightarrow{}^2P_{3/2}$; $^2D_{3/2}\rightarrow{}^2P_{1/2}$

24.29. Which noble gas completes the $6p$ subshell?   *Ans.* $_{86}$Rn $\equiv$ Rn $(Z = 86)$

24.30. Give the electron configuration for the alkali metal built on the Xe noble gas core.
Ans. $1s^2\,2s^2\,2p^6\,3s^2\,3p^6\,4s^2\,3d^{10}\,4p^6\,5s^2\,4d^{10}\,5p^6\,6s^1$ $(_{55}$Cs$)$

24.31. The *alkaline earths* have two electrons more than a noble gas. Give the electron configuration of the first four.

Ans.   Be $(Z = 4)$: $1s^2\,2s^2$
Mg $(Z = 12)$: $1s^2\,2s^2\,2p^6\,3s^2$
Ca $(Z = 20)$: $1s^2\,2s^2\,2p^6\,3s^2\,3p^6\,4s^2$
Sr $(Z = 38)$: $1s^2\,2s^2\,2p^6\,3s^2\,3p^6\,4s^2\,3d^{10}\,4p^6\,5s^2$

24.32. Give the electron configuration for the halogen with one electron less than the noble gas Xe.
Ans. $1s^2\,2s^2\,2p^6\,3s^2\,3p^6\,4s^2\,3d^{10}\,4p^6\,5s^2\,4d^{10}\,5p^5$ $(_{53}$I$)$

24.33. What transition element has the electron configuration $1s^2\,2s^2\,2p^6\,3s^2\,3p^6\,3d^6\,4s^2$?   *Ans.* $_{26}$Fe

24.34. Give in spectral notation the possible states of an atom which has a closed core plus one d electron.
Ans. $^2D_{5/2}$; $^2D_{3/2}$

24.35. List the possible states (M_L, M_S) of two equivalent nd electrons.
Ans.   The 45 allowable combinations of quantum numbers (10 things taken 2 at a time) yield 23 states: $(\pm 4, 0)$, $(\pm 3, 1)$, $(\pm 3, 0)$, $(\pm 3, -1)$, $(\pm 2, 1)$, $(\pm 2, 0)$, $(\pm 2, -1)$, $(\pm 1, 1)$, $(\pm 1, 0)$, $(\pm 1, -1)$, $(0, 1)$, $(0, 0)$, $(0, -1)$.

24.36. For a field of 2 T, calculate the Zeeman energy splitting of the $^2P_{1/2}$ and $^2S_{1/2}$ states in Na.
Ans. $\pm 3.86 \times 10^{-5}$ eV; $\pm 11.58 \times 10^{-5}$ eV

24.37. The $^2P_{1/2} \rightarrow {}^2S_{1/2}$ transition in Na has a wavelength of 5895.9 Å. Calculate the wavelength changes seen in a 2 T magnetic field.   *Ans.* ± 0.216 Å; ± 0.433 Å

24.38. Assuming the $\mathbf{L} \cdot \mathbf{S}$ interaction to be much larger than the interaction with an external magnetic field, calculate the anomalous Zeeman splitting of the $^2D_{3/2}$ and $^2D_{5/2}$ states in hydrogen in a field of 0.05 T.
Ans. for $^2D_{3/2}$: $\pm 3.47 \times 10^{-6}$ eV, $\pm 1.16 \times 10^{-6}$ eV; for $^2D_{5/2}$: $\pm 8.68 \times 10^{-6}$ eV, $\pm 5.21 \times 10^{-6}$ eV,
   $\pm 1.74 \times 10^{-6}$ eV

Chapter 25

Inner-Electron Transitions: X-Rays

25.1 X-RAY APPARATUS

X-rays, discovered by Wilhelm Roentgen in 1895, are high-energy photons (1–100 keV) with wavelengths of the order of 1 Å. They are usually produced by bombarding a target with high-energy electrons, as shown in Fig. 25-1. The kinetic energy of the electrons at the cathode is negligible, so that when they strike the target the electrons have kinetic energy $K = eV$.

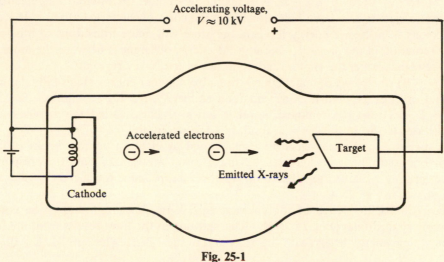

Fig. 25-1

25.2 PRODUCTION OF BREMSSTRAHLUNG

The bombarding electrons can interact with the atoms of the target in a number of different ways. In one type of interaction the electrons are accelerated by the positively charged nuclei, as shown in Fig. 25-2. Whenever a charge is accelerated it produces radiation which, according to the quantum

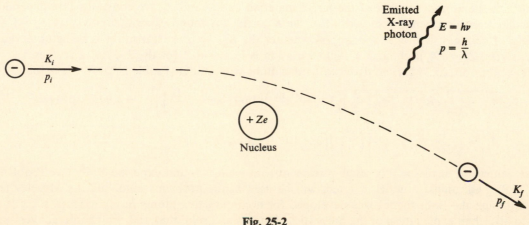

Fig. 25-2

picture, will be in the form of a photon with energy $h\nu$ equal to the change in the electron's kinetic energy, i.e. $h\nu = K_i - K_f$. Radiation produced in this fashion is called *bremsstrahlung* after the German word for "braking" or "slowing down" radiation.

An electron may produce a number of such photons before coming to rest. Clearly, the most energetic photon possible will occur when an electron loses all of its initial kinetic energy in a single interaction, producing a single photon with a maximum frequency or minimum wavelength given by

$$h\nu_{\max} = \frac{hc}{\lambda_{\min}} = eV \tag{25.1}$$

Thus the bremsstrahlung process will produce radiation with a continuous spectrum that has a cutoff frequency or wavelength which depends on the accelerating voltage according to (25.1).

25.3 PRODUCTION OF CHARACTERISTIC X-RAY SPECTRA

It is also possible for the incident electrons to excite the electrons in the atoms of the target of the X-ray tube. Moreover, because of the large accelerating voltage, the bombarding electrons can have sufficient energy to eject tightly bound electrons from the cores of the target atoms. If a core electron is ejected, electrons from higher energy levels in the atom will make transitions to this lower vacated state, emitting radiation in the process. Because the energy differences between the inner levels of the target atoms are rather large, the emitted radiation lies in the X-ray region.

If K-shell ($n = 1$) electrons are removed, electrons from higher energy states falling into the K shell produce a series of lines denoted in X-ray notation as the K_α, K_β, K_γ, . . . , lines. See Fig. 25-3(a). If L-shell ($n = 2$) electrons are removed, another series of lines, called the L series, is produced. Similarly, transitions to the M shell ($n = 3$) result in an M series, etc.

When the spectrum of a many-electron atom excited by electron bombardment is observed, one sees a smooth bremsstrahlung background having a lower-wavelength cutoff corresponding to the maximum accelerating voltage, together with intense sharp lines produced by the K_α, K_β, etc., transitions, as shown in Fig. 25-4.

Upon closer observation each of the characteristic X-ray lines is found to be composed of a number of closely spaced lines [Fig. 25-3(b)]. This splitting in the lines results from the fine-structure splitting of the atomic energy levels (with the exception of the K shell where $n = 1$, $L = 0$), as described in Chapter 22.

25.4 THE MOSELEY RELATION

In 1913 H. Moseley found that the observed frequencies ν for the K and L X-ray series could be fitted by the relation

$$\nu^{1/2} = A(Z - Z_0) \tag{25.2}$$

where Z is the atomic number of the target material, and A and Z_0 are constants that depend on the particular transition being observed. For the K series it is found experimentally that $Z_0 = 1$ and the value of A changes slightly depending on whether the K_α, K_β, . . . , transition is being observed. For the L series, $Z_0 = 7.4$, and again a slight variation is found in A for the L_α, L_β, . . . , lines.

The form of (25.2) can be deduced from a Bohr-type model (Problem 25.6). It is found that

$$A_{K_\alpha} = \left(\frac{3}{4} cR_\infty\right)^{1/2} = \left[\left(\frac{3}{4}\right)\left(3 \times 10^8 \ \frac{m}{s}\right)\left(1.097 \times 10^7 \ \frac{1}{m}\right)\right]^{1/2} = 4.97 \times 10^7 \ Hz^{1/2}$$

$$A_{L_\alpha} = \left(\frac{5}{36} cR_\infty\right)^{1/2} = \left[\left(\frac{5}{36}\right)\left(3 \times 10^8 \ \frac{m}{s}\right)\left(1.097 \times 10^7 \ \frac{1}{m}\right)\right]^{1/2} = 2.14 \times 10^7 \ Hz^{1/2}$$

These values are in reasonably good agreement with what is found experimentally (see Problems 25.7 and 25.8), and, unless otherwise stated, will be used in the problems involving the Moseley relation.

Although the Bohr theory was developed for noninteracting atoms in the gaseous state, it is seen also to afford an explanation of the properties of atoms in a solid material, which interact very strongly

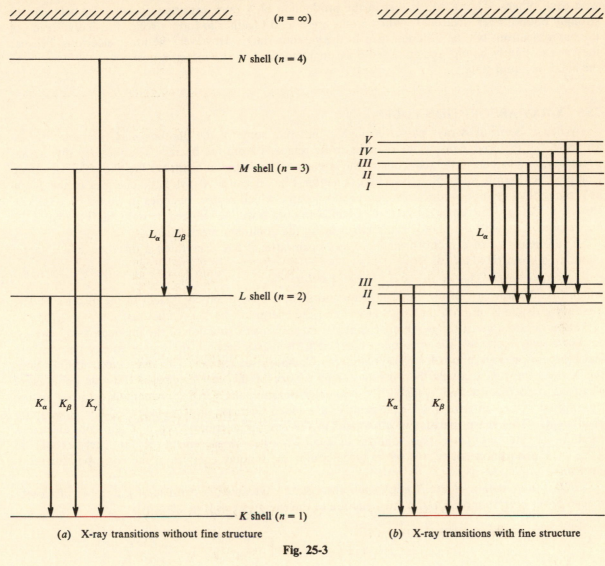

(a) X-ray transitions without fine structure

(b) X-ray transitions with fine structure

Fig. 25-3

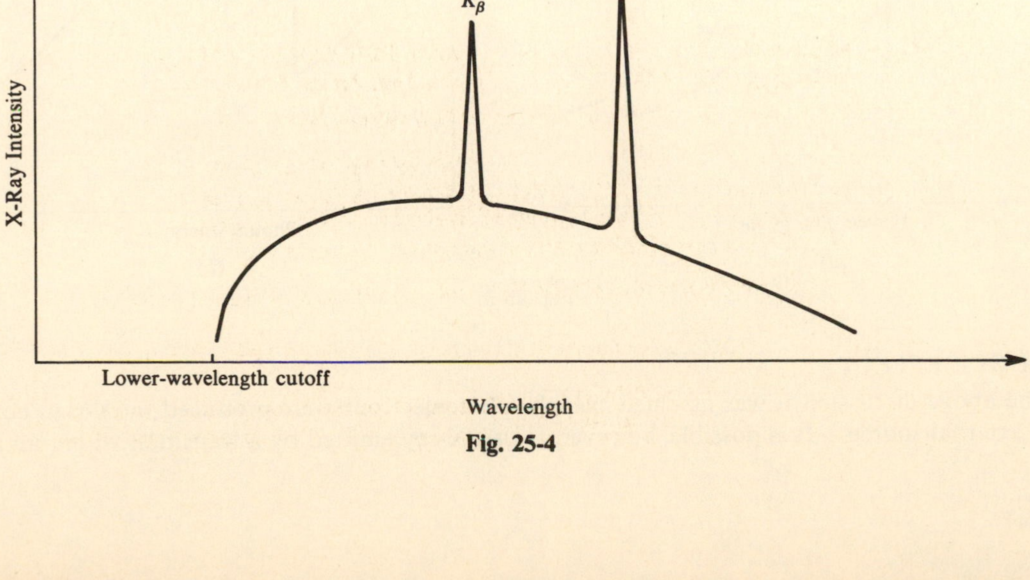

Fig. 25-4

with each other.   The reason is that in the production of X-rays transitions occur only between the strongly bound inner electrons.   When atoms are bound together to form a solid, the energy levels of the outer electrons will be different than in the gaseous state.   However, the inner electrons, because they are so tightly bound, remain essentially unchanged when the material goes from the gaseous to the solid or liquid state.

25.5   X-RAY ABSORPTION EDGES

When a beam of X-rays passes through a material, some of the photons may interact with the atoms of the material, causing the photons to be removed from the beam.   The primary interaction processes responsible for the reduction of the intensity of any photon beam are the photoelectric effect, Compton scattering, and pair production (Chapter 14).   Because X-rays have energies in the 1–100 keV range they cannot produce electron-positron pairs, which require energies in excess of 1000 keV (Problem 13.12).   Therefore the intensity of an X-ray beam will be reduced by only the first two of the above processes, with the photoelectric effect being the dominant mechanism.

The intensity I of a monochromatic X-ray beam after it has penetrated a distance x in a target material is given by (14.1) as

$$I = I_0 e^{-\mu x}$$

where I_0 is the intensity of the incident beam and μ is the absorption coefficient of the material.   The quantity μ depends on both the target atoms and the energy of the X-ray photons.

Suppose, for a given target material, that μ is measured as a function of the incident X-ray energy. As this energy increases, the absorption coefficient decreases because the higher-energy photons are less likely to produce photoelectrons or undergo Compton scattering.   This decrease continues until the X-ray energy just equals the binding energy of one of the core electrons.   At this point more electrons suddenly become available for photoelectric emission, causing a marked decrease in the transmitted X-ray intensity, or equivalently, a sudden increase in the value of the absorption coefficient.   This sharp increase in μ happens at the binding energies of each of the core electrons, resulting in the *absorption edges* shown in Fig. 25-5(a).   Measurement of the energies of the $K, L, \ldots$, absorption edges thus serves to determine the binding energies of the corresponding core electrons.

With the exception of the K edge, each absorption edge actually consists of a number of closely spaced peaks corresponding to the fine structure of the energy levels [Fig. 25-5(b)].

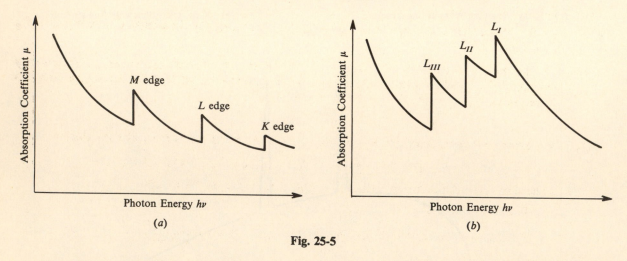

(a)   (b)

Fig. 25-5

25.6   AUGER EFFECT

In the above discussion it was assumed that the photoelectrons were produced by X-rays coming from an external source.   It is possible, however, for an X-ray emitted by a transition within an atom to be absorbed by an electron in the same atom, causing its ejection. Photoelectrons produced by such internal conversions of X-rays are called *Auger* (pronounced OH·ZHAY) *electrons*.

25.7 X-RAY FLUORESCENCE

X-ray photons can be used to excite or eject core electrons. The resulting downward transitions as the atom returns to its ground state will then produce additional X-ray photons of smaller energy than the incident X-ray. This phenomenon is known as *X-ray fluorescence*.

Solved Problems

25.1. A TV tube operates with a 20 kV accelerating potential. What are the maximum-energy X-rays from the TV set?

The electrons in the TV tube have an energy of 20 keV, and if these electrons are brought to rest by a collision in which one X-ray photon is emitted, the photon energy is 20 keV. The corresponding wavelength is

$$\lambda = \frac{c}{\nu} = \frac{hc}{h\nu} = \frac{12.4 \text{ keV} \cdot \text{Å}}{20 \text{ keV}} = 0.62 \text{ Å}$$

25.2. Determine the wavelength of the K_α line for molybdenum ($Z = 42$).

From the Moseley relation we have

$$\nu^{1/2} = A(Z-1) = (4.97 \times 10^7 \text{ Hz}^{1/2})(42-1) = 2.04 \times 10^9 \text{ Hz}^{1/2}$$

$$\lambda = \frac{c}{\nu} = \frac{3 \times 10^8 \text{ m/s}}{(2.04 \times 10^9 \text{ Hz}^{1/2})^2} = 0.721 \times 10^{-10} \text{ m} = 0.721 \text{ Å}$$

This compares well with the experimental value of 0.709 Å.

25.3. In the Moseley relation, which will have the greater value for the constant A, a K_α or a K_β transition?

The Moseley relation for K transitions is $\nu^{1/2} = A(Z-1)$. The K_β transitions are larger-energy than the K_α transitions; therefore higher-frequency photons are emitted in the K_β transitions, and A is larger for the K_β transitions than for the K_α transitions.

25.4. An experiment measuring the K_α lines for various elements yields the following data:

Fe: 1.94 Å   Co: 1.79 Å   Ni: 1.66 Å   Cu: 1.54 Å

Determine the atomic number of each of the elements from these data.

The Moseley relation gives

$$\nu^{1/2} = (4.97 \times 10^7 \text{ Hz}^{1/2})(Z-1) \quad \text{or} \quad Z = 1 + \frac{\nu^{1/2}}{4.97 \times 10^7 \text{ Hz}^{1/2}}$$

and using $\nu = c/\lambda$ we obtain

$$Z = 1 + \frac{c^{1/2}}{\lambda^{1/2}} \left(\frac{1}{4.97 \times 10^7 \text{ Hz}^{1/2}} \right) = 1 + \frac{34.85}{\lambda^{1/2}} \quad (\lambda \text{ in Å})$$

The results are given in Table 25-1.

Table 25-1

Element	λ, Å	Z
Fe	1.94	$26.02 \approx 26$
Co	1.79	$27.04 \approx 27$
Ni	1.66	$28.04 \approx 28$
Cu	1.54	$29.08 \approx 29$

Before Moseley's work Ni, whose atomic weight is 58.69, was listed in the periodic table before Co, whose atomic weight is 58.94, and it was believed that the atomic numbers for Ni and Co were 27 and 28, respectively. By using the above experimental data, Moseley showed that this ordering and the corresponding atomic numbers should be reversed.

25.5. A beam of 100 keV electrons is incident on a Mo ($Z = 42$) target. The binding energies of the core electrons in Mo are given in Table 25-2. Calculate the wavelengths of the transitions that occur.

Table 25-2

Shell	K	L_I	L_{II}	L_{III}	M_I	M_{II}	M_{III}	M_{IV}	M_V
Electron	$1s$	$2s$	$2p$	$2p$	$3s$	$3p$	$3p$	$3d$	$3d$
Binding Energy, keV	20.000	2.866	2.625	2.520	0.505	0.410	0.393	0.230	0.227

Only transitions with $\Delta l = \pm 1$ are allowed. These are shown in Fig. 25-6. The wavelengths are found from

$$\lambda = \frac{c}{\nu} = \frac{hc}{h\nu} = \frac{hc}{E_u - E_l} = \frac{12.4 \text{ keV} \cdot \text{Å}}{E_u - E_l}$$

resulting in Table 25-3.

Table 25-3

Transition		E_u, keV	E_l, keV	λ, Å
K_α	1	− 2.625	− 20.000	0.7137
	2	− 2.520	− 20.000	0.7094
K_β	1	− 0.410	− 20.000	0.6330
	2	− 0.393	− 20.000	0.6324
L_α	1	− 0.505	− 2.520	6.1538
	2	− 0.505	− 2.625	5.8491
	3	− 0.410	− 2.866	5.0489
	4	− 0.393	− 2.866	5.0142
	5	− 0.230	− 2.520	5.4148
	6	− 0.230	− 2.625	5.1775
	7	− 0.227	− 2.520	5.4078
	8	− 0.227	− 2.625	5.1710

This problem illustrates the fine structure of characteristic X-rays.

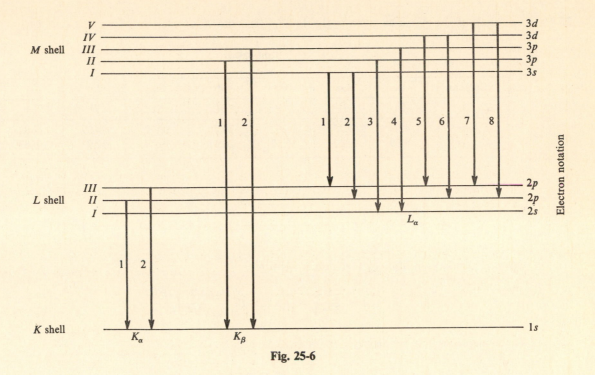

Fig. 25-6

25.6.   Using a simple Bohr picture, calculate the value of A in Moseley's equation for the K_α and L_α series of transitions.

For one-electron atoms we know that (Section 19.3)

$$\frac{1}{\lambda} = \frac{\nu}{c} = R_\infty Z^{*2}\left(\frac{1}{n_l^2} - \frac{1}{n_u^2}\right) \qquad \text{or} \qquad \nu^{1/2} = \left[\left(\frac{1}{n_l^2} - \frac{1}{n_u^2}\right)cR_\infty\right]^{1/2} Z^*$$

where R_∞ is the Rydberg constant, n_u and n_l the principal quantum numbers of the upper and lower states for the electron transition, and Z^*e the net positive charge acting on the electron. For K_α and L_α transitions we have respectively $n_u = 2$, $n_l = 1$ and $n_u = 3$, $n_l = 2$, so that

$$\nu_{K_\alpha}^{1/2} = \left[\left(\frac{1}{1^2} - \frac{1}{2^2}\right)cR_\infty\right]^{1/2} Z^* = \left(\frac{3}{4}cR_\infty\right)^{1/2} Z^* = (4.97 \times 10^7 \text{ Hz}^{1/2})Z^*$$

$$\nu_{L_\alpha}^{1/2} = \left[\left(\frac{1}{2^2} - \frac{1}{3^2}\right)cR_\infty\right]^{1/2} Z^* = \left(\frac{5}{36}cR_\infty\right)^{1/2} Z^* = (2.14 \times 10^7 \text{ Hz}^{1/2})Z^*$$

Therefore, $A_{K_\alpha} = 4.97 \times 10^7 \text{ Hz}^{1/2}$ and $A_{L_\alpha} = 2.14 \times 10^7 \text{ Hz}^{1/2}$. If it is assumed in a K_α transition that the inner electrons are not affected by the outer electrons in the atom, the L electron before the transition will see an effective charge of $(Z - 1)e$, because the remaining K electron shields the atomic nucleus, whose charge is Ze. Using $Z^* = Z - 1$ in the above expression for the K_α transition, we obtain

$$\nu_{K_\alpha}^{1/2} = (4.97 \times 10^7 \text{ Hz}^{1/2})(Z - 1)$$

In transitions involving electrons from shells farther out than the L shell the shielding effect becomes more complicated and the constant Z_0 in the Moseley relation is no longer equal to the number of screening electrons.

25.7. For each of the K_α lines given in Table 25-4 find the value of A in Moseley's relation.

Table 25-4

Element	Sc	Ga	Nb	Sb	Pm	Lu	Tl
Z	21	31	41	51	61	71	81
$K_{\alpha1}$, keV	4.09	9.25	16.62	26.36	38.72	54.07	72.87
$K_{\alpha2}$, keV	4.09	9.22	16.52	26.11	38.17	52.97	70.83

Using

$$\nu^{1/2} = \left(\frac{E}{h}\right)^{1/2} = \left(\frac{E}{4.136 \times 10^{-18} \text{ keV} \cdot \text{s}}\right)^{1/2} = \left(49.17 \times 10^7 \frac{\text{Hz}^{1/2}}{\text{keV}^{1/2}}\right)E^{1/2}$$

where E is in keV, in Moseley's relation, we obtain

$$A = \frac{\nu^{1/2}}{Z-1} = \left(49.17 \times 10^7 \frac{\text{Hz}^{1/2}}{\text{keV}^{1/2}}\right)\frac{E^{1/2}}{Z-1}$$

The numerical results are displayed in Table 25-5.

Table 25-5

Z	21	31	41	51	61	71	81
A_1, $\text{Hz}^{1/2} \times 10^7$	4.97	4.98	5.01	5.05	5.10	5.17	5.25
A_2, $\text{Hz}^{1/2} \times 10^7$	4.97	4.98	5.00	5.02	5.06	5.11	5.17

These values of A agree reasonably well with the value $A = 4.97 \times 10^7$ Hz obtained in Problem 25.6.

25.8. The energies of the L and M shells in W $(Z = 74)$ are

Shell	L_I	L_{II}	L_{III}	M_I	M_{II}	M_{III}	M_{IV}	M_V
Energy, keV	12.099	11.542	10.205	2.820	2.575	2.281	1.872	1.810

Determine the values of A in the Moseley relation for the largest- and smallest-frequency L_α lines and compare them with the result of Problem 25.6.

The L_α lines arise from transitions from the M to the L shell. The largest and smallest allowed frequencies arise respectively from the $M_{III} \to L_I$ and $M_I \to L_{III}$ transitions (see Problem 25.5). The corresponding frequencies are

$$\nu_{\max} = \frac{E_{L_I} - E_{M_{III}}}{h} = \frac{12.099 \text{ keV} - 2.281 \text{ keV}}{4.136 \times 10^{-18} \text{ keV} \cdot \text{s}} = 2.374 \times 10^{18} \text{ Hz}$$

$$\nu_{\min} = \frac{E_{L_{III}} - E_{M_I}}{h} = \frac{10.205 \text{ keV} - 2.820 \text{ keV}}{4.136 \times 10^{-18} \text{ keV} \cdot \text{s}} = 1.786 \times 10^{18} \text{ Hz}$$

Substituting these values in the Moseley relation, we find

$$A_{max} = \frac{\nu_{max}^{1/2}}{Z - 7.4} = \frac{(2.374 \times 10^{18} \text{ Hz})^{1/2}}{74 - 7.4} = 2.31 \times 10^7 \text{ Hz}^{1/2}$$

$$A_{min} = \frac{\nu_{min}^{1/2}}{Z - 7.4} = \frac{(1.786 \times 10^{18} \text{ Hz})^{1/2}}{74 - 7.4} = 2.01 \times 10^7 \text{ Hz}^{1/2}$$

These values straddle the value $A = 2.14 \times 10^7 \text{ Hz}^{1/2}$ obtained in Problem 25.6.

25.9. Assume the following model for the two K electrons in an atom: the total energy of each electron is given by the Bohr energy, for an "effective" nuclear charge Z^*e, plus the Coulomb interaction energy, where it is assumed the two electrons are maximally separated at a distance equal to twice the Bohr radius, r_B. Find the nuclear shielding factor $Z_0 = Z - Z^*$ for the elements whose K-shell binding energies are given in Table 25-6.

Table 25-6

Element	Ni	Zr	Sb	Gd	Ta
Z	28	40	51	64	73
K-Shell Binding Energy, keV	8.333	17.998	30.491	50.239	67.417

By (19.5) with $n = 1$, the Bohr energy is $-Z^{*2}E_1^\circ$; and by Problem (24.10), the Coulomb interaction energy is $ke^2/2r_B$. The binding energy of each electron is the negative of its total energy; hence

$$\text{BE} = -\left(-Z^{*2}E_1^\circ + \frac{ke^2}{2r_B}\right) = E_1^\circ Z^{*2} - \frac{ke^2}{2r_B} \qquad (1)$$

in which $E_1^\circ = 13.6 \text{ eV} = 0.0136 \text{ keV}$ and, by (19.4) with $n = 1$,

$$r_B = \frac{r_1^\circ}{Z^*} = \frac{1}{Z^*}\frac{h^2}{4\pi^2 kme^2} = \frac{1}{Z^*}\frac{ke^2}{2E_1^\circ}$$

or

$$\frac{ke^2}{2r_B} = E_1^\circ Z^*$$

Substituting these values in (1) yields the quadratic equation

$$Z^{*2} - Z^* - \frac{\text{BE}}{E_1^\circ} = 0$$

whose positive root is Z^*. The numerical results are displayed in Table 25-7.

Table 25-7

Z	Z^*	Z_0
28	25.3	2.7
40	36.9	3.1
51	47.9	3.1
64	61.3	2.7
73	70.9	3.1

The fact that the shielding factors Z_0 are all approximately equal shows that the Bohr model is a reasonably good approximation for the K-shell electrons.

25.10. In uranium ($Z = 92$) the K absorption edge is 0.107 Å and the K_α line is 0.126 Å. Determine the wavelength of the L absorption edge.

From Fig. 25-7 it is seen that

$$E_K = -\frac{hc}{\lambda_K} = -\frac{12.4 \text{ keV} \cdot \text{Å}}{0.107 \text{ Å}} = -115.9 \text{ keV}$$

and

$$E_L - E_K = \frac{hc}{\lambda_{K_\alpha}} = \frac{12.4 \text{ keV} \cdot \text{Å}}{0.126 \text{ Å}} = 98.4 \text{ keV}$$

Therefore $E_L = 98.4 - 115.9 = -17.5$ keV, and

$$\lambda_L = \frac{hc}{-E_L} = \frac{12.4 \text{ keV} \cdot \text{Å}}{17.5 \text{ keV}} = 0.709 \text{ Å}$$

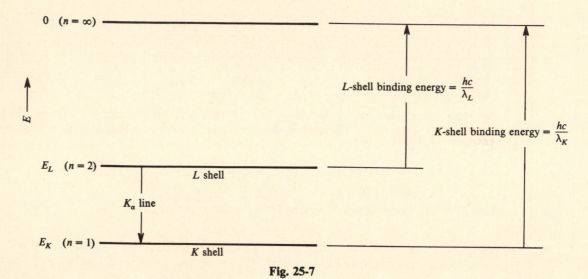

Fig. 25-7

25.11. The K absorption edge for Y ($Z = 39$) is 0.7277 Å. In order to produce emission of the K series from Y an accelerating potential of at least 17.039 kV is required. Determine h/e from these data.

The energy of the 0.7277 Å photon that will just remove an electron from the K shell is

$$E_\nu = h\frac{c}{\lambda} = h\frac{2.998 \times 10^8 \text{ m/s}}{0.7277 \times 10^{-10} \text{ m}} = (4.120 \times 10^{18} \text{ s}^{-1})h$$

The energy of the bombarding electron that will just remove an electron from the K shell is

$$E_e = eV = e(17.039 \times 10^3 \text{ V})$$

Since both of these energies must be equal, we have

$$(4.120 \times 10^{18} \text{ s}^{-1})h = e(17.039 \times 10^3 \text{ V})$$

$$\frac{h}{e} = 4.136 \times 10^{-15} \text{ V} \cdot \text{s} = 4.136 \times 10^{-15} \frac{\text{J} \cdot \text{s}}{\text{C}}$$

25.12. A material whose K absorption edge is 0.15 Å is irradiated with 0.10 Å X-rays. What is the maximum kinetic energy of photoelectrons that are emitted from the K shell?

The K-shell binding energy is

$$|E_K| = \frac{hc}{\lambda_K} = \frac{12.4 \text{ keV} \cdot \text{Å}}{0.15 \text{ Å}} = 82.7 \text{ keV}$$

The energy of the incident photon is

$$E_\nu = \frac{hc}{\lambda} = \frac{12.4 \text{ keV} \cdot \text{Å}}{0.10 \text{ Å}} = 124 \text{ keV}$$

The maximum kinetic energy is the difference between these two values.

$$K_{\max} = E_\nu - |E_K| = 124 \text{ keV} - 82.7 \text{ keV} = 41.3 \text{ keV}$$

25.13. When 0.50 Å X-rays strike a material, the photoelectrons from the K shell are observed to move in a circle of radius 23 mm in a magnetic field of 2×10^{-2} T. What is the binding energy of K-shell electrons?

The velocity of the photoelectrons is found from $F = ma$:

$$evB = m\frac{v^2}{R} \qquad \text{or} \qquad v = \frac{e}{m}BR$$

The kinetic energy of the photoelectrons is then

$$K = \frac{1}{2}mv^2 = \frac{1}{2}\frac{e^2B^2R^2}{m} = \frac{1}{2}\frac{(1.6 \times 10^{-19} \text{ C})^2(2 \times 10^{-2} \text{ T})^2(23 \times 10^{-3} \text{ m})^2}{(9.11 \times 10^{-31} \text{ kg})} = 2.97 \times 10^{-15} \text{ J}$$

or

$$K = (2.97 \times 10^{-15} \text{ J})\frac{1 \text{ keV}}{1.6 \times 10^{-16} \text{ J}} = 18.6 \text{ keV}$$

The energy of the incident photon is

$$E_\nu = \frac{hc}{\lambda} = \frac{12.4 \text{ keV} \cdot \text{Å}}{0.50 \text{ Å}} = 24.8 \text{ keV}$$

The binding energy is the difference between these two values.

$$\text{BE} = E_\nu - K = 24.8 \text{ keV} - 18.6 \text{ keV} = 6.2 \text{ keV}$$

25.14. Stopping potentials of 24, 100, 110, and 115 kV are measured for photoelectrons emitted from a certain element when it is irradiated with monochromatic X-rays. If this element is used as a target in an X-ray tube, what will be the wavelength of the K_α line?

The stopping potential energy, eV_s, is equal to the difference between the energy of the incident photon and the binding energy of the electron in a particular shell.

$$eV_s = E_p - E_B$$

The different stopping potentials arise from electrons being emitted from different shells, with the smallest value (24 kV) corresponding to ejection of a K-shell electron. Subtracting the expression for the two smallest stopping potentials, we obtain

$$eV_{sL} - eV_{sK} = (E_p - E_{BL}) - (E_p - E_{BK}) = E_{BK} - E_{BL}$$

or

$$100 \text{ keV} - 24 \text{ keV} = E_{BK} - E_{BL}$$

This difference, 76 keV, is the energy of the K_α line. The corresponding wavelength is

$$\lambda = \frac{hc}{E_{BK} - E_{BL}} = \frac{12.4 \text{ keV} \cdot \text{Å}}{76 \text{ keV}} = 0.163 \text{ Å}$$

25.15. In Zn ($Z = 30$) the ionization (binding) energies of the K and L shells are, respectively, 9.659 keV and 1.021 keV. Determine the kinetic energy of an Auger electron emitted from the L shell by a K_α X-ray.

The energy of the K_α X-ray is

$$E_{K_\alpha} = E_L - E_K = -1.021 \text{ keV} - (-9.659 \text{ keV}) = 8.638 \text{ keV}$$

The kinetic energy of the Auger electron will be equal to the difference between the K_α photon energy and the binding energy of the electron in the L shell:

$$K = E_{K_\alpha} - E_{BL} = 8.638 \text{ keV} - 1.021 \text{ keV} = 7.617 \text{ keV}$$

25.16. The K, L, and M energy levels for Cu, Ni, and Co are given in Table 25-8. It is desired to filter the K_β line from the K_α and K_β radiation emitted from Cu. Which will be the better filter, Ni or Co?

Table 25-8

Element	Z	E_K, keV	E_L, keV	E_M, keV
Cu	29	-8.979	-0.931	-0.074
Ni	28	-8.333	-0.855	-0.068
Co	27	-7.709	-0.779	-0.060

The energies of the X-rays emitted from Cu are equal to the differences between the various energy levels in Cu. Thus,

$$E_{K_\alpha} = -0.931 \text{ keV} - (-8.979 \text{ keV}) = 8.048 \text{ keV}$$

$$E_{K_\beta} = -0.074 \text{ keV} - (-8.979 \text{ keV}) = 8.905 \text{ keV}$$

A material will filter out X-rays if the energy of the X-rays is larger than the energy required to eject electrons from the atoms of the material. Otherwise the incident X-rays will not interact appreciably with the material and will pass through unfiltered. It is seen that the K absorption edge of Ni (8.333 keV) lies between the K_α (8.048 keV) and K_β (8.905 keV) X-ray energies from Cu, so that the K_β photons will interact much more with Ni than will the K_α photons. On the other hand, both K_α and K_β photons can cause photoelectric emission from the K shell (7.709 keV) of Co. Thus Ni will be the better filter.

25.17. Electrons are accelerated between a filament and a grid through mercury vapor by a variable potential V, as shown in Fig. 25-8(a). A small retarding potential, $V_R \approx 0.5$ V is maintained between the grid and the collector plate. When the current I in the collector is measured as a function of the accelerating voltage, the curve of Fig. 25-8(b) is obtained. Determine the first excitation energy of mercury and the wavelength of the light emitted by mercury in the experiment.

In order to reach the collector the electrons must have a kinetic energy greater than the retarding potential energy of about 0.5 eV between the grid and the collector. As the accelerating potential is increased, the electrons acquire larger and larger kinetic energies and hence more and more reach the collector, resulting in an increasing current. Eventually, however, the electrons acquire an energy equal to the first excited state of the mercury atoms. At this point the electrons can excite the mercury atoms into this state, thereby, of course, losing kinetic energy. Thus fewer electrons will have sufficient energy to overcome the retarding potential V_R, resulting in the observed dip in the collector current. In addition, the mercury vapor, previously dark, will emit radiation as the atoms return to their ground state.

Upon further increase in V the current will again begin to increase because the electrons can acquire additional kinetic energy after they excite a mercury atom. At still greater accelerating potentials electrons will have sufficient energy to excite two mercury atoms, resulting in a second dip in I, and so on. (We are neglecting the possibility that an electron might put a mercury atom into a higher excited

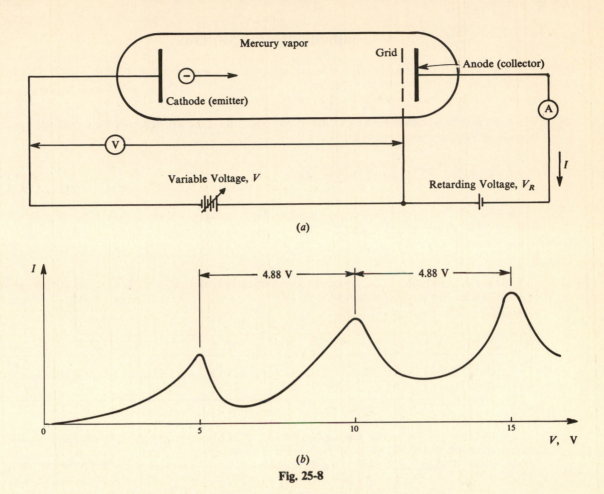

Fig. 25-8

state. This could happen, but special potential variations across the vapor tube would be required.) The voltage difference between the various current peaks is thus seen to correspond to the energy required to excite mercury into its first excited state, so that

$$\Delta E = e \, \Delta V = 4.88 \text{ eV}$$

(The potential of the first peak cannot be used because of the existence of V_R and various contact potentials.) The wavelength of the photons emitted when the excited mercury atoms return to their ground state is

$$\lambda = \frac{c}{\nu} = \frac{hc}{h\nu} = \frac{hc}{\Delta E} = \frac{12.4 \times 10^3 \text{ eV} \cdot \text{Å}}{4.88 \text{ eV}} = 2540 \text{ Å}$$

This experiment was first performed by J. Franck and G. Hertz in 1914, and was the first experiment to demonstrate the existence of stationary states in atoms, further confirming Bohr's emerging quantum hypothesis. In addition it showed that atoms can be excited by interacting with energetic electrons.

Supplementary Problems

25.18. If a K_α X-ray from a certain element is measured to have a wavelength of 0.786 Å, what is the element?
Ans. $_{40}Zr$

25.19. An electron is accelerated through a 10^5 V potential.   Find the smallest possible wavelength produced when this electron interacts with a heavy target.   *Ans.*   0.124 Å

25.20. How many lines are in the K_α fine structure?   *Ans.*   2

25.21. Determine the constant A in Moseley's equation for L_α transitions if the L_α lines have the values 30.1 Å in Ca ($Z = 20$) and 11.2 Å in Zn ($Z = 30$).
Ans.   for Ca, $A = 2.51 \times 10^7$ Hz$^{1/2}$; for Zn, $A = 2.29 \times 10^7$ Hz$^{1/2}$

25.22. From the data in Problem 25.7 determine for each line the value of Z_0 in Moseley's relation if the value of A for each line is taken to be 4.97×10^7 Hz$^{1/2}$.
Ans.

Element	Sc	Ga	Nb	Sb	Pm	Lu	Tl
Z_0 for $K_{\alpha 1}$	0.99	0.91	0.67	0.21	− 0.56	− 1.75	− 3.46
Z_0 for $K_{\alpha 2}$	0.99	0.96	0.79	0.45	− 0.12	− 1.00	− 2.26

25.23. In a NaCl crystal the lattice spacing is 2.820 Å.   If a first-order Bragg reflection for a K_α X-ray is observed from a principal plane at 15.8°, what is its wavelength?   (This problem shows how one can measure wavelengths of X-rays.)   *Ans.*   1.54 Å

25.24. X-rays from copper ($K_\alpha = 1.54$ Å, $K_\beta = 1.39$ Å, $K_{abs} = 1.38$ Å) are passed through a sheet of nickel ($K_\alpha = 1.66$ Å, $K_\beta = 1.50$ Å, $K_{abs} = 1.49$ Å).   What intense wavelengths emerge?   *Ans.*   8.05 Å

25.25. In a given element which is larger, the K absorption energy or the energy of a K_α X-ray?
Ans. K absorption energy

25.26. In Os ($Z = 76$) the K and L absorption edges have respective wavelengths of 0.168 Å and 1.17 Å.   What is the wavelength of the K_α line?   *Ans.*   0.196 Å

25.27. In Ta ($Z = 73$) the K_α line is 0.216 Å and the L absorption edge is 1.25 Å.   What is the wavelength of the K absorption edge?   *Ans.*   0.184 Å

25.28. The stopping potential for the photoelectrons emitted from the L shell of a material when it is irradiated with 0.257 Å X-rays is found to be 8.20 V.   Find the wavelength of the L absorption edge.
Ans.   0.310 Å

25.29. The kinetic energy of an Auger electron emitted by a K_α X-ray from the L shell of a material with a K absorption edge of 0.827 Å is measured as 10.2 keV.   Find the energy of the K_α X-ray and the wavelength of the L absorption edge.   *Ans.*   12.6 keV; 5.17 Å

Chapter 26

Nucleon and Deuteron Properties

26.1 THE NUCLEONS

All nuclei are composed of two types of particles, positively charged *protons* and neutral *neutrons*, referred to collectively as *nucleons*. Their principal properties are listed in Table 26-1.

Table 26-1

	Proton	Neutron
Charge	$+1.6 \times 10^{-19}$ C	0 C
Rest Mass	1.67252×10^{-27} kg 938.256 MeV 1.007277 u	1.67482×10^{-27} kg 939.550 MeV 1.008665 u
Spin	1/2	1/2
Magnetic Moment	$+2.7928\beta_n$	$-1.9128\beta_n$

The *atomic mass unit*, u, is defined such that a $^{12}_{6}$C atom has a rest mass of exactly 12 u; the *nuclear magneton*, β_n, is given by

$$\beta_n = \frac{e\hbar}{2m_p} = \frac{(1.6 \times 10^{-19}\text{ C})(6.58 \times 10^{-16}\text{ eV}\cdot\text{s})}{2(1.673 \times 10^{-27}\text{ kg})} = 3.15 \times 10^{-8}\ \frac{\text{eV}}{\text{T}}$$

where m_p is the proton rest mass. The positive or negative sign for the magnetic moment indicates that the magnetic moment and spin vectors are in the same or opposite directions, respectively. It is interesting to note that even though the neutron has no charge, it still possesses a magnetic moment.

Protons have an infinite half-life; left alone they will not decay. Neutrons, on the other hand, have a half-life of 12 minutes; if a group is left alone, half the number, on the average, will decay every 12 minutes.

26.2 NUCLEON FORCES

When nucleons are brought close together (on the order of 10^{-15} m = 1 fm), it is found that they exhibit a strong attractive force which has a short range; i.e. at distances greater than a few femtometers the nucleon force is essentially zero. The attractive force is found to be independent of the charge of the nucleons; this means that the proton-proton, neutron-neutron, and proton-neutron forces are all approximately equal.

The force between two nucleons is made up of several different parts. Besides the normal central component, there is a spin-dependent term which is different when the nucleon spins are aligned and when they are antialigned. In addition there is a noncentral component that does not point along the line joining the two nucleons. This noncentral component depends upon how the nucleon spins are oriented relative to the line joining the nucleons.

At distances very much smaller than 1 fm the nucleon force changes in character from being attractive to being repulsive. This behavior is usually referred to as the *repulsive nuclear core*.

163

26.3   THE DEUTERON

A *deuteron* or *deuterium* is a bound system composed of a proton and a neutron, and as such represents the simplest nucleus having more than one nucleon. The properties of a deuteron are: charge, $+1.6 \times 10^{-19}$ C; mass, 1875.5803 MeV or 2.013553 u; spin, $S = 1$ (this is the sum of the neutron and proton spins); magnetic moment, $+0.8574 \beta_n$; total angular momentum, $J = 1$.

It is possible to assign to an atom a single quantized orbital angular momentum **L**. Because the proton-neutron interaction is noncentral, however, it is found that a deuteron does not possess a definite orbital angular momentum. Instead, a deuteron in its ground state has a 96% probability of being in an S ($L = 0$) state and a 4% probability of being in a D ($L = 2$) state.

A deuteron is also found not to be spherical. A quantity that measures a charged body's deviation from sphericity is its *electric quadrupole moment* (see Problem 26.5). If a body is spherical its quadrupole moment is zero. The quadrupole moment of a deuteron is found experimentally to be $+0.282$ fm^2.

The above discussion shows that even though a deuteron is composed of only two nucleons, its structure is quite complex, giving an indication of the complications to be expected when heavier nuclei are investigated.

Solved Problems

26.1. If an electron is confined within a nucleus whose diameter is 10^{-14} m, estimate its minimum kinetic energy.

The de Broglie wavelength of a minimum-energy electron confined inside the nucleus would be approximately twice the nuclear diameter (one-half a wavelength would fit into the diameter). Therefore the electron's momentum would be of the order of magnitude

$$p = \frac{h}{\lambda} = \frac{hc}{\lambda c} = \frac{12.4 \times 10^3 \text{ eV} \cdot \text{Å}}{(2 \times 10^{-4} \text{ Å})c} = 62 \times 10^6 \frac{\text{eV}}{c} = 62 \frac{\text{MeV}}{c}$$

corresponding to a kinetic energy of

$$K = \sqrt{(pc)^2 + E_0^2} - E_0 = \sqrt{\left(62 \frac{\text{MeV}}{c} \times c\right)^2 - (0.511 \text{ MeV})^2} - 0.511 \text{ MeV} = 61 \text{ MeV}$$

26.2. For the nucleus of Problem 26.1 estimate the Coulomb energy of the electron.

The nucleon number A is roughly (see Section 27.3)

$$A = \left(\frac{R}{r_0}\right)^3 = \left(\frac{0.5 \times 10^{-14} \text{ m}}{1.4 \times 10^{-15} \text{ m}}\right)^3 \approx 46$$

For nuclei of this size the number of protons is $Z \approx A/2 = 23$. If the electron is assumed to be at the edge of the nucleus, the Coulomb energy is given by

$$E_C = -\frac{ke^2 Z}{R} = -\frac{(1.44 \text{ MeV} \cdot \text{fm})(23)}{5 \text{ fm}} = -6.6 \text{ MeV}$$

This correction is negligible compared to the 61 MeV electron kinetic energy found in Problem 26.1. Electrons emitted from nuclei have kinetic energies of a few MeV, and not ≈ 54 MeV as predicted by this and the previous problem. Furthermore, some type of positive barrier must exist so that electrons could be bound in the nucleus with a positive energy. Neither of these effects can be produced by a Coulomb interaction, and a reasonable conclusion is that electrons are not nuclear building blocks.

An alternate proposal, made in 1920 by E. Rutherford, was that neutral particles of mass approximately equal to that of a proton but of charge zero were contained in the nucleus. In 1932, J. Chadwick's experiments verified the existence of these *neutrons* (see Problems 30.14 through 30.16), thereby establishing that a nucleus is composed of Z protons and $N = A - Z$ neutrons, for a total of A nucleons.

26.3. Calculate the binding energy of the deuteron.

The binding energy of the deuteron is the amount of energy needed to separate the deuteron into a proton and a neutron.

$$BE = (m_p + m_n - M_d)c^2$$

$$= 938.256 \text{ MeV} + 939.550 \text{ MeV} - 1875.5803 \text{ MeV} = 2.226 \text{ MeV}$$

26.4. Calculate the difference between the deuteron magnetic moment and the sum of the neutron and proton magnetic moments.

Proton magnetic moment	$2.793\beta_n$
Neutron magnetic moment	$-1.913\beta_n$
SUM	$0.880\beta_n$
Deuteron magnetic moment	$0.857\beta_n$
DIFFERENCE	$0.023\beta_n$

The deuteron magnetic moment does not equal the sum of the proton and neutron magnetic moments because the deuteron is not always in an S $(L = 0)$ state but can also be found in a D $(L = 2)$ state 4% of the time.

26.5. The electric quadrupole moment of a nuclear charge distribution which is symmetric about the z-axis is given by

$$\mathcal{Q} = \frac{1}{e} \int_V (3z^2 - r^2)\rho(x, y, z)\, d\tau \tag{1}$$

with $\rho(x, y, z)$ the charge density and $r^2 = x^2 + y^2 + z^2$. For the uniformly charged ellipsoid of revolution defined by the equation

$$\frac{x^2 + y^2}{a^2} + \frac{z^2}{b^2} = 1 \tag{2}$$

the electric quadrupole moment reduces to

$$\mathcal{Q} = \frac{2Z}{5}(b^2 - a^2) \tag{3}$$

where Ze is the total nuclear charge. If the average nuclear radius is taken to be $R_0^3 = a^2 b$ (the volume of the ellipsoid is $\frac{4}{3}\pi a^2 b$), with $R_0 + \delta R_0 = b$, show that the electric quadrupole moment is

$$\mathcal{Q} = \frac{6Z}{5}R_0^2\left(\frac{\delta R_0}{R_0}\right)$$

If $b = R_0 + \delta R_0$ and $R_0^3 = a^2 b$, then, for $\delta R_0 \ll R_0$,

$$a^2 = \frac{R_0^3}{R_0 + \delta R_0} = \frac{R_0^2}{1 + \dfrac{\delta R_0}{R_0}} \approx R_0^2\left(1 - \frac{\delta R_0}{R_0}\right)$$

and

$$b^2 - a^2 \approx R_0^2 \left[1 + 2\left(\frac{\delta R_0}{R_0} \right) + \left(\frac{\delta R_0}{R_0} \right)^2 \right] - R_0^2 \left(1 - \frac{\delta R_0}{R_0} \right)$$

$$= R_0^2 \left[3\left(\frac{\delta R_0}{R_0} \right) + \left(\frac{\delta R_0}{R_0} \right)^2 \right] \approx 3 R_0^2 \left(\frac{\delta R_0}{R_0} \right)$$

Hence,

$$\mathcal{Q} = \frac{2Z}{5} (b^2 - a^2) = \frac{6Z}{5} R_0^2 \left(\frac{\delta R_0}{R_0} \right)$$

26.6.   For $^{155}_{64}$Gd the quadrupole moment is 130 fm². If R_0 is given by $R_0 = (1.4 \text{ fm})A^{1/3}$, find $\delta R_0 / R_0$.

The average radius is

$$R_0 = (1.4 \text{ fm})A^{1/3} = (1.4 \text{ fm})(155)^{1/3} = (1.4 \text{ fm})(5.37) = 7.52 \text{ fm}$$

From Problem 26.5,

$$\mathcal{Q} = \frac{6Z}{5} R_0^2 \left(\frac{\delta R_0}{R_0} \right)$$

$$130 \text{ fm}^2 = \frac{6(64)}{5} (7.52 \text{ fm})^2 \left(\frac{\delta R_0}{R_0} \right)$$

$$\frac{\delta R_0}{R_0} = 2.99 \times 10^{-2} = 2.99\%$$

This shows that for $^{155}_{64}$Gd the nucleus is almost spherical, deviating from sphericity by only 2.99% of the average radius.

26.7.   Determine the possible states of a deuteron if its total angular momentum has quantum number $J = 1$.

The total angular momentum (**J**) of the deuteron is the vector sum of the orbital angular momentum for the neutron-proton bound system (**L**) and the total intrinsic spin of the neutron-proton system (**S**). Since both neutron and proton have spin $S = 1/2$, the total intrinsic spin is 0 (singlet state) or 1 (triplet state). Since $\mathbf{J} = \mathbf{L} + \mathbf{S}$ and $J = 1$ and $S = 0, 1$, the only possible values for L are 0, 1 and 2. In the spectroscopic notation of Section 24.3, the possible deuteron states are $^3S_1$, $^3P_1$, $^1P_1$ and $^3D_1$.

The ground state of the deuteron is a mixture of $^3S_1$ and $^3D_1$.

Supplementary Problems

26.8.   Evaluate the nuclear magneton in units of J/T.   *Ans.* 5.03×10^{-27} J/T

26.9.   What is the ratio of the nuclear magneton to the Bohr magneton of an electron?   *Ans.* 5.45×10^{-4}

26.10.   Find the ratio of the nuclear to the atomic density for hydrogen.   (Assume the nuclear radius to be 1 fm.)   *Ans.* 0.15×10^{15}

26.11.   A 6 MeV γ-ray is absorbed and dissociates a deuteron into a proton and neutron.   If the neutron makes an angle of 90° with the direction of the γ-ray, determine the kinetic energies of the proton and neutron. *Ans.* $K_p = 1.91$ MeV; $K_n = 1.86$ MeV

26.12.   For Problem 26.11 find the angle the proton makes with the γ-ray.   *Ans.*   84°

26.13.   Refer to Problem 26.5.   Derive (*3*) from (*1*) and (*2*) and the fact that the ellipsoid is uniformly charged.

26.14.   The electric quadrupole moment of $^{165}_{67}$Ho is 300 fm^2.   If $R_0 = (1.4)A^{1/3}$ find $\delta R_0 / R_0$.   *Ans.*   6.33%

Chapter 27

Properties of Nuclei

Of all known nuclei about 270 are stable, while roughly four and a half times that number are unstable. The following is a description of some basic properties of nuclei in their ground (lowest-energy) states. Part of this discussion is also applicable to nuclei in excited states.

27.1 DESIGNATION OF NUCLEI

Each nucleus is identified by the atomic number Z, an integer equal to the number of protons in the nucleus; an integer N, equal to its number of neutrons, and a mass number $A = N + Z$, which is the total number of nucleons. Nuclei are designated by giving the symbol X of the chemical element, with the Z value as a pre-subscript and the A value as a pre-superscript; thus, $_Z^A X$. For example, $_{11}^{23}\text{Na}$ has 11 protons, 23 nucleons, and $23 - 11 = 12$ neutrons.

Nuclei are grouped into three categories. *Isotopes* are nuclei with the same atomic (proton) number Z, e.g. $_8^{16}\text{O}$ and $_8^{17}\text{O}$. *Isotones* are nuclei with the same neutron number N, e.g. $_6^{13}\text{C}$ and $_7^{14}\text{N}$. *Isobars* are nuclei with the same mass number A, e.g. $_6^{14}\text{C}$ and $_7^{14}\text{N}$.

27.2 RELATIVE NUMBER OF PROTONS AND NEUTRONS

In light nuclei the number of neutrons is about equal to the number of protons ($N \approx Z$). As the number of nucleons increases, however, it is found that for stable nuclei the number of neutrons becomes greater than the number of protons ($N > Z$), following roughly the curve shown in Fig. 27-1. The neutron excess occurs because the protons' repulsive Coulomb force keeps them farther apart. Therefore proton matter is less dense than neutron matter, and as the number of nucleons increases there are fewer protons than neutrons in a given nuclear volume.

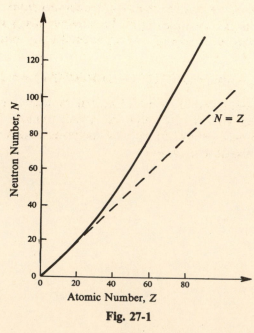

Fig. 27-1

168

27.3   THE NUCLEUS AS A SPHERE

If the density of nuclear matter is assumed to be constant, the volume of a nucleus will be directly proportional to the number of nucleons, A, in it.   For spherical symmetry we then have $V = (\frac{4}{3} \pi r_0^3)A$, giving the nuclear radius R as

$$R = r_0 A^{1/3}$$

Several experiments have been performed to check this relation and obtain r_0.   It is found that the value of r_0 depends upon the nuclear property being measured.   For the size of the mass distribution, $r_0 = 1.4$ fm; while for the size of the charge distribution, $r_0 = 1.2$ fm.   Unless specified otherwise, we will use the value $r_0 = 1.4$ fm in the following discussion and problems.

From the picture of a nucleus as a sphere with uniformly distributed charge Ze it follows that the nucleus will have an electrostatic energy of (see Problem 27.7)

$$E_C = \frac{3}{5} \frac{kZ(Z-1)e^2}{R}$$

$$\approx \frac{3}{5} \frac{kZ^2e^2}{R} \qquad \text{(for large } Z\text{)}$$

This relationship provides a method for determining the size of nuclear charge distributions.

27.4   NUCLEAR BINDING ENERGY

It is found that the rest mass of a stable nucleus is less than the sum of the rest masses of its constituent nucleons.   The mass decrease arises because negative energy is required to hold the nucleons together in the nucleus (see Problem 26.3).   The total nuclear binding energy, BE, is given by the difference between the rest energies of the constituent nucleons and the rest energy of the final nucleus:

$$\text{BE} = (Zm_p)c^2 + (Nm_n)c^2 - M_{\text{nuc}}c^2$$

with m_p, m_n, and M_{nuc} being respectively the proton, neutron, and nuclear rest masses.   The "liquid drop" model (Section 28.1) can be used to calculate the binding energies of stable nuclei.

Usually tables list the *atomic* masses rather than the *nuclear* masses of elements.   In order to find the nuclear mass one must subtract the mass of the atom's electrons from the atomic mass.   (Strictly speaking, one should also add the mass equivalent of the binding energy of the electrons, but this is usually negligible compared to the rest masses.)   As an example, ^6_3Li, which has an atomic mass of 6.015125 u, has a nuclear mass of

$$M_{\text{nuc}} = M_{\text{atom}} - Zm_e = 6.015125 \text{ u} - 3(0.000549 \text{ u}) = 6.013478 \text{ u}$$

Unless stated otherwise, the masses given in the problems will be atomic masses.   Where masses are not supplied, the reader should consult the table of atomic masses in the Appendix.   Where applicable, the mass of hydrogen will be used in place of the proton mass in the expression for BE to compensate for the electrons in the atomic masses.

Solved Problems

27.1. Determine the radii of a ^{16}O and a ^{208}Pb nucleus.

From $R = r_0 A^{1/3} = (1.4 \text{ fm}) A^{1/3}$,

$$R_O = (1.4 \text{ fm})(16)^{1/3} = 3.53 \text{ fm}$$

$$R_{Pb} = (1.4 \text{ fm})(208)^{1/3} = 8.29 \text{ fm}$$

27.2. Determine the approximate density of a nucleus.

If the nucleus is treated as a uniform sphere,

$$\text{density} = \frac{\text{mass}}{\text{volume}} \approx \frac{A \times (\text{mass of a nucleon})}{\frac{4}{3}\pi R^3} = \frac{A(1.7 \times 10^{-27} \text{ kg})}{\frac{4}{3}\pi (1.4 \times 10^{-15} A^{1/3} \text{ m})^3} = 1.5 \times 10^{17} \frac{\text{kg}}{\text{m}^3}$$

A cubic inch of nuclear material would weigh about 1 billion tons!

27.3. Determine the stable nucleus that has a radius 1/3 that of ^{189}Os.

Since $R \propto A^{1/3}$,

$$\frac{1}{3} = \frac{R}{R_{Os}} = \left(\frac{A}{A_{Os}}\right)^{1/3} = \left(\frac{A}{189}\right)^{1/3}$$

giving

$$A = \frac{189}{27} = 7$$

corresponding to ^{7}Li.

27.4. A nucleus with $A = 235$ splits into two new nuclei whose mass numbers are in the ratio $2:1$. Find the radii of the new nuclei.

The new mass numbers are

$$A_1 = \frac{1}{3}(235) \qquad A_2 = \frac{2}{3}(235)$$

so that

$$r_1 = (1.4 \text{ fm})A_1^{1/3} = (1.4 \text{ fm})\left(\frac{235}{3}\right)^{1/3} = 5.99 \text{ fm}$$

$$r_2 = (1.4 \text{ fm})A_2^{1/3} = (1.4 \text{ fm})\left[\frac{2}{3}(235)\right]^{1/3} = 7.55 \text{ fm}$$

27.5. Calculate the binding energy of $^{126}_{52}$Te.

The binding energy is given by

$$\text{BE} = (Zm_p)c^2 + (Nm_n)c^2 - M_{nuc}c^2$$

$$= (52 \times 1.007825 \text{ u} + 74 \times 1.008665 \text{ u} - 125.903322 \text{ u}) \times 931.5 \text{ MeV/u} = 1.066 \times 10^3 \text{ MeV}$$

or 1.066 GeV.

27.6. What is the energy required to remove the least tightly bound neutron from $^{40}_{20}$Ca?

From conservation of mass-energy,

$$M_{^{40}Ca}c^2 + E = (M_{^{39}Ca} + m_n)c^2$$

$$(39.962589 \text{ u})(931.5 \text{ MeV/u}) + E = (38.970691 \text{ u} + 1.008665 \text{ u})(931.5 \text{ MeV/u})$$

$$E = 15.6 \text{ MeV}$$

27.7. Determine the electrical potential energy of the protons in a nucleus if it is assumed that the charge is uniformly spherically distributed.

Consider a thin spherical shell of charge,

$$dq = \rho\, dV = \rho(4\pi r^2\, dr)$$

that is added to a sphere that has the same charge density and has total charge

$$q = \rho V = \rho\left(\frac{4}{3}\pi r^3\right)$$

The electrical potential energy dE of the thin shell is then

$$dE = \frac{kq}{r}\, dq = \frac{k}{r}\left(\frac{4}{3}\rho\pi r^3\right)(4\pi\rho r^2\, dr) = 3k\left(\frac{4}{3}\rho\pi\right)^2 r^4\, dr$$

The total electrical potential energy of the sphere is found by integrating dE from $r = 0$ to $r = R$, the final radius of the sphere.

$$E = \int_0^R dE = 3k\left(\frac{4}{3}\rho\pi\right)^2\int_0^R r^4\, dr = \frac{3}{5}k\left(\frac{4}{3}\rho\pi\right)^2 R^5 = \frac{3k}{5R}\left(\frac{4}{3}\rho\pi R^3\right)^2$$

Since $\frac{4}{3}\rho\pi R^3 = \rho V = Q = Ze$, we have

$$E = \frac{3}{5}\frac{kZ^2 e^2}{R}$$

The charges forming a nucleus are actually not continuous but must be brought in discrete amounts. For $Z = 1$ the Coulomb energy should be zero, but the above expression gives a finite answer. To correct the above relationship Z^2 should be changed to $Z(Z - 1)$. For large values of Z this is a minor correction, but not for small values of Z. The correct Coulomb energy is

$$E_C = \frac{3}{5}\frac{kZ(Z-1)e^2}{R}$$

27.8. Calculate the Coulomb energy of $^{73}_{32}$Ge.

Using the result of Problem 27.7, we have

$$E_C = \frac{3}{5}\frac{kZ(Z-1)e^2}{R} = \frac{3}{5}\frac{ke^2 Z(Z-1)}{r_0 A^{1/3}} = \frac{3}{5}\frac{(1.44\ \text{MeV}\cdot\text{fm})}{(1.4\ \text{fm})A^{1/3}}Z(Z-1)$$

$$= (0.617\ \text{MeV})\frac{Z(Z-1)}{A^{1/3}} = (0.617\ \text{MeV})\frac{32(31)}{(73)^{1/3}} = 146\ \text{MeV}$$

Supplementary Problems

Atomic masses are tabulated in the Appendix.

27.9. Using standard notation, give the symbols for neon with 20 nucleons and yttrium with 89 nucleons.
Ans. $^{20}_{10}$Ne; $^{89}_{39}$Y

27.10. The radius of Ge is measured to be twice the radius of $^{9}_{4}$Be.   From this information, how many nucleons are in Ge?   *Ans.* 72

27.11. What is the energy required to remove the least tightly bound proton from $^{40}_{20}$Ca?   Compare this answer with that found in Problem 27.6.   *Ans.* 8.33 MeV

27.12. Calculate the ratio of the nuclear radius of $^{208}_{82}$Pb to the radius of its innermost electrons as calculated from Bohr theory.   *Ans.*   1/77.8

27.13. Determine the value of Z for which the correct and approximate expressions for the Coulomb energy, given in Problem 27.7, differ by 5%.   *Ans.*   21

27.14. Calculate the Coulomb energies of $^{18}_{8}$O and $^{175}_{71}$Lu.   *Ans.*   13.2 MeV; 548 MeV

27.15. Calculate the binding energy of $^{39}_{19}$K.   *Ans.*   333.7 MeV

Chapter 28

Nuclear Models

At present there exists no fundamental theory that will explain all the observed properties of nuclei. In lieu of a theory different models have been developed, each of which successfully explains some, but not all, nuclear properties.

28.1 LIQUID DROP MODEL

C. v. Weiszäcker in 1935 recognized that the nuclear properties connected with size, mass, and binding energy resemble what is found in a drop of liquid. In a liquid drop the density is constant, the size is proportional to the number of particles, or molecules, in the drop, and the heat of vaporization, or binding energy, of the drop is directly proportional to the mass or number of particles forming the drop.

As we shall now demonstrate, a liquid drop model for a nucleus leads to the following expression, known as the *semiempirical mass formula*, for the dependence of the mass of a nucleus on A and Z:

$$M = Zm_p + (A - Z)m_n - b_1A + b_2A^{2/3} + b_3Z^2A^{-1/3} + b_4(A - 2Z)^2A^{-1} + b_5A^{-3/4} \quad (28.1)$$

The constants in (28.1) are determined from experimental data; their values (in energy units) can be taken as

$$b_1 = 14.0 \text{ MeV} \qquad b_3 = 0.58 \text{ MeV}$$

$$b_2 = 13.0 \text{ MeV} \qquad b_4 = 19.3 \text{ MeV}$$

and b_5 is given according to the following scheme:

A	Z	b_5
Even	Even	-33.5 MeV
Odd		0
Even	Odd	$+33.5$ MeV

The various terms in (28.1) are obtained by a series of successive corrections, in the following manner.

With binding energy neglected, the first estimate of the mass of a nucleus composed of Z protons and $N = A - Z$ neutrons would be $Zm_p + (A - Z)m_n$.

Next, this estimate of the mass is corrected to account for the binding energy of the nucleons. Because the nuclear force is attractive, this binding energy will be positive (positive work must be done to separate the nucleons), so that the mass of the nucleus will be smaller than the mass of the separate nucleons. From the liquid drop model the heat of vaporization (binding energy) will be directly proportional to the number of nucleons A, resulting in a correction of $-b_1A$ ($b_1 > 0$).

The assumption made in the first correction, that the binding energy is b_1 per nucleon, is tantamount to assuming that all nucleons are equally surrounded by other nucleons. This is, of course, not true for nucleons on the nuclear surface, which are more weakly bound. Thus, too much was subtracted in the first correction, and a mass correction proportional to the nuclear surface area, $b_2A^{2/3}$, must be added to account for this "surface" effect.

The positive Coulomb energy between the protons, E_C, (which is equivalent to binding energy $-E_C$) increases the mass of the nucleus by an amount E_C/c^2. By Problem 27.7, for large Z,

$$E_C \propto Z^2 R^{-1} = Z^2 \left(r_0 A^{1/3}\right)^{-1} \propto Z^2 A^{-1/3}$$

which accounts for the term $b_3 Z^2 A^{-1/3}$.

To this point all the terms in the nuclear mass expression have been obtained from analogy to a charged incompressible liquid drop. In addition, because of quantum mechanical effects, two extra terms are usually added as follows.

It is found that if there are more neutrons than protons (or vice versa) in a nucleus, its energy, and correspondingly its mass, will be increased because of the Pauli exclusion principle. The correction term for this effect depends on the neutron (or proton) excess according to

$$b_4 (N - Z)^2 A^{-1} = b_4 (A - 2Z)^2 A^{-1}$$

(see Problem 28.16).

Nucleons in nuclei also tend to "pair," that is, neutrons or protons group together with opposite spins. Because of this effect it is found that a pairing energy exists that varies as $A^{-3/4}$ and increases with the number of unpaired nucleons. This number is determined as follows:

A	Z	Number of Unpaired Nucleons
Even	Even	0
Odd		1
Even	Odd	2 (1 neutron and 1 proton)

The inclusion of this pairing energy term then gives the final expression, (28.1), for the nuclear mass.

The *average binding energy per nucleon* is obtained from (28.1) by taking the difference between the nuclear mass-energy and the mass-energies of its constituent nucleons and dividing by the number of nucleons:

$$\text{BE}/A = \frac{\left[Z m_p + (A - Z)m_n - M\right]c^2}{A} = b_1 - b_2 A^{-1/3} - b_3 Z^2 A^{-4/3} - b_4 (A - 2Z)^2 A^{-2} - b_5 A^{-7/4}$$

$$(28.2)$$

(It should be noted that the BE/A is not the same as the energy required to remove a single nucleon from a given nucleus.) A plot of this equation is shown in Fig. 28-1. It is seen that for large A the value of BE/A is approximately constant at 8 MeV.

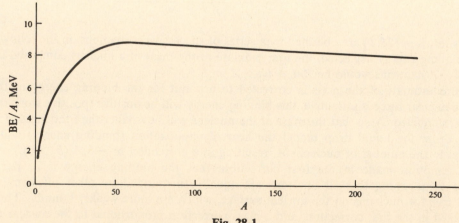

Fig. 28-1

It should be emphasized that (28.1) or (28.2) does not give exact values but predicts only approximate values, with the accuracy being different for different nuclei, as demonstrated in the Solved Problems.

28.2 SHELL MODEL

In the liquid drop model the nucleons are not treated individually, but instead their effects are averaged out over the nucleus. This model is successful in explaining some nuclear properties like the average binding energy per nucleon. However, other nuclear properties, such as the energies of excited states and nuclear magnetic moments, require a microscopic model that takes into account the behavior of the individual nucleons.

As nuclear data were accumulated it became evident that gross changes in nuclear properties occurred in nuclei with N or Z equal to 2, 8, 20, 28, 50, 82, or 126, usually referred to as "magic numbers." At these magic numbers nuclei are found to be particularly stable and numerous, and the last or magic nucleons that complete these "shells" have high binding energies. In addition, the energies of the first excited states are found to be larger than for nearby nuclei that do not have magic numbers. As an example, tin, with the magic number $Z = 50$, has 10 stable isotopes (same Z, different A), the energy required to remove a proton is about 11 MeV, and the first excited states of the even-even isotopes (i.e. both N and Z even) are about 1.2 MeV above the ground state. In contrast, for the nearby tellurium isotopes, with $Z = 52$, the energy required to remove a proton is about 7 MeV and for the even-even isotopes the first excited state has an energy of about 0.60 MeV.

We recall that similar fluctuations in behavior are observed in atoms, as the electrons completely fill the various atomic shells (Chapter 24). This similarity in behavior suggests that some nuclear properties might be explainable in terms of a *nuclear shell model*.

The *atomic* shell structure is obtained by a series of successive approximations. It is first assumed that the energy levels for a nucleus of charge Ze are successively filled with Z electrons as if they do not interact with each other, and then corrections are made to account for the various interaction effects. These corrections, however, are small; the main effect, resulting in the first approximation to the shell levels, is that on the average the electrons move independently in the Coulomb field of the nucleus.

If the same approach is taken to develop a shell picture for the nucleus, a different potential must be used to represent the short-range nuclear forces. One approach is to assume that the nucleons move in an average harmonic oscillator potential

$$V = \frac{1}{2} kR^2 = \frac{1}{2} m\omega^2 R^2$$

A quantum mechanical treatment then shows that the energy levels are given by

$$E = \left(\mathfrak{N} + \frac{3}{2} \right)\hbar\omega \qquad (28.3)$$

with $\mathfrak{N} = 2(n - 1) + l$. The quantity l is the orbital angular momentum quantum number and takes on the values 0, 1, 2, 3, . . . ; it is related to the orbital angular momentum vector in the usual fashion by $|\mathbf{l}| = \sqrt{l(l + 1)}\,\hbar$. (For nucleons, both quantized vectors and quantum numbers will be represented by lowercase letters.) The quantity n is an integer, taking on the values 1, 2, 3, 4, ... In contrast with the hydrogen atom solution, however, the value of l is not limited by n.

Nucleon orbital angular momentum states are indicated in spectroscopic notation:

value of l:	0	1	2	3	4	5	. . .
letter symbol:	s	p	d	f	g	h	. . .

By prefixing the letter symbol with the value of n, the order (with respect to increasing energy) of a given l-state is shown. (For fixed l, $\mathfrak{N}$ increases with n.) Thus, the $2d$ state is the next-to-lowest $l = 2$ state.

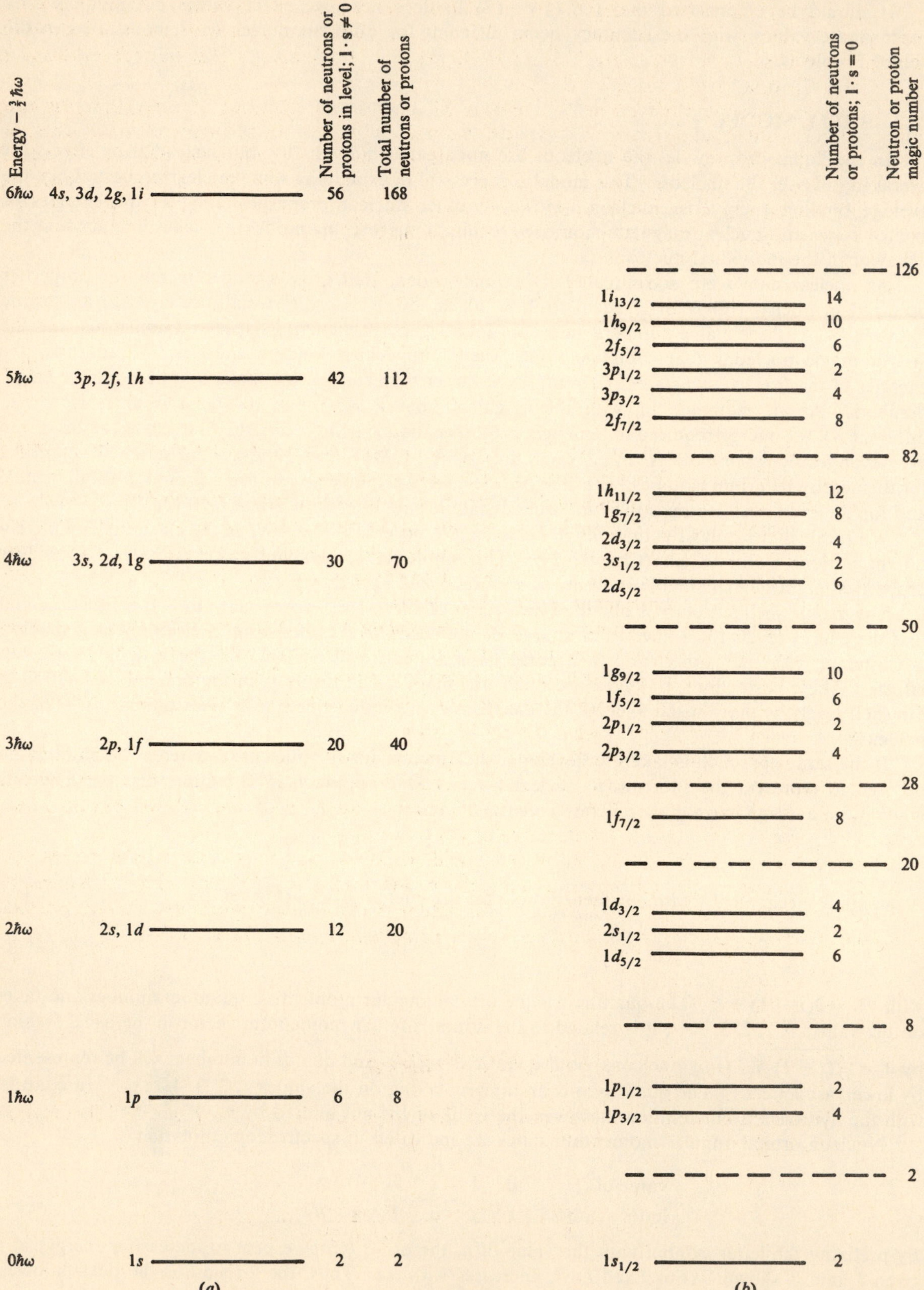

(a)

Energy $-\frac{3}{2}\hbar\omega$		Number of neutrons or protons in level; $l \cdot s \neq 0$	Total number of neutrons or protons
$6\hbar\omega$	$4s, 3d, 2g, 1i$	56	168
$5\hbar\omega$	$3p, 2f, 1h$	42	112
$4\hbar\omega$	$3s, 2d, 1g$	30	70
$3\hbar\omega$	$2p, 1f$	20	40
$2\hbar\omega$	$2s, 1d$	12	20
$1\hbar\omega$	$1p$	6	8
$0\hbar\omega$	$1s$	2	2

(b)

	Number of neutrons or protons; $l \cdot s = 0$	Neutron or proton magic number
		126
$1i_{13/2}$	14	
$1h_{9/2}$	10	
$2f_{5/2}$	6	
$3p_{1/2}$	2	
$3p_{3/2}$	4	
$2f_{7/2}$	8	
		82
$1h_{11/2}$	12	
$1g_{7/2}$	8	
$2d_{3/2}$	4	
$3s_{1/2}$	2	
$2d_{5/2}$	6	
		50
$1g_{9/2}$	10	
$1f_{5/2}$	6	
$2p_{1/2}$	2	
$2p_{3/2}$	4	
		28
$1f_{7/2}$	8	
		20
$1d_{3/2}$	4	
$2s_{1/2}$	2	
$1d_{5/2}$	6	
		8
$1p_{1/2}$	2	
$1p_{3/2}$	4	
		2
$1s_{1/2}$	2	

Fig. 28-2

Figure 28-2(a) shows the energy levels predicted from a harmonic oscillator potential, together with the maximum number of nucleons in each energy level consistent with the Pauli exclusion principle. It is seen that the energy level closings are at 2, 8, 20, 40, 70, 112, and 168 nucleons, of which only the first three are magic numbers.

To account for the observed magic numbers M. Mayer and J. Jensen, in 1949, independently proposed the existence of a spin-orbit ($\mathbf{l} \cdot \mathbf{s}$) interaction in addition to the harmonic oscillator potential. Because nucleons have the single value $s = \frac{1}{2}$ for their spin quantum number, the spin-orbit effect will cause each orbital angular momentum state with $l > 0$ to split into two *orbits* (or *orbitals*), according to whether the total angular momentum quantum number j is $j = l + s$ or $j = l - s$ [see (22.6)]. The relative energy splitting is found by evaluating $\mathbf{l} \cdot \mathbf{s}$ (Problem 22.1):

$$\mathbf{l} \cdot \mathbf{s} = \frac{1}{2}\left[j(j+1) - l(l+1) - s(s+1)\right]\hbar^2$$

$$= \begin{cases} \dfrac{l}{2}\,\hbar^2 & j = l + \frac{1}{2} \\[2mm] -\dfrac{l+1}{2}\,\hbar^2 & j = l - \frac{1}{2} \end{cases} \qquad (28.4)$$

Subtraction of these two expressions shows that the energy separation between the two orbits is proportional to $2l + 1$ and therefore becomes larger as l increases.

Orbits are designated by appending the value of j as a post-subscript to the symbol for the orbital angular momentum state. For example, $1d_{3/2}$ stands for the combination of quantum numbers $n = 1$, $l = 2, j = l - s = 3/2$. For nuclei, it is convenient to rewrite the Pauli exclusion principle as follows: no two nucleons may have the same set of quantum numbers (n, l, j, m_j). It follows (Problem 28.10) that an orbit may contain a maximum of $2j + 1$ nucleons.

In atoms the spin-orbit splitting is a small effect giving rise to the "fine" structure. In nuclei, however, the spin-orbit interaction is rather strong and gives rise to energy splittings comparable to the separation between the harmonic oscillator energy levels. Another difference between $\mathbf{l} \cdot \mathbf{s}$ splitting in nuclei and in atoms is that in nuclei the energy of the $j = l + \frac{1}{2}$ orbit is *lower* than that of the $j = l - \frac{1}{2}$ orbit, which is just the opposite of what is found in atoms.

It is not possible to predict whether a spin-orbit splitting will or will not result in "crossovers" of the initial harmonic oscillator levels. The final ordering of the orbits is determined from experimental evidence, and is shown in Fig. 28-2(b). The *shell closings*—the total number of nucleons up to each large energy gap—correspond to the magic numbers.

Protons (and neutrons) in the same orbit tend to pair to states of zero angular momentum. Therefore, even-even nuclei will have a total angular momentum, $\mathbf{J} = \Sigma \mathbf{j}$, of zero, while if a nucleus has an odd proton or neutron its total angular momentum is the angular momentum of the last (odd) nucleon. For odd-odd nuclei the situation is more complicated (see Problem 28.13).

Solved Problems

28.1. What is the Coulomb repulsion energy of the two protons in ^3_2He if it is assumed that they are separated by a nuclear radius?

The Coulomb energy is

$$E_C = \frac{ke^2}{R} = \frac{ke^2}{(r_0 A^{1/3})} = \frac{1.44\ \text{MeV} \cdot \text{fm}}{(1.4\ \text{fm})(3^{1/3})} = 0.71\ \text{MeV}$$

28.2. What is the difference between the binding energies of ^3_2He and ^3_1H?

The binding energy for $^3_2$He is

$$BE_{He} = (Zm_p + Nm_n - M)c^2 = [2(1.007825 \text{ u}) + 1.008665 \text{ u} - 3.016030 \text{ u}](931.5 \text{ MeV/u}) = 7.72 \text{ MeV}$$

The binding energy for $^3_1$H is

$$BE_H = [1.007825 \text{ u} + 2(1.008665 \text{ u}) - 3.016050 \text{ u}](931.5 \text{ MeV/u}) = 8.48 \text{ MeV}$$

Note that the binding energy of $^3_2$He is lower than that of $^3_1$H by an amount (0.76 MeV) that is approximately equal to the Coulomb repulsion energy of $^3_2$He as estimated in Problem 28.1.

28.3. Calculate the binding energy per nucleon for $^{98}_{42}$Mo.

$$BE/A = \frac{(Zm_p + Nm_n - M_{nuc})c^2}{A}$$

where the atomic masses are used for m_p and M_{nuc} (so that the electron masses cancel).

$$BE/A = \frac{42(1.007825 \text{ u}) + 56(1.008665 \text{ u}) - 97.905409 \text{ u}}{98} \left(931.5 \frac{\text{MeV}}{\text{u}}\right) = 8.64 \text{ MeV}$$

28.4. Compare the minimum energies required to remove a neutron from $^{41}_{20}$Ca, $^{42}_{20}$Ca and $^{43}_{20}$Ca.

For $^{41}_{20}$Ca the energy needed to remove a neutron is obtained from the process

$$^{41}_{20}\text{Ca} + E \rightarrow\, ^{40}_{20}\text{Ca} + n$$

so

$$E = (M_{^{40}\text{Ca}} + m_n - M_{^{41}\text{Ca}})c^2 = (39.962589 \text{ u} + 1.008665 \text{ u} - 40.962275 \text{ u})(931.5 \text{ MeV/u}) = 8.36 \text{ MeV}$$

For $^{42}_{20}$Ca,

$$E = (M_{^{41}\text{Ca}} + m_n - M_{^{42}\text{Ca}})c^2 = (40.962275 \text{ u} + 1.008665 \text{ u} - 41.958625 \text{ u})(931.5 \text{ MeV/u}) = 11.47 \text{ MeV}$$

For $^{43}_{20}$Ca,

$$E = (M_{^{42}\text{Ca}} + m_n - M_{^{43}\text{Ca}})c^2 = (41.958625 \text{ u} + 1.008665 \text{ u} - 42.958780 \text{ u})(931.5 \text{ MeV/u}) = 7.93 \text{ MeV}$$

The energy needed to remove a neutron from $^{42}_{20}$Ca is 3.11 MeV larger than the energy needed to remove a neutron from $^{41}_{20}$Ca and 3.54 MeV larger than the energy needed to remove a neutron from $^{43}_{20}$Ca, even though all the neutrons are in the $1f_{7/2}$ shell (which consists of the $1f_{7/2}$ orbit). The reason for this difference is that neutrons in the same orbit tend to pair. Therefore, in $^{42}_{20}$Ca with 22 neutrons, one needs not only the normal neutron binding energy but an additional energy to break the $1f_{7/2}$ neutron pair. In $^{41}_{20}$Ca and $^{43}_{20}$Ca there is an unpaired neutron available, so less energy is needed. It is interesting to note that $^{44}_{20}$Ca requires 11.14 MeV to remove a neutron (it has two neutron pairs).

28.5. "Mirror" nuclei have the same odd value of A, but the values of N and Z are interchanged. Determine the mass difference between two mirror nuclei which have N and Z differing by one unit.

The term $A - 2Z$ in the semiempirical mass formula can be written as

$$A - 2Z = N + Z - 2Z = N - Z$$

so that if N and Z differ by one unit, $A - 2Z = \pm 1$. If we now subtract the two masses M_{Z+1} and M_Z from each other, the $(A - 2Z)^2$ term will cancel, leaving, for constant $A = 2Z + 1$,

$$M_{Z+1} - M_Z = (m_p - m_n)[(Z + 1) - Z] + b_3 A^{-1/3}[(Z + 1)^2 - Z^2] = m_p - m_n + b_3 A^{2/3}$$

28.6. The masses of $^{23}_{11}$Na and $^{23}_{12}$Mg are 22.989771 u and 22.994125 u, respectively. From these data determine the constant b_3 in the semiempirical mass formula.

The two nuclei are mirror nuclei.   From Problem 28.5,

$$M_{Z+1} - M_Z = m_p - m_n + b_3 A^{2/3}$$

$$22.994125 \text{ u} - 22.989771 \text{ u} = 1.007825 \text{ u} - 1.008665 \text{ u} + b_3 (23)^{2/3}$$

$$b_3 = 6.42 \times 10^{-4} \text{ u} = 0.598 \text{ MeV}$$

28.7.   From Problem 27.7, the Coulomb energy of a nucleus is, for large Z,

$$E_C = \frac{3}{5} \frac{kZ^2 e^2}{R}$$

Calculate b_3 in the semiempirical mass formula, taking $r_0 = 1.5$ fm.

For a nucleus, $R = r_0 A^{1/3}$, and the Coulomb energy is

$$E_C = \frac{3}{5} \frac{kZ^2 e^2}{r_0 A^{1/3}} = b_3 \frac{Z^2}{A^{1/3}}$$

Therefore,

$$b_3 = \frac{3}{5} \frac{ke^2}{r_0} = \frac{3(1.44 \text{ MeV} \cdot \text{fm})}{5(1.5 \text{ fm})} = 0.58 \text{ MeV}$$

If r_0 is taken as 1.4 fm, the value of b_3 becomes 0.62 MeV.   These answers agree reasonably well with the value of b_3 found in Problem 28.6.

28.8.   Using the liquid drop model, find the most stable isobar for a given odd A.

For odd A, $b_5 = 0$ in the semiempirical mass formula and the binding energy is found to be

$$\text{BE} = b_1 A - b_2 A^{2/3} - b_3 Z^2 A^{-1/3} - b_4 (A - 2Z)^2 A^{-1}$$

The most stable isobar ($A = $ constant) is the one with the maximum binding energy.   This is found by setting $d(\text{BE})/dZ = 0$.

$$\frac{d(\text{BE})}{dZ} = -2b_3 Z A^{-1/3} + 4b_4 (A - 2Z) A^{-1} = 0$$

$$Z = \frac{4b_4}{2b_3 A^{-1/3} + 8b_4 A^{-1}} = \frac{A}{\dfrac{b_3}{2b_4} A^{2/3} + 2}$$

Using $b_3 = 0.58$ MeV and $b_4 = 19.3$ MeV gives

$$Z = \frac{A}{0.015 \, A^{2/3} + 2}$$

28.9.   For $A = 25, 43, 77$, find the most stable nuclei.

The result of Problem 28.8 gives, for $A = 25$,

$$Z = \frac{A}{0.015 \, A^{2/3} + 2} = \frac{25}{(0.015)(25)^{2/3} + 2} = 11.7 \approx 12$$

and $^{25}_{12}\text{Mg}$ is in fact stable.   It is found experimentally that $^{25}_{13}\text{Al}$ and $^{25}_{11}\text{Na}$ are not stable.
For $A = 43$,

$$Z = \frac{43}{(0.015)(43)^{2/3} + 2} = 19.7 \approx 20$$

and it is found experimentally that $^{43}_{20}\text{Ca}$ is stable, while $^{43}_{19}\text{K}$ and $^{43}_{21}\text{Sc}$ are unstable.
For $A = 77$,

$$Z = \frac{77}{(0.015)(77)^{2/3} + 2} = 33.9 \approx 34$$

and it is found experimentally that $^{77}_{34}\text{Se}$ is stable, while $^{77}_{33}\text{As}$ and $^{77}_{35}\text{Br}$ are unstable.

28.10. Show that in an orbit of given j there may be at most $2j + 1$ nucleons. Demonstrate that for p states ($l = 1$) this is consistent with the fact that the Pauli principle allows $2(2l + 1) = 6$ nucleons.

For given j,

$$m_j = j, j - 1, \ldots, -(j - 1), -j$$

a total of $2j + 1$ values. Therefore the Pauli principle allows $2j + 1$ nucleons in the orbit.

A p state is split into a $p_{3/2}$ orbit, which may contain $2j + 1 = 2(3/2) + 1 = 4$ nucleons, and a $p_{1/2}$ orbit, which may contain $2j + 1 = 2(1/2) + 1 = 2$ nucleons. The total is 6 nucleons.

28.11. For $A = 50$ the known masses are: $^{50}_{21}\text{Sc}$, 49.951730 u; $^{50}_{22}\text{Ti}$, 49.944786 u; $^{50}_{23}\text{V}$, 49.947164 u; $^{50}_{24}\text{Cr}$, 49.946055 u; $^{50}_{25}\text{Mn}$, 49.954215 u. From these data estimate the constant b_5, the strength of the pairing term, in the semiempirical mass formula.

For fixed even A and for Z odd (whence $Z + 1$ is even, etc.) the semiempirical mass formula can be written as:

$$M(Z) = a_1 Z^2 \qquad\quad + a_2 Z \qquad\;\; + a_3 + b_5 A^{-3/4}$$
$$M(Z + 1) = a_1(Z + 1)^2 + a_2(Z + 1) + a_3 - b_5 A^{-3/4}$$
$$M(Z + 2) = a_1(Z + 2)^2 + a_2(Z + 2) + a_3 + b_5 A^{-3/4}$$
$$M(Z + 3) = a_1(Z + 3)^2 + a_2(Z + 3) + a_3 - b_5 A^{-3/4}$$

where a_1, a_2, and a_3 are constants. Taking the *third difference*:

$$M(Z + 3) - 3M(Z + 2) + 3M(Z + 1) - M(Z) = -8b_5 A^{-3/4}$$

Applying this to $Z = 21$ gives (after reducing the data by 49 u)

$$0.946055\text{ u} - 3(0.947164\text{ u}) + 3(0.944786\text{ u}) - 0.951730\text{ u} = -8(50^{-3/4})b_5$$

from which

$$b_5 = \frac{50}{8}^{3/4}(0.012809\text{ u}) = 0.0301\text{ u} = 28.0\text{ MeV}$$

If instead we take another third difference,

$$M(Z + 4) - 3M(Z + 3) + 3M(Z + 2) - M(Z + 1) = +8b_5 A^{-3/4}$$

we obtain

$$b_5 = \frac{50}{8}^{3/4}(0.012756\text{ u}) = 0.0300\text{ u} = 27.9\text{ MeV}$$

and the two estimates for b_5 are seen to be very close together. The accepted value is 33.5 MeV.

28.12. Find the ground-state angular momentum of (a) $^{15}_8\text{O}$, (b) $^{39}_{19}\text{K}$, (c) $^{20}_{10}\text{Ne}$.

The ground-state configurations, as given by the shell model, are as shown in Fig. 28-3.
(a) All nucleons are paired except the $1p_{1/2}$ neutron; therefore, the total angular momentum in the ground state is $J = 1/2$.
(b) All nucleons are paired except a $1d_{3/2}$ proton; the total angular momentum is $J = 3/2$.
(c) All nucleons are paired; $J = 0$.

28.13. What are the possible values of the ground-state angular momentum for $^{32}_{15}\text{P}$?

The shell model ground-state description of $^{32}_{15}\text{P}$, assuming all the lowest energy levels are totally filled, is shown in Fig. 28-4. All the nucleons in this model of $^{32}_{15}\text{P}$ are paired to zero angular momentum except for the $2s_{1/2}$ proton and the $1d_{3/2}$ neutron. The ground-state angular momentum of $^{32}_{15}\text{P}$ in this picture must be the vector sum of the angular momenta of the $j = 1/2$ and $j = 3/2$ particles. For the proton we have as the possible values of m_j

$$m_{1/2} = \frac{1}{2},\ -\frac{1}{2}$$

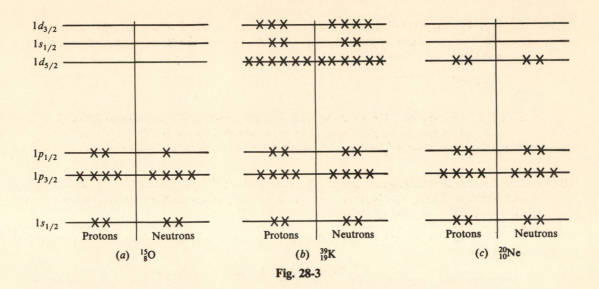

Fig. 28-3

and for the neutron,

$$m_{3/2} = \frac{3}{2}, \ \frac{1}{2}, \ -\frac{1}{2}, \ -\frac{3}{2}$$

Then

$$M_J = m_{1/2} + m_{3/2} = 2, \ \begin{Bmatrix} 1 \\ 1 \end{Bmatrix}, \ \begin{Bmatrix} 0 \\ 0 \end{Bmatrix}, \ \begin{Bmatrix} -1 \\ -1 \end{Bmatrix}, \ -2$$

The upper line of M_J-values corresponds to $J = 2$; the lower line, to $J = 1$.

The experimental value of the ground state of $^{32}_{15}$P is $J = 1$, $J = 2$ corresponding to the first excited state.

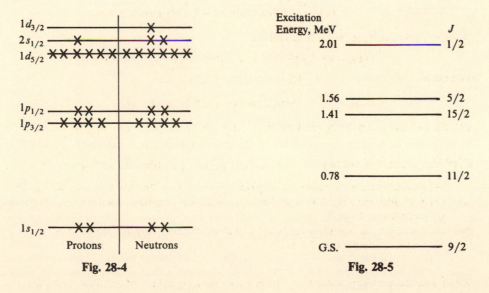

Fig. 28-4   Fig. 28-5

28.14. The first four excited states of $^{209}_{82}$Pb are as shown in Fig. 28-5. Explain the spectrum as a single-particle excitation, using the shell model.

The isotope $^{208}_{82}$Pb, with 82 protons and 126 neutrons, is doubly magic. As shown in Fig. 28-6(a), $^{209}_{82}$Pb has a $^{208}_{82}$Pb closed core and one additional neutron in the $2g_{9/2}$ orbit. The 11/2, 15/2, 5/2 and 1/2 excited states correspond to excitations of the neutron to the $1i_{11/2}$, $1j_{15/2}$, $3d_{5/2}$ and $4s_{1/2}$ shell-model orbits [Fig. 28-6(b) through (e)].

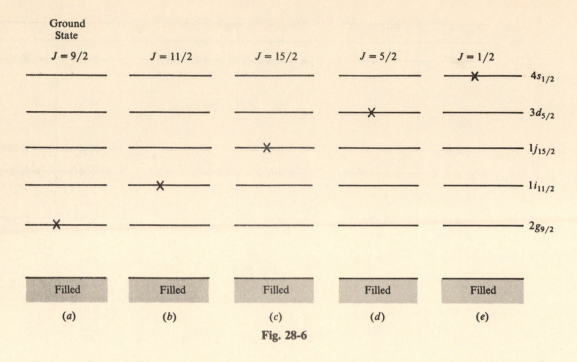

Fig. 28-6

28.15. The masses of $^{40}_{20}Ca$, $^{41}_{20}Ca$, and $^{39}_{20}Ca$ are 39.962589 u, 40.962275 u, and 38.970691 u, respectively. Calculate the energy difference between the $1d_{3/2}$ and the $1f_{7/2}$ neutron shells (i.e. the energy gap corresponding to neutron magic number 20).

From the shell model, $^{39}_{20}Ca$ has one neutron missing in the $1d_{3/2}$ shell, $^{40}_{20}Ca$ completes this shell and $^{41}_{20}Ca$ adds a neutron to the $1f_{7/2}$ shell.

The binding energy of a $1d_{3/2}$ neutron in ^{40}Ca is

$$BE_1 = (M_{^{39}Ca} + m_n - M_{^{40}Ca})c^2$$

$$= (38.970691\ u + 1.008665\ u - 39.962589\ u)(931.5\ MeV/u) = 15.62\ MeV$$

while the binding energy of the $1f_{7/2}$ neutron in ^{41}Ca is

$$BE_2 = (M_{^{40}Ca} + m_n - M_{^{41}Ca})c^2$$

$$= (39.962589\ u + 1.008665\ u - 40.962275\ u)(931.5\ MeV/u) = 8.36\ MeV$$

The difference in binding energies must be the energy separation, δ, of the $1f_{7/2}$ and $1d_{3/2}$ shells:

$$\delta = BE_1 - BE_2 = 15.62\ MeV - 8.36\ MeV = 7.26\ MeV$$

28.16. Consider a shell model in which the nucleons are in pairs in equally spaced energy levels. If one starts with the same number of neutrons and protons, calculate the energy needed to change n proton pairs to neutrons and move them to neutron orbits.

The problem is illustrated in Fig. 28-7. If the final nucleus is to have N neutrons and Z protons, we define the nucleon difference by $n \equiv N - Z$, which is *twice* the number of nucleon pairs to be moved from proton levels to neutron levels. If all levels are separated by an energy δ, the total energy needed to create the final nucleus is

$$E = (2\delta)(1) + (2\delta)(3) + (2\delta)(5) + \cdots + (2\delta)(n-1) = 2\delta[1 + 3 + 5 + \cdots + (n-1)]$$

$$= 2\delta\left(\frac{n^2}{4}\right) = \frac{\delta}{2}(N-Z)^2 = \frac{\delta}{2}(A-2Z)^2$$

This term is directly related to the $b_4(A-2Z)^2 A^{-1}$ term in the semiempirical mass formula, which is an expression for the neutron or proton excess energy.

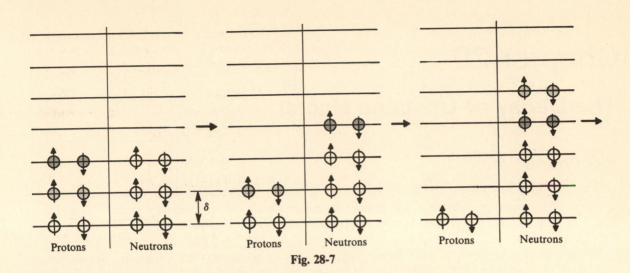

Fig. 28-7

Supplementary Problems

Atomic masses are tabulated in the Appendix.

28.17. Calculate the binding energy per nucleon for (a) $^4_2$He, (b) $^{12}_6$C, (c) $^{40}_{20}$Ca, (d) $^{202}_{80}$Hg.
Ans. (a) 7.07 MeV; (b) 7.68 MeV; (c) 8.55 MeV; (d) 7.90 MeV

28.18. Calculate the energy needed to remove the least tightly bound neutron in $^{17}_8$O. *Ans.* 4.14 MeV

28.19. Determine the value of b_3 in the semiempirical mass formula from the masses of $^{22}_{10}$Ne (21.991385 u) and $^{22}_{11}$Na (21.994437 u). Compare your value with that found in Problem 28.6. *Ans.* 0.484 MeV

28.20. For $A = 57$ find the most stable nucleus. *Ans.* $Z = 26$ ($^{57}_{26}$Fe)

28.21. Find the ground-state angular momentum of (a) $^{41}_{20}$Ca, (b) $^{80}_{36}$Kr, (c) $^{91}_{40}$Zr.
Ans. (a) 7/2; (b) 0; (c) 5/2

28.22. Give the expected total angular momentum for the following states in $^{13}_6$C: (a) the ground state, with all neutron and proton orbits filled through the $1p_{3/2}$ and the extra neutron in the $1p_{1/2}$ orbit; (b) an excited state which is the same as (a) except that the extra neutron is excited to the $2s_{1/2}$ orbit; (c) an excited state which is the same as (a) except that the extra neutron is in the $1d_{5/2}$ orbit; (d) an excited state with the proton orbits filled through the $1p_{3/2}$, and with two neutrons in the $1s_{1/2}$ orbit, three neutrons in the $1p_{3/2}$ orbit and two neutrons in the $1p_{1/2}$ orbit. *Ans.* (a) 1/2; (b) 1/2; (c) 5/2; (d) 3/2

28.23. The *mass excesses*,

$$\delta \equiv [M \text{ (in u)} - A(\times 1 \text{ u})] \times (931.5 \text{ MeV/u})$$

for $^{15}_8$O, $^{16}_8$O and $^{17}_8$O are respectively 2859.9 keV, -4736.6 keV and -807.7 keV. From this data calculate the difference in energy between the $1p_{1/2}$ and $1d_{5/2}$ neutron shells (i.e. the energy gap corresponding to neutron magic number 8). *Ans.* 11.53 MeV

Chapter 29

The Decay of Unstable Nuclei

29.1 INTRODUCTION

As discussed in Chapter 28, nuclei have excited states. These excited states can decay by the emission of high-energy photons to the ground state, directly or via lower energy states. In addition, nuclei in both excited and ground states can spontaneously emit other particles to reach lower-energy configurations.

When nuclear decay was first investigated, the decay products were given the names γ-*rays*, *α-particles*, *β⁻-particles* and *β⁺-particles*. It was not until later that it was recognized that the decay products were not new entities, but that γ-rays are high-energy photons, α-particles are helium nuclei, β⁻-particles are electrons, and β⁺-particles are positrons.

In the various reactions the usual conservation laws of mass-energy, charge, and linear and angular momenta always apply. In nuclear decay, however, it is found that a law of conservation of nucleons also holds: the number of nucleons before and after a decay must be equal.

29.2 THE STATISTICAL RADIOACTIVE DECAY LAW

In a typical radioactive decay an initially unstable nucleus, called the *parent*, emits a particle and decays into a nucleus called the *daughter*; effectively, the birth of the daughter arises from the death of the parent. The daughter may be either the same nucleus in a lower energy state, as in the case of γ-decay, or an entirely new nucleus, as arises from α- and β-decays. No matter what types of particles are emitted, all nuclear decays follow the same *radioactive decay law*. If there are initially N_0 unstable parent nuclei present, the number N of parents that will be left after a time t is (Problem 29.1)

$$N = N_0 e^{-\lambda t} \qquad (29.1)$$

The constant λ is called the *decay constant* or *disintegration constant* and depends on the particular decay process.

Equation (29.1) is a statistical, not a deterministic, law; it gives the *expected number* N of parents that survive after a time t. However, if N_0 is very large (as it always is in applications), the actual number of survivors and the expected number will almost certainly differ by no more than an insignificant fraction of N_0.

The rapidity of decay of a particular radioactive sample is usually measured by the *half-life*, $T_{1/2}$, defined as the time interval in which the number of parent nuclei at the beginning of the interval is reduced by a factor of one-half. The half-life is readily obtained in terms of λ as

$$T_{1/2} = \frac{\ln 2}{\lambda} = \frac{0.693}{\lambda} \qquad (29.2)$$

Thus, starting initially with N_0 nuclei, $N_0/2$ will be left after a time $T_{1/2}$, $N_0/4$ will remain after a time $2T_{1/2}$, etc.

Another quantity that measures how fast a sample decays is the *average* or *mean lifetime* of a nucleus, T_m, given by

$$T_m = \frac{1}{\lambda} = \frac{T_{1/2}}{\ln 2} \qquad (29.3)$$

(see Problem 29.7).

The law describing the increase in daughter nuclei, assuming they are stable, is obtained from (29.1) as

$$N_D = N_0 - N = N_0(1 - e^{-\lambda t}) \qquad (29.4)$$

In many decays the daughter nucleus is also unstable and decays further into a granddaughter; this situation will be treated in the Solved Problems.

The *activity* of a sample is defined as the value of the disintegration rate:

$$\text{activity} = \left| \frac{dN}{dt} \right| = \lambda N_0 e^{-\lambda t} = \lambda N \qquad (29.5)$$

The unit of measure of the disintegration rate, or activity, is the *curie*, defined as $1 \text{ Ci} = 3.700 \times 10^{10}$ disintegrations per second.

29.3 GAMMA DECAY

In a gamma decay a nucleus initially in an excited state makes a transition to a lower energy state and in the process emits a photon, called a γ-ray. It is found that the γ-rays emerge with discrete energies, which shows that nuclei possess discrete energy levels. The energy of the γ-ray photon is given by the usual expression

$$h\nu = E_u - E_l \qquad (29.6)$$

In contrast to photons emitted in atomic transitions, where the energies are of the order of a few eV, the energies of γ-rays range from tens of keV's to MeV's.

Because γ-ray photons carry no charge or mass, the charge and atomic number of the nucleus do not change in gamma decay. If the excited nucleus is designated by $(Z^A)^*$, a gamma decay to the ground state can be written symbolically as

$$(Z^A)^* \rightarrow Z^A + \gamma$$

Most excited nuclei that undergo gamma decay have immeasurably small half-lives of the order of 10^{-14} s, much shorter than the half-life of excited electronic states. The excited states of some nuclei, however, are very long and their half-lives are readily measurable. These excited nuclei are called *isomers* and the excited states are referred to as *isomeric states*.

29.4 ALPHA DECAY

In alpha decay an α-particle is ejected from a nucleus. Inasmuch as an α-particle is a helium nucleus, the parent nucleus loses two protons and two neutrons. Therefore its atomic number Z decreases by two units and its mass number A decreases by four units, so that the daughter, D, and parent, P, are different chemical elements. Applying conservation of charge and nucleons, we can write alpha decay symbolically as

$$^A_Z P \rightarrow ^{A-4}_{Z-2} D + ^4_2 He$$

For example,

$$^{238}_{92}U \rightarrow ^{234}_{90}Th + ^4_2 He$$

In a system where the parent is at rest, we find from conservation of energy that

$$M_P c^2 = M_D c^2 + M_\alpha c^2 + K_D + K_\alpha \qquad (29.7)$$

where K_D and K_α are the respective kinetic energies of the daughter and α-particle, and M_P, M_D and M_α are the rest masses of the parent, daughter and α-particle, respectively. Because kinetic energy can never be negative, alpha decay cannot occur unless

$$M_P \geqslant M_D + M_\alpha \qquad (29.8)$$

In addition to energy, momentum must be conserved. Since only two particles result from alpha decay, the two conservation conditions fix uniquely the kinetic energies (and momenta) of the α-particle and daughter nucleus. If a parent nucleus of mass number A decays at rest, the kinetic energy of the α-particle is given by (see Problem 29.15)

$$K_\alpha = \left(\frac{A - 4}{A} \right) Q \tag{29.9}$$

where the *disintegration energy* Q is the total energy released in the reaction:

$$Q = (M_P - M_D - M_\alpha)c^2 \tag{29.10}$$

The quantity Q is a constant for any alpha decay and has the same value for all observers. In the rest frame of the parent nucleus,

$$Q = K_D + K_\alpha \tag{29.11}$$

29.5 BETA DECAY AND THE NEUTRINO

It is possible for a nuclear process to occur where the charge Ze of a nucleus changes, but the number of nucleons, A, remains unchanged. This can happen with a nucleus emitting an electron (β^- decay), emitting a positron (β^+ decay), or capturing an inner atomic electron (*electron capture*). In each of these processes either a proton is converted into a neutron or vice versa.

It is also found that in each of these processes an extra particle, called a *neutrino* (ν), appears as one of the decay products. The properties of a neutrino are: electric charge, 0; rest mass, 0; intrinsic spin, 1/2; and, as with all massless particles, it has speed c (speed of light).

The existence of a neutrino was first postulated by W. Pauli in 1930 in order to preserve conservation of energy and momentum in beta decays. For example, neutron beta decay is

$$n \rightarrow p + e^- + \bar{\nu} \tag{29.12}$$

If the neutrino were not part of the decay products, it would follow from conservation of energy and momentum that for the two-body decay the electrons would be ejected with one single energy, as was described above in alpha decay. Experimentally, however, it had been found that the ejected electrons have a distribution of energies ranging from zero up to a maximum energy, as shown in Fig. 29-1. Moreover, since originally there is a single particle with spin $\frac{1}{2}$, the creation of only two particles, each with spin $\frac{1}{2}$, would violate conservation of angular momentum. Actual observation of a neutrino did not take place until the experiments of C. L. Cowan and F. Reines in 1956.

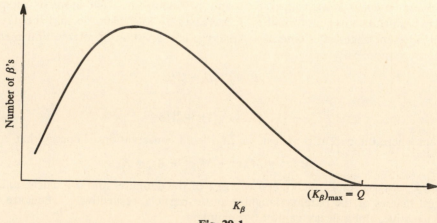

Fig. 29-1

In (29.12) the emitted neutrino has been designated by $\bar{\nu}$ rather than by ν. This is done because there are actually two different types of neutrinos, the "neutrino" (ν) and the "antineutrino" ($\bar{\nu}$). The antineutrino arises in β^- decay, while the neutrino occurs in the other beta processes. Antiparticles will be discussed further in Chapter 32.

In general, a β^- decay can be expressed as

$$_Z^A P \rightarrow _{Z+1}^A D + e^- + \bar{\nu}$$

a typical example being

$$_5^{12} B \rightarrow _6^{12} C + e^- + \bar{\nu}$$

Thus in β^- decay a neutron is converted into a proton. For β^+ decay, where a positron is emitted,

$$_Z^A P \rightarrow _{Z-1}^A D + e^+ + \nu$$

so that a proton is converted into a neutron. An example of β^+ decay is

$$_7^{12} N \rightarrow _6^{12} C + e^+ + \nu$$

From conservation of energy, in a system where the parent nucleus is at rest, we have for both β^- and β^+ decays (the electron and positron masses are equal)

$$M_P c^2 = M_D c^2 + m_e c^2 + K_{\text{total}}$$

giving a disintegration energy Q of

$$Q = K_{\text{total}} = (M_P - M_D - m_e)c^2$$

In electron capture, where an inner atomic electron (usually a K electron) is captured by a nucleus, no charged particle is emitted. Instead, electron capture is accompanied by the emission of a neutrino, followed by the emission of characteristic X-ray photons as the outer electrons make transitions to the vacant inner energy levels. In electron capture a proton is converted into a neutron. Moreover, the emitted X-rays are characteristic of the daughter atom and not the parent atom, since they are produced after the electron capture has taken place. An electron capture process may be written as

$$e^- + _Z^A P \rightarrow _{Z-1}^A D + \nu$$

where an example is

$$e^- + _4^7 Be \rightarrow _3^7 Li + \nu$$

It must be emphasized that in beta decay or electron capture the electrons or positrons do not exist inside the nucleus. The nucleus is composed only of protons and neutrons. The creation or absorption of the electrons or positrons results in the rearrangement of the nucleus into a state of lower energy by the transformation of a proton into a neutron or vice versa.

In the following problems, unless otherwise stated, all given mass values are atomic masses.

Solved Problems

29.1. Derive the decay law $N = N_0 e^{-\lambda t}$.

The number of nuclei dN that decay in a time interval dt will be proportional to that time interval and proportional to the number of nuclei N that are present. Thus

$$dN = -\lambda N \, dt$$

where λ is the proportionality constant and the minus sign is introduced because N decreases. Integration of this expression yields the decay law.

An alternative derivation will shed light on the statistical nature of the decay law. Consider the lifetime T (a random variable) of a single parent nucleus. Suppose that the probability that the nucleus will decay within the next τ seconds (τ arbitrary) is independent of how long the nucleus has already lived. Then it can be shown that T has an *exponential distribution*:

$$\text{Prob}\{t < T \leqslant t + dt\} = \lambda e^{-\lambda t}\, dt \qquad (1)$$

where λ is a positive constant. It follows from (1) that the nucleus has probability

$$p = \int_t^\infty \lambda e^{-\lambda t}\, dt = e^{-\lambda t} \qquad (2)$$

of living at least t seconds.

Now imagine that we have an initial sample consisting of N_0 independent nuclei, and that we observe the sample over a period of t seconds. The decay process can be modeled by tossing a coin N_0 times, where the probability of "heads" (survival over the period) is p on each toss. The expected (average) number of heads, N, is just

$$N = N_0 p = N_0 e^{-\lambda t}$$

which is the radioactive decay law.

29.2. What is the activity of one gram of $^{226}_{88}\text{Ra}$, whose half-life is 1622 years?

The number of atoms in 1 g of radium is

$$N = (1 \text{ g})\left(\frac{1 \text{ g-mole}}{226 \text{ g}}\right)\left(6.025 \times 10^{23}\, \frac{\text{atoms}}{\text{g-mole}}\right) = 2.666 \times 10^{21}$$

The decay constant is related to the half-life by

$$\lambda = \frac{0.693}{T_{1/2}} = \left(\frac{0.693}{1622 \text{ y}}\right)\left(\frac{1 \text{ y}}{365 \text{ d}}\right)\left(\frac{1 \text{ d}}{8.64 \times 10^4 \text{ s}}\right) = 1.355 \times 10^{-11}\, \text{s}^{-1}$$

The activity is then found from

$$\text{activity} = \lambda N = (1.355 \times 10^{-11}\, \text{s}^{-1})(2.666 \times 10^{21}) = 3.612 \times 10^{10}\, \text{disintegrations/s}$$

The definition of the curie is $1 \text{ Ci} = 3.700 \times 10^{10}$ disintegrations/s. This is approximately equal to the value found above.

29.3. Over what distance in free space will the intensity of a 5 eV neutron beam be reduced by a factor of one-half? ($T_{1/2} = 12.8$ min.)

The speed of the neutrons in the beam is found from $\frac{1}{2} mv^2 = K$:

$$\frac{1}{2}(1.67 \times 10^{-27}\text{ kg})v^2 = (5 \text{ eV})\left(\frac{1.6 \times 10^{-19}\text{ J}}{1 \text{ eV}}\right)$$

$$v = 31.0 \text{ km/s}$$

During a time of $T_{1/2} = 12.8$ min, half the neutrons will have decayed from the beam. The distance traveled by the undecayed neutrons during this time is

$$d = vt = (31.0 \text{ km/s})(12.8 \text{ min})(60 \text{ s/min}) = 23{,}800 \text{ km}$$

or about 2 earth diameters.

29.4. How much time is required for 5 mg of ^{22}Na ($T_{1/2} = 2.60$ y) to reduce to 1 mg?

Since the mass of a sample will be proportional to the number of atoms in the sample, we may write

$$m = m_0 e^{-\lambda t} = m_0 e^{-(0.693/T_{1/2})t}$$

$$1 \text{ mg} = (5 \text{ mg})e^{-(0.693/2.60 \text{ y})t}$$

$$e^{(0.693/2.60 \text{ y})t} = 5$$

Taking logarithms of both sides, we find

$$\frac{0.693t}{2.60 \text{ y}} = \ln 5 = 1.61 \qquad \text{or} \qquad t = 6.04 \text{ y}$$

29.5. If 3×10^{-9} kg of radioactive $^{200}_{79}$Au has an activity of 58.9 Ci, what is its half-life?

The number of atoms in 3×10^{-9} kg of $^{200}_{79}$Au is

$$N = (3 \times 10^{-9} \text{ kg})\left(\frac{1 \text{ kmol}}{200 \text{ kg}}\right)\left(6.025 \times 10^{26} \frac{\text{atoms}}{\text{kmol}}\right) = 9.04 \times 10^{15} \text{ atoms}$$

The activity is

$$\text{activity} = (58.9 \text{ Ci})\left(\frac{3.7 \times 10^{10} \text{ disintegration/s}}{1 \text{ Ci}}\right) = 2.18 \times 10^{12} \frac{\text{disintegrations}}{\text{s}}$$

The decay constant is found from

$$\text{activity} = \lambda N$$

$$\lambda = \frac{2.18 \times 10^{12} \text{ s}^{-1}}{9.04 \times 10^{15}} = 2.41 \times 10^{-4} \text{ s}^{-1}$$

Finally,

$$T_{1/2} = \frac{\ln 2}{\lambda} = \frac{0.693}{2.41 \times 10^{-4} \text{ s}^{-1}} = 2.88 \times 10^{3} \text{ s} = 48 \text{ min}$$

29.6. The activity of a sample of $^{55}_{24}$Cr at the end of 5-min intervals is found to be 19.2, 7.13, 2.65, 0.99, and 0.37 millicuries. What is the half-life of $^{55}_{24}$Cr?

$$\text{activity} = \lambda N = \lambda N_0 e^{-\lambda t}$$

Taking the natural logarithms of both sides, we have

$$\ln(\text{activity}) = \ln\left(\lambda N_0 e^{-\lambda t}\right) = \ln(\lambda N_0) - \lambda t$$

Thus $\ln(\text{activity})$ varies linearly with the time t, with slope $-\lambda$. Plotting the data

Time, min	0	5	10	15	20
Activity, mCi	19.2	7.13	2.65	0.99	0.37
ln (activity)	2.95	1.96	0.974	−0.010	−0.994

we obtain the curve shown in Fig. 29-2. From the curve we find

$$|\text{slope}| = \lambda = 0.197 \text{ min}^{-1}$$

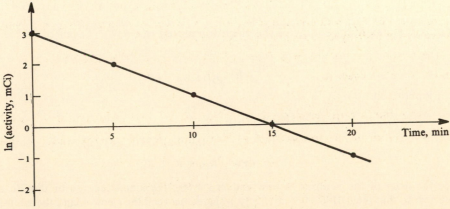

Fig. 29-2

Finally,

$$T_{1/2} = \frac{\ln 2}{\lambda} = \frac{0.693}{0.197 \text{ min}^{-1}} = 3.52 \text{ min}$$

29.7. Show that the average lifetime of a radioactive nucleus is $T_m = 1/\lambda$.

If a sample starts out with N_0 nuclei, the average lifetime as it decays to zero nuclei is given by

$$T_m = \frac{\int_{N_0}^{0} t \, dN}{\int_{N_0}^{0} dN} = \frac{1}{-N_0} \int_{N_0}^{0} t \, dN$$

From $N = N_0 e^{-\lambda t}$ we have $dN = -\lambda N_0 e^{-\lambda t} \, dt$, and the limits N_0, 0 change to 0, ∞ in terms of the variable t. Thus

$$T_m = \frac{1}{-N_0} \int_{0}^{\infty} t\left(-\lambda N_0 e^{-\lambda t} \, dt\right) = \lambda \int_{0}^{\infty} t e^{-\lambda t} \, dt = \lambda\left(\frac{1}{\lambda^2}\right) = \frac{1}{\lambda}$$

Alternatively, we have from the exponential distribution, (1) of Problem 29.1,

$$T_m = \int_{0}^{\infty} t\lambda e^{-\lambda t} \, dt = \frac{1}{\lambda}$$

29.8. In terms of the parent and daughter rest masses determine the Q-values for β^- decay, β^+ decay, and electron capture.

The three reactions are (P = parent, D = daughter):

$$^A_Z P \rightarrow _{Z+1}^{A} D + e^- + \bar{\nu} \quad (\beta^- \text{ decay})$$

$$^A_Z P \rightarrow _{Z-1}^{A} D + e^+ + \nu \quad (\beta^+ \text{ decay})$$

$$^A_Z P + e^- \rightarrow _{Z-1}^{A} D + \nu \quad \text{(electron capture)}$$

The corresponding mass-energy relations are, after subtracting the electron masses from the atomic masses to obtain the nuclear masses,

$$\left.\begin{array}{l} (M_P - Zm_e)c^2 = [M_D - (Z+1)m_e]c^2 + m_e c^2 + Q \\ Q = (M_P - M_D)c^2 \end{array}\right\} (\beta^- \text{ decay})$$

$$\left.\begin{array}{l} (M_P - Zm_e)c^2 = [M_D - (Z-1)m_e]c^2 + m_e c^2 + Q \\ Q = (M_P - M_D - 2m_e)c^2 \end{array}\right\} (\beta^+ \text{ decay})$$

$$\left.\begin{array}{l} (M_P - Zm_e)c^2 + m_e c^2 = [M_D - (Z-1)m_e]c^2 + Q \\ Q = (M_P - M_D)c^2 \end{array}\right\} \text{(electron capture)}$$

29.9. What is the maximum energy of the electron emitted in the β^- decay of $^3_1 H$?

The reaction is

$$^3_1 H \rightarrow _2^3 He + e^- + \bar{\nu}$$

From Problem 29.8,

$$Q = (M_H - M_{He})c^2$$

$$= (3.016050 \text{ u} - 3.016030 \text{ u})(931.5 \text{ MeV/u})$$

$$= 0.0186 \text{ MeV} = K_{He} + K_e + K_\nu$$

Since the mass of the neutrino is zero and $M_{He} \gg m_e$, the kinetic energy of the He nucleus can be neglected, so that the 0.0186 MeV of energy is shared between the electron and the neutrino. When the energy of the neutrino is zero, kinetic energy of the electron will have its maximum value, 0.0186 MeV.

29.10. Determine the minimum energy of an antineutrino to produce the reaction $\bar{\nu} + p \rightarrow n + e^+$.

From conservation of mass-energy,

$$E_\nu + m_p c^2 = m_n c^2 + m_e c^2 + K_n + K_e$$

The required neutrino energy will be minimum when the neutron and positron are both emitted with zero kinetic energy:

$$E_{\nu_{min}} + 938.2 \text{ MeV} = 939.5 \text{ MeV} + 0.5 \text{ MeV} \qquad \text{or} \qquad E_{\nu_{min}} = 1.8 \text{ MeV}$$

29.11. Determine the energy and momentum of the daughter and the neutrino that are produced when ${}^7_4\text{Be}$ undergoes electron capture at rest.

The electron capture reaction is

$$ {}^7_4\text{Be} + e^- \rightarrow {}^7_3\text{Li} + \nu $$

From Problem 29.8,

$$ Q = (M_\text{Be} - M_\text{Li})c^2 $$
$$ = (7.016929 \text{ u} - 7.016004 \text{ u})(931.5 \text{ MeV/u}) = 0.862 \text{ MeV} $$

This energy is split between the neutrino and the ${}^7_3\text{Li}$ nucleus. However, because of the large mass of the ${}^7_3\text{Li}$ nucleus and the zero rest mass of the neutrino, almost all the energy is carried by the neutrino, so that

$$ E_\nu \approx 0.862 \text{ MeV} $$

Assuming that the ${}^7_4\text{Be}$ nucleus was initially at rest, the magnitudes of the momenta of the neutrino and ${}^7_3\text{Li}$ nucleus must be equal. Using $p_\nu = E_\nu / c$, we then have

$$ p_\nu = p_\text{Li} = 0.862 \text{ MeV}/c $$

The kinetic energy of the ${}^7_3\text{Li}$ nucleus can now be found from

$$ K_\text{Li} = \frac{p_\text{Li}^2}{2M_\text{Li}} = \frac{(p_\text{Li}c)^2}{2M_\text{Li}c^2} = \frac{(0.862 \text{ MeV})^2}{2(7.02 \text{ u} \times 931.5 \text{ MeV/u})} = 5.68 \times 10^{-5} \text{ MeV} = 56.8 \text{ eV} $$

29.12. ${}^{20}_9\text{F}$ decays to the ground state of ${}^{20}_{10}\text{Ne}$ as follows:

$$ {}^{20}_9\text{F} \rightarrow ({}^{20}_{10}\text{Ne})^* + e^- + \bar{\nu} $$
$$ \hookrightarrow {}^{20}_{10}\text{Ne} + \gamma $$

where $({}^{20}_{10}\text{Ne})^*$ is an excited state of ${}^{20}_{10}\text{Ne}$. If the maximum kinetic energy of the emitted electrons is 5.4 MeV and the γ-ray energy is 1.6 MeV, determine the mass of ${}^{20}_9\text{F}$ ($M_\text{Ne} = 19.99244 \text{ u}$).

Conservation of mass-energy applied to each of the reactions yields (the energy of the neutrino is zero in the limiting case and the recoil energies of $({}^{20}_{10}\text{Ne})^*$ and ${}^{20}_{10}\text{Ne}$ are negligible)

$$ (M_\text{F} - 9m_e)c^2 = (M_{\text{Ne}^*} - 10m_e)c^2 + m_e c^2 + K_e \qquad \text{or} \qquad M_\text{F}c^2 = M_{\text{Ne}^*}c^2 + K_e $$
$$ M_{\text{Ne}^*}c^2 = M_\text{Ne}c^2 + E_\gamma $$

Rearranging these two expressions, we have

$$ M_\text{F}c^2 = M_\text{Ne}c^2 + E_\gamma + K_e $$
$$ M_\text{F} = 19.99244 \text{ u} + (1.6 \text{ MeV} + 5.4 \text{ MeV})\left(\frac{1 \text{ u}}{931.5 \text{ MeV}} \right) = 20.000 \text{ u} $$

29.13. From the β^+ decay of ${}^{13}_7\text{N}$ find the value of r_0 in the expression $R = r_0 A^{1/3}$ (Section 27.3). The maximum energy of a β^+ is found to be 1.19 MeV.

From conservation of energy for the β^+ decay of ${}^{13}_7\text{N}$,

$$ {}^{13}_7\text{N} \rightarrow {}^{13}_6\text{C} + e^+ + \nu $$

one has

$$M_{13_N}c^2 = M_{13_C}c^2 + m_e c^2 + K_e + K_\nu$$

Substituting into this expression the masses of the odd-A nuclei obtained from the liquid drop model, with the Coulomb term shown explicitly (Problem 27.7), one has

$$7m_p c^2 + 6m_n c^2 - b_1 A + b_2 A^{2/3} + \frac{3}{5} ke^2 \frac{(7)(6)}{R} + \frac{b_4}{A}$$

$$= 6m_p c^2 + 7m_n c^2 - b_1 A + b_2 A^{2/3} + \frac{3}{5} ke^2 \frac{(6)(5)}{R} + \frac{b_4}{A} + m_e c^2$$

$$+ K_e + K_\nu$$

$$\frac{3}{5} ke^2 \frac{(42 - 30)}{R} = (m_n - m_p + m_e)c^2 + K_e + K_\nu$$

If the kinetic energy of the β^+ is to be a maximum, the kinetic energy of the neutrino must be zero. Therefore,

$$\frac{3}{5} ke^2 \frac{12}{R} = (m_n - m_p + m_e)c^2 + K_e$$

$$\frac{3}{5} (1.44 \text{ MeV} \cdot \text{fm}) \frac{12}{R} = 1.80 \text{ MeV} + 1.19 \text{ MeV}$$

$$R = 3.47 \text{ fm}$$

If we take $R = r_0 A^{1/3} = r_0 (13)^{1/3} = 2.35 r_0$, then

$$r_0 = \frac{3.47 \text{ fm}}{2.35} = 1.48 \text{ fm}$$

This value is in good agreement with the value $r_0 = 1.4$ fm given in Section 27.3.

29.14. For $A = 104$ show that the graph of mass versus atomic number Z predicts the stable isobars to be $^{104}_{44}\text{Ru}$ and $^{104}_{46}\text{Pd}$.

As seen from Fig. 29-3, $^{104}_{42}\text{Mo}$ and $^{104}_{43}\text{Tc}$ undergo β^- decay ending with $^{104}_{44}\text{Ru}$. This is energetically possible because $^{104}_{42}\text{Mo}$ is heavier than $^{104}_{43}\text{Tc}$, which is heavier than $^{104}_{44}\text{Ru}$. (If $^{104}_{44}\text{Ru}$ were to β^- decay, it would become $^{104}_{45}\text{Rh}$, which is heavier than $^{104}_{44}\text{Ru}$; hence that decay is forbidden.) Also it is shown that $^{104}_{45}\text{Rh}$ decays to $^{104}_{44}\text{Ru}$ and $^{104}_{46}\text{Pd}$; while $^{104}_{48}\text{Cd}$ decays to $^{104}_{47}\text{Ag}$, which then decays to $^{104}_{46}\text{Pd}$. These processes are all energetically possible and show $^{104}_{46}\text{Pd}$ to be stable. Note that the mass of $^{104}_{44}\text{Ru}$ is larger than that of $^{104}_{46}\text{Pd}$, but this decay process is forbidden because it must form $^{104}_{45}\text{Rh}$ as an intermediate nucleus, and, as mentioned before, this is forbidden. It is seen that the data follow very closely the parabolic shape predicted from the liquid drop model for $A = \text{const.}$

29.15. Determine the kinetic energy of the α-particles emitted in an alpha decay in terms of the Q of the reaction.

An alpha decay reaction has the form

$$^A_Z P \rightarrow ^{A-4}_{Z-2} D + ^4_2 \text{He}$$

Assuming that the parent is initially at rest, we obtain from conservation of momentum $p_D = p_\alpha$. Because the kinetic energies will be very small compared to the rest energy of the parent, we can use the nonrelativistic relation $K = p^2/2M$ to obtain

$$\frac{K_D}{K_\alpha} = \frac{M_\alpha}{M_D} \approx \frac{4}{A-4}$$

where A is the mass number of the parent. The Q of the reaction is

$$Q = K_D + K_\alpha = \frac{4}{A-4} K_\alpha + K_\alpha = \frac{A K_\alpha}{A-4}$$

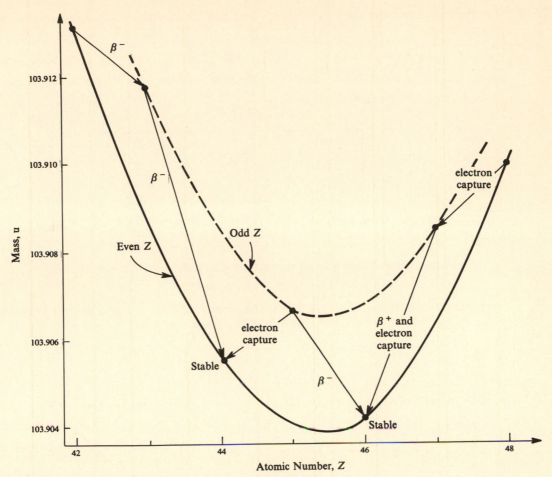

Fig. 29-3

so

$$K_\alpha = \left(\frac{A-4}{A} \right) Q$$

Since Q has a precise value, so does K_α; so that in this two-body decay the α-particles are monoenergetic. The kinetic energy of the daughter nucleus is

$$K_D = \frac{4}{A-4} \left(\frac{A-4}{A} \, Q \right) = \frac{4Q}{A}$$

Note that the larger A is, the more nearly is K_α equal to the total available energy Q and the smaller is K_D.

29.16. Show that $^{236}_{94}$Pu is unstable and will α-decay.

For $^{236}_{94}$Pu to spontaneously α-decay,

$$^{236}_{94}\text{Pu} \rightarrow ^{232}_{92}\text{U} + ^4_2\text{He} + Q$$

the value of Q must be positive. Solving for Q gives

$$Q = (M_{\text{Pu}} - M_{\text{U}} - M_{\text{He}})c^2$$

$$= (236.046071 \text{ u} - 232.037168 \text{ u} - 4.002603 \text{ u})(931.5 \text{ MeV/u})$$

$$= 5.87 \text{ MeV}$$

Therefore $^{236}_{94}$Pu can, and in fact does, spontaneously α-decay.

29.17. An analysis shows that the kinetic energy and half-life in alpha decay for even-even isotopes are related by

$$T_{1/2} = Be^{b/K_\alpha^{1/2}}$$

where B and b are constants.   Show that the data in Table 29-1 satisfy this expression.

Table 29-1

Isotope	$T_{1/2}$	K_α, MeV
$^{210}_{84}\text{Po}$	138.4 d	5.30
$^{212}_{84}\text{Po}$	3×10^{-7} s	8.78
$^{214}_{84}\text{Po}$	1.64×10^{-4} s	7.68
$^{216}_{84}\text{Po}$	0.15 s	6.78
$^{218}_{84}\text{Po}$	3.05 min	6.00

Taking logarithms of both sides of the expression, one obtains

$$\ln T_{1/2} = \ln B + bK_\alpha^{-1/2}$$

according to which $\ln T_{1/2}$ should be a linear function of $K_\alpha^{-1/2}$.   The data give the values shown in Table 29-2 and plotted in Fig. 29-4.   The fit is good if $b \approx 3 \times 10^2$ MeV$^{1/2}$, $B \approx 1 \times 10^{-56}$ s.

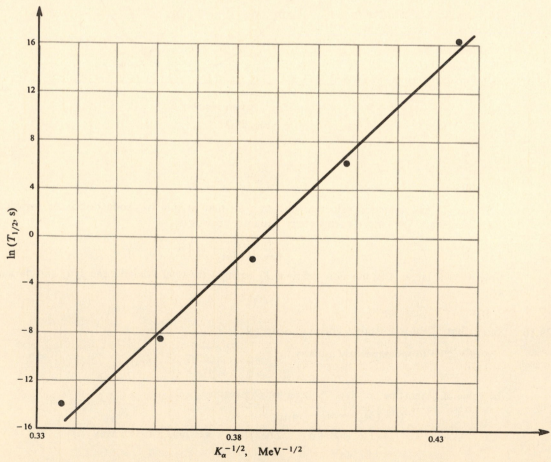

Fig. 29-4

Table 29-2

Isotope	$T_{1/2}$, s	$\ln T_{1/2}$	$K_\alpha^{-1/2}$, MeV$^{-1/2}$
$^{210}_{84}$Po	1.20×10^7	16.30	0.434
$^{212}_{84}$Po	3×10^{-7}	-15.02	0.337
$^{214}_{84}$Po	1.64×10^{-4}	-8.72	0.361
$^{216}_{84}$Po	0.15	-1.90	0.384
$^{218}_{84}$Po	1.83×10^2	5.21	0.408

29.18. An unstable element is produced in a nuclear reactor at a constant rate R. If its half-life for β^- decay is $T_{1/2}$, how much time, in terms of $T_{1/2}$, is required to produce 50% of the equilibrium quantity?

We have:

$$\text{rate of increase of element} = \frac{\text{number of nuclei produced by reactor}}{\text{s}} - \frac{\text{number of nuclei decaying}}{\text{s}}$$

$$\frac{dN}{dt} = R - \lambda N$$

or

$$\frac{dN}{dt} + \lambda N = R$$

The solution to this is the sum of the homogeneous solution, $N_h = ce^{-\lambda t}$, where c is a constant, and a particular solution, $N_p = R/\lambda$.

$$N = N_h + N_p = ce^{-\lambda t} + \frac{R}{\lambda}$$

The constant c is obtained from the requirement that the initial number of nuclei be zero:

$$N(0) = 0 = c + \frac{R}{\lambda} \qquad \text{or} \qquad c = -\frac{R}{\lambda}$$

so that

$$N = \frac{R}{\lambda}(1 - e^{-\lambda t})$$

The equilibrium value is $N(\infty) = R/\lambda$. Setting N equal to $1/2$ of this value gives

$$\frac{1}{2}\left(\frac{R}{\lambda}\right) = \frac{R}{\lambda}(1 - e^{-\lambda t})$$

$$e^{-\lambda t} = \frac{1}{2}$$

$$t = \frac{\ln 2}{\lambda} = T_{1/2}$$

The result is independent of R.

29.19. Radioactive material a (decay constant λ_a) decays into a material b (decay constant λ_b) which is also radioactive. Determine the amount of material b remaining after a time t.

$$\text{rate of increase of } b \text{ nuclei} = \frac{\text{number of } b \text{ nuclei produced by } a}{\text{s}} - \frac{\text{number of } b \text{ nuclei decaying}}{\text{s}}$$

For every a nucleus that decays one b nucleus is formed, so that b nuclei are formed at the rate of

$$-\frac{dN_a}{dt} = \lambda_a N_a$$

giving

$$\frac{dN_b}{dt} = \lambda_a N_a - \lambda_b N_b = \lambda_a N_{a0} e^{-\lambda_a t} - \lambda_b N_b$$

or

$$\frac{dN_b}{dt} + \lambda_b N_b = \lambda_a N_{a0} e^{\lambda_a t} \tag{1}$$

This is a first-order linear differential equation solvable by conventional techniques. The homogeneous equation has the solution

$$(N_b)_h = c e^{-\lambda_b t}$$

where c is an arbitrary constant. A particular solution is obtained by trying $(N_b)_p = D e^{-\lambda_a t}$ in (1):

$$(-\lambda_a + \lambda_b) D e^{-\lambda_a t} = \lambda_a N_{a0} e^{-\lambda_a t}$$

$$D = \frac{\lambda_a N_{a0}}{\lambda_b - \lambda_a}$$

The complete solution is then

$$N_b = (N_b)_h + (N_b)_p = c e^{-\lambda_b t} + \frac{\lambda_a N_{a0}}{\lambda_b - \lambda_a} e^{-\lambda_a t} \tag{2}$$

The constant c is evaluated by requiring $N_b = N_{b0}$ at $t = 0$,

$$N_{b0} = c + \frac{\lambda_a N_{a0}}{\lambda_b - \lambda_a}$$

giving finally

$$N_b = N_{b0} e^{-\lambda_b t} + \frac{\lambda_a N_{a0}}{\lambda_b - \lambda_a} (e^{-\lambda_a t} - e^{-\lambda_b t}) \tag{3}$$

29.20. If it is assumed in Problem 29.19 that $N_{b0} = 0$, find the time at which N_b (the number of daughter nuclei) is a maximum.

With $N_{b0} = 0$, we have

$$N_b = \frac{\lambda_a N_{a0}}{\lambda_b - \lambda_a} (e^{-\lambda_a t} - e^{-\lambda_b t})$$

from which, for a maximum,

$$\frac{dN_b}{dt} = \frac{\lambda_a N_{a0}}{\lambda_b - \lambda_a} (-\lambda_a e^{-\lambda_a t} + \lambda_b e^{-\lambda_b t}) = 0$$

Solving for t gives

$$t = \frac{1}{\lambda_a - \lambda_b} \ln \left(\frac{\lambda_a}{\lambda_b} \right)$$

29.21. Refer to Problem 29.19. If the material b decays into a stable substance c, determine how the amount of c varies with time, assuming $N_{b0} = 0$.

The total number of nuclei present at any time will be N_{a0}, so

$$N_c = N_{a0} - N_a - N_b$$

$$= N_{a0} - N_{a0} e^{-\lambda_a t} - \frac{\lambda_a N_{a0}}{\lambda_b - \lambda_a} (e^{-\lambda_a t} - e^{-\lambda_b t})$$

$$= N_{a0} \left(1 - \frac{\lambda_b}{\lambda_b - \lambda_a} e^{-\lambda_a t} + \frac{\lambda_a}{\lambda_b - \lambda_a} e^{-\lambda_b t} \right)$$

Supplementary Problems

Atomic masses are tabulated in the Appendix.

29.22. Determine the Q-values of alpha, proton, neutron, and deuteron decays of $^{232}_{92}U$.
Ans.   5.42 MeV; -6.09 MeV; -7.23 MeV; -10.59 MeV

29.23. What is the kinetic energy of the α-particles emitted in the alpha decay of $^{232}_{92}U$?   Assume that the $^{232}_{92}U$ nucleus decays at rest.   Ans.   5.33 MeV

29.24. Which of the following are possible decay modes for $^{40}_{19}K$: β^- decay, β^+ decay, alpha decay, electron capture, neutron emission?   Ans. β^- decay; β^+ decay; electron capture

29.25. The maximum kinetic energy of the β^--particle emitted in a tritium (^3_1H) decay is 19 keV.   If the mass of tritium is 3.0160504 u, what is the mass of the decay product?   Ans.   3.016030 u

29.26. Determine the energy of the neutrino emitted in electron capture for $^{41}_{20}Ca$.   Ans.   0.41 MeV

29.27. Element a ($T_{1/2} = 2.1$ h) decays into element b ($T_{1/2} = 4.6$ h), which then decays into element c.   If the initial amount of element b is zero, find the value of N_b/N_{a0} after 2 h.
Ans.   0.41

29.28. Determine the disintegration constant of $^{90}_{38}Sr$ ($T_{1/2} = 28$ y).   Ans.   0.0247 y^{-1}

29.29. What is the energy of the α-particle emitted in the alpha decay of $^{226}_{88}Ra$, if the recoil energy of the radium nucleus is neglected?   Ans.   4.87 MeV

29.30. Solve Problem 29.29 taking into account the recoil energy of the radium nucleus.
Ans.   4.78 MeV

29.31. Determine the maximum possible speed of the daughter in a β^- decay of a ^6_2He nucleus which is initially at rest.   Ans. 1.0×10^5 m/s

29.32. What is the mass of a sample of $^{14}_6C$ ($T_{1/2} = 5570$ y) that has an activity of 5 Ci?   Ans.   1.09 g

29.33. What is the activity of 5×10^{-7} kg of $^{230}_{92}U$, whose half-life is 0.180×10^7 s?   Ans.   14.9 Ci

29.34. How much time is required for an amount of $^{90}_{38}Sr$ ($T_{1/2} = 28$ y) to be reduced by 75%?   Ans.   56 y

29.35. Determine the energies of the α-particle and daughter nucleus in the decay $^{144}_{60}Nd \rightarrow ^{140}_{58}Ce + \alpha$.
Ans.   1.85 MeV; 0.53 MeV

29.36. The α-particles emitted in the alpha decay of $^{243}_{95}Am$ have an energy of 5.3 MeV.   Assuming that the α-particles have the same kinetic energy inside the nucleus, determine the number of collisions per second that the α-particles make with the walls of the nucleus.   Ans. 9.2×10^{20}

29.37. A substance with atomic number A undergoes alpha decay by emitting two groups of α-particles with kinetic energies $K_{\alpha 1}$ and $K_{\alpha 2}$.   Show that the energy of the accompanying γ-ray is

$$E_\gamma = \frac{A}{A-4}(K_{\alpha 1} - K_{\alpha 2})$$

29.38. The maximum distance traveled, or *range R* (in cm), of an α-particle in a bubble chamber is related to its kinetic energy K (in MeV) by the empirical equation $R = 0.318K^{3/2}$. The α-particles emitted in the decay of $^{219}_{86}$Ru are measured in a bubble chamber to have ranges of 5.66 cm, 5.33 cm, and 5.18 cm. What are their energies?   *Ans.*   6.82 MeV; 6.55 MeV; 6.43 MeV

29.39. Refer to Problem 29.38. If the daughter nucleus is produced in the ground and two excited states, determine the energies of the emitted γ-rays.   *Ans.*   0.28 MeV; 0.40 MeV; 0.12 MeV

Chapter 30

Nuclear Reactions

30.1 INTRODUCTION

A great deal of the nuclear data now available has come from the analysis of reaction experiments. In these experiments nuclei are bombarded with known projectiles and the final products observed. Isotopes of nuclei with atomic numbers as high as $Z = 18$ are used as projectiles, but only the following incident particles will be considered in this chapter:

Particle	Notation
neutron	n
proton	p, ^1_1H
deuteron	d, ^2_1H
triton	t, ^3_1H
helium-3	h, ^3_2He
helium-4 (alpha particle)	α, ^4_2He

Normally the reaction results in a final *residual nucleus* (which is usually not observed) plus another particle which is detected experimentally. (Sometimes both final particles are observed.)

Nuclear reactions are indicated in equation form,

$$\text{PROJECTILE} + \begin{array}{c}\text{TARGET}\\\text{NUCLEUS}\end{array} \rightarrow \begin{array}{c}\text{RESIDUAL}\\\text{NUCLEUS}\end{array} + \begin{array}{c}\text{DETECTED}\\\text{PARTICLE}\end{array}$$

or in condensed form,

$$\text{TARGET(PROJECTILE, DETECTED PARTICLE)RESIDUAL NUCLEUS}$$

In the equation for any nuclear reaction the total charge (total Z) and the total number of nucleons (total A) must be the same on the left-hand and right-hand sides.

For example, the first nuclear reaction observed (by Rutherford in 1919) was

$$^{14}_7\text{N} + ^4_2\text{He} \rightarrow ^1_1\text{H} + ^{17}_8\text{O} \qquad \text{or} \qquad ^{14}_7\text{N}(\alpha, p)^{17}_8\text{O}$$

30.2 CLASSIFICATION OF NUCLEAR REACTIONS

Reactions are classified according to the projectile, detected particle and residual nucleus. If the projectile and detected particle are the same, one has a *scattering reaction*. If the residual nucleus is left in its lowest or ground state, the scattering is *elastic*; when the residual nucleus is left in an excited state, the scattering is called *inelastic*.

Processes in which the bombarding projectile gains nucleons from, or loses nucleons to, the target are referred to respectively as *pickup* and *stripping reactions*. Two examples of pickup reactions are:

$$^{16}_8\text{O}(d, t)^{15}_8\text{O} \qquad \text{or} \qquad ^{16}_8\text{O} + d \rightarrow ^{15}_8\text{O} + t$$

$$^{41}_{20}\text{Ca}(h, \alpha)^{40}_{20}\text{Ca} \qquad \text{or} \qquad ^{41}_{20}\text{Ca} + h \rightarrow ^{40}_{20}\text{Ca} + \alpha$$

and two stripping reactions are:

$$^{90}_{40}\text{Zr}(d, p)^{91}_{40}\text{Zr} \qquad \text{or} \qquad ^{90}_{40}\text{Zr} + d \rightarrow ^{91}_{40}\text{Zr} + p$$

$$^{23}_{11}\text{Na}(h, d)^{24}_{12}\text{Mg} \qquad \text{or} \qquad ^{23}_{11}\text{Na} + h \rightarrow ^{24}_{12}\text{Mg} + d$$

Pickup and stripping reactions are often observed at high enough energies that one can assume the reaction to be *direct*. In a direct stripping or pickup reaction it is assumed that the nucleon taking part in the process enters or leaves a definite shell-model orbit of the target without disturbing the other nucleons in the target.

Quite the opposite type of reaction is one in which the incident projectile and target form a new nucleus, referred to as a *compound nucleus*, which lives for a short time in an excited state and then decays. The lifetime of a typical compound nucleus is of the order of 10^{-16} s. Although 10^{-16} s is so short that the compound nucleus cannot be observed directly, it is much longer than the time a bombarding particle takes to traverse a nuclear distance, which is of the order of 10^{-21} s. It is therefore assumed that the decay of a compound nucleus does not depend on how it was formed; the compound nucleus does not "remember" how it was formed.

There are usually several different reactions that will give rise to the same compound nucleus, and also several different modes or channels in which this compound nucleus can decay. As an example, for the compound nucleus $^{20}_{10}\text{Ne}$ formed in an excited state, $[^{20}_{10}\text{Ne}]^*$, we can have the reactions shown in Fig. 30-1.

$$
\left.
\begin{array}{l}
^{19}_{9}\text{F} + p \\
^{17}_{8}\text{O} + h \\
^{16}_{8}\text{O} + \alpha \\
^{14}_{7}\text{N} + ^{6}_{3}\text{Li} \\
^{12}_{6}\text{C} + ^{8}_{4}\text{Be} \\
^{10}_{5}\text{B} + ^{10}_{5}\text{B}
\end{array}
\right\}
\rightarrow
[^{20}_{10}\text{Ne}]^*
\rightarrow
\left\{
\begin{array}{l}
^{19}_{9}\text{F} + p \\
^{19}_{10}\text{Ne} + n \\
^{20}_{10}\text{Ne} + \gamma \\
^{18}_{9}\text{F} + d \\
^{17}_{9}\text{F} + t \\
^{17}_{8}\text{O} + h \\
^{16}_{8}\text{O} + \alpha \\
^{14}_{7}\text{N} + ^{6}_{3}\text{Li} \\
^{13}_{7}\text{N} + ^{7}_{3}\text{Li} \\
^{12}_{6}\text{C} + ^{8}_{4}\text{Be} \\
^{11}_{6}\text{C} + ^{9}_{4}\text{Be} \\
^{10}_{5}\text{B} + ^{10}_{5}\text{B} \\
^{9}_{5}\text{B} + ^{11}_{5}\text{B}
\end{array}
\right.
$$

Fig. 30-1

30.3 LABORATORY AND CENTER-OF-MASS SYSTEMS

Experimental nuclear reactions are often analyzed in what is called the *center-of-mass system*. This system moves at constant velocity with respect to the laboratory system in such a manner that in it the colliding particles (and final particles) have *zero total momentum*.

If the target nucleus is at rest in the laboratory system, the velocity V_{cm} of the center-of-mass system will be along the direction of the incident bombarding particle. Therefore, with respect to the center of mass, the magnitudes of the velocity of the target nucleus (V') and the incident particle (v') are respectively (in a nonrelativistic treatment)

$$V' = V_{\text{cm}} \qquad v' = v - V_{\text{cm}} \tag{30.1}$$

with the directions as shown in Fig. 30-2(a). Here v is the velocity of the incident particle as measured in the laboratory. Requiring the sum of the momenta of the target nucleus (mass M_i) and incident particle (mass m_i) to be zero in the center-of-mass system, we obtain the center-of-mass velocity from

$$-M_i V' + m_i v' = 0$$

$$-M_i V_{\text{cm}} + m_i (v - V_{\text{cm}}) = 0$$

$$(M_i + m_i) V_{\text{cm}} = m_i v \tag{30.2}$$

where $(M_i + m_i) V_{\text{cm}}$ is the momentum of the center of mass in the laboratory system. From (30.1) and (30.2), the target nucleus and incident particle have, before the reaction takes place, respective velocities in the center-of-mass system of

$$V' = \frac{m_i}{m_i + M_i} v \qquad v' = \frac{M_i}{m_i + M_i} v \qquad (30.3)$$

After a reaction the final particles (if there are only two) must move in opposite directions with equal momenta in the center-of-mass system, because the initial total momentum in this system was zero [see Fig. 30-2(b)].

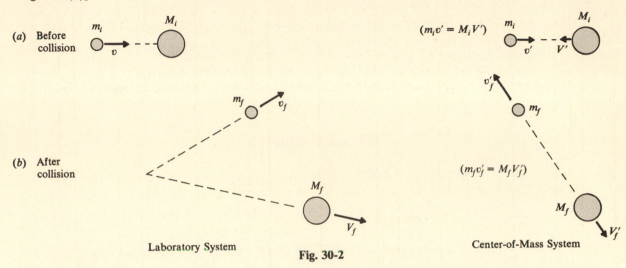

Fig. 30-2

30.4 ENERGETICS OF NUCLEAR REACTIONS

Energy is often released or absorbed in a nuclear reaction. To say that a reaction "releases energy" means that the kinetic energy of the particles after the reaction is greater than the kinetic energy of the particles before the reaction, the increase resulting from the transformation of rest mass into kinetic energy. The amount of energy released is measured by the Q-value of the nuclear reaction, defined as the difference between the final and the initial kinetic energies:

$$Q \equiv K_{\text{after}} - K_{\text{before}}$$

Since total energy, $E = E_0 + K$, is conserved, we also have

$$Q = E_{0\,\text{before}} - E_{0\,\text{after}}$$

In other words, Q/c^2 is the difference between the total initial and final rest masses; this is exactly the definition given in Chapter 29 for the special class of decay reactions.

A reaction with $Q > 0$, so that energy is released, is called an *exothermic* or *exoergic* reaction; the reaction can occur even if both initial particles are at rest. If $Q < 0$ energy is absorbed or consumed and the reaction is called *endothermic* or *endoergic*; the reaction cannot occur unless the bombarding particle has a certain threshold kinetic energy (see Problem 30.8). If $Q = 0$ and if the particles are the same before and after the reaction, we have an *elastic collision*.

30.5 NUCLEAR CROSS SECTIONS

When a target material is bombarded with incident particles to produce a nuclear reaction, there is no guarantee that a particular bombarding projectile will interact with a target nucleus to bring about the reaction. A *cross section*, σ, is a quantity that measures the probability that a nuclear reaction will occur in a given region of target material. It is defined as:

$$\sigma \equiv \frac{\text{number of reactions per second per nucleus}}{\text{number of projectiles incident per second per area}}$$

The larger the value of σ, the more probable will it be for a particular reaction to occur. A cross section has the dimensions of area and is usually measured in terms of a unit called the *barn*, where

$$1 \text{ barn} = 10^{-28} \text{ m}^2$$

so that one barn is of the order of the square of a nuclear radius.

If the number of target nuclei per unit volume in a material is n, the number N_{sc} of particles scattered when a beam of N_0 projectiles is incident on a thickness T of the material is (see Problem 30.12)

$$N_{sc} = N_0(1 - e^{-n\sigma T})$$

Cross sections will be different for different reactions, and for a given reaction will vary with the energy of the bombarding particle. If the reaction is endothermic, the cross section will be zero if the energy is below the threshold value.

Solved Problems

30.1. When ^6_3Li is bombarded with 4 MeV deuterons, one reaction that is observed is the formation of two alpha particles, each with 13.2 MeV of energy. Find the Q-value for this reaction.

$$Q = (K_{\alpha 1} + K_{\alpha 2}) - K_d = (13.2 \text{ MeV} + 13.2 \text{ MeV}) - 4 \text{ MeV} = 22.4 \text{ MeV}$$

30.2. Determine the unknown particle in the following nuclear reactions: (a) $^{18}_8\text{O}(d, p)X$, (b) $X(p, \alpha)^{87}_{39}\text{Y}$, (c) $^{122}_{52}\text{Te}(X, d)^{124}_{53}\text{I}$.

(a) In the process $^{18}_8\text{O}(d, p)X$ a neutron is added to $^{18}_8\text{O}$ to form X, which is $^{19}_8\text{O}$.

(b) In the process $X(p, \alpha)^{87}_{39}\text{Y}$ a proton and two neutrons have been removed from X to form $^{87}_{39}\text{Y}$, so X is $^{90}_{40}\text{Zr}$.

(c) In the process $^{122}_{52}\text{Te}(X, d)^{124}_{53}\text{I}$ a deuteron (^2_1H) and $^{124}_{53}\text{I}$ have been formed from $^{122}_{52}\text{Te}$ and X. Therefore X must have two protons and a total of four nucleons, and is ^4_2He.

30.3. Determine the compound nucleus and some of the possible reaction products when α-particles are incident on $^{19}_9\text{F}$.

The compound nucleus has $Z = Z_1 + Z_2 = 2 + 9 = 11$ and $A = A_1 + A_2 = 4 + 19 = 23$. Therefore we have

$$^4_2\text{He} + ^{19}_9\text{F} \rightarrow [^{23}_{11}\text{Na}]^*$$

This nucleus can then decay to many products, such as

$$[^{23}_{11}Na]^* \rightarrow \begin{cases} ^{23}_{11}\text{Na} + \gamma \\ ^{22}_{10}\text{Ne} + p \\ ^{21}_{10}\text{Ne} + d \\ \vdots \end{cases}$$

30.4. Calculate the Q-values for the reactions (a) $^{16}_8\text{O}(\gamma, p)^{15}_7\text{N}$, (b) $^{150}_{62}\text{Sm}(p, \alpha)^{147}_{61}\text{Pm}$.

For the reaction $M_i(m_i, m_f)M_f$, the Q-value is

$$Q = [M_i + m_i - (M_f + m_f)]c^2$$

(a) $Q = [15.994915 \text{ u} + 0 \text{ u} - (15.000108 \text{ u} + 1.007825 \text{ u})](931.5 \text{ MeV/u}) = -12.13 \text{ MeV}$

(b) $Q = [149.917276 \text{ u} + 1.007825 \text{ u} - (146.915108 \text{ u} + 4.002603 \text{ u})](931.5 \text{ MeV/u}) = 6.88 \text{ MeV}$

30.5. Calculate the mass excess (Problem 28.23) for (a) $^{42}_{20}$Ca, (b) $^{130}_{52}$Te.

(a) $\delta = 41.958625\ u - 42\ u = -0.041375\ u = -38.540\ MeV$

(b) $\delta = 129.906238\ u - 130\ u = -0.093762\ u = -87.337\ MeV$

Nuclear data are often given in terms of mass excess rather than atomic weight.

30.6. Using the data

Nucleus	Mass Excess
$^{192}_{76}$Os	$-0.038550\ u$
$^{191}_{76}$Os	-0.039030
d	$+0.014102$
t	$+0.016050$

find the Q-value for the reaction $^{192}_{76}$Os$(d, t)^{191}_{76}$Os.

Since total A is conserved in any reaction, we can replace rest masses by mass excesses in calculating Q. Thus

$$Q = \left[M_i + m_i - (M_f + m_f) \right] c^2$$
$$= \left[-0.038550\ u + 0.014102\ u - (-0.039030\ u + 0.016050\ u) \right] (931.5\ MeV/u) = -1.37\ MeV$$

30.7. As observed in the laboratory system, a 6 MeV proton is incident on a stationary ^{12}C target. Find the velocity of the center-of-mass system. Take the mass of the proton to be 1 u.

Using a nonrelativistic treatment, the proton velocity is found from $K_i = \frac{1}{2} m_i v^2$:

$$v = \sqrt{\frac{2K_i}{m_i}} = c\sqrt{\frac{2K_i}{m_i c^2}} = (3 \times 10^8\ m/s)\sqrt{\frac{2(6\ MeV)}{(1\ u)(931.5\ MeV/u)}} = 3.41 \times 10^7\ m/s$$

By (30.2),

$$V_{cm} = \frac{m_i}{M_i + m_i} v = \frac{1\ u}{12\ u + 1\ u} (3.41 \times 10^7\ m/s) = 2.62 \times 10^6\ m/s$$

in the direction of the proton.

30.8. For the endothermic reaction $M_i(m_i, m_f)M_f$, determine how the Q-value is related to the threshold energy of the incoming particle. Use a nonrelativistic treatment.

The desired answer is obtained most easily by first doing the calculation in the center-of-mass system, where the total momentum is zero, and then transforming the results to the laboratory system. Using the notation of Fig. 30-2, we have for the total initial kinetic energy in the center-of-mass system

$$K_{i\ cm} = \frac{1}{2} m_i v'^2 + \frac{1}{2} M_i V'^2$$

Transforming to the laboratory system via (30.3), and recalling that the target particle is at rest in the laboratory system,

$$K_{i\ cm} = \frac{1}{2} m_i \left(\frac{M_i}{m_i + M_i} v \right)^2 + \frac{1}{2} M_i \left(\frac{m_i}{m_i + M_i} v \right)^2$$

$$= \left(\frac{1}{2} m_i v^2 \right)\left[\frac{M_i(M_i + m_i)}{(m_i + M_i)^2} \right]$$

$$= K_{i\ lab}\left(\frac{M_i}{m_i + M_i} \right) \tag{1}$$

Equation (1) is the general relation between the initial kinetic energies measured in the laboratory and center-of-mass systems.

The Q-value of the reaction, which depends only on rest masses, is the same in both systems:

$$K_{f\,\text{lab}} - K_{i\,\text{lab}} = K_{f\,\text{cm}} - K_{i\,\text{cm}} = Q \tag{2}$$

The threshold energy, $K_{\text{th cm}}$, in the center-of-mass system is the initial kinetic energy that will produce the final two particles at rest ($K_{f\,\text{cm}} = 0$), so that

$$K_{\text{th cm}} = -Q \tag{3}$$

The corresponding energy in the laboratory system is then obtained from (1) as

$$K_{\text{th lab}} = -Q\left(\frac{M_i + m_i}{M_i}\right) \tag{4}$$

A more revealing expression for $K_{\text{th lab}}$ can be obtained by considering the kinetic energy, K^*, *of the center of mass* (in the laboratory system) when the incident particle has the threshold energy. We have, using (30.2) and (4),

$$K^* = \tfrac{1}{2}(M_i + m_i)V_{\text{cm th}}^2 = \frac{m_i}{M_i + m_i}\left(\tfrac{1}{2}m_i v_{\text{th}}^2\right) = \frac{m_i}{M_i + m_i}K_{\text{th lab}}$$

$$= \left(1 - \frac{M_i}{M_i + m_i}\right)K_{\text{th lab}} = K_{\text{th lab}} + Q$$

or

$$K_{\text{th lab}} = -Q + K^* \tag{5}$$

Equation (5) states that the incident particle must have sufficient energy to start the endothermic reaction $(-Q)$ *and* to account for the gross motion of the system (K^*, which remains unchanged in the reaction).

30.9.   Find the Coulomb barriers of $^{16}_{8}\text{O}$, $^{93}_{41}\text{Nb}$ and $^{209}_{83}\text{Bi}$ as seen by a proton.

    The Coulomb barrier is the energy needed to bring the proton to the edge of the nucleus (Fig. 30-3). If we define $\Delta \equiv R + r = r_0(A^{1/3} + 1)$, then

$$E_C = k\,\frac{(Ze)e}{\Delta} = k\,\frac{Ze^2}{r_0(A^{1/3} + 1)} = \left(\frac{1.44\text{ MeV}\cdot\text{fm}}{1.4\text{ fm}}\right)\left(\frac{Z}{A^{1/3} + 1}\right) = (1.03\text{ MeV})\left(\frac{Z}{A^{1/3} + 1}\right)$$

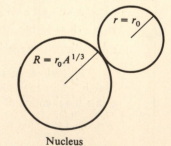

For $^{16}_{8}\text{O}$,

$$E_C = (1.03\text{ MeV})\left(\frac{8}{16^{1/3} + 1}\right) = 2.34\text{ MeV}$$

For $^{93}_{41}\text{Nb}$,

$$E_C = (1.03\text{ MeV})\left(\frac{41}{93^{1/3} + 1}\right) = 7.64\text{ MeV}$$

For $^{209}_{83}\text{Bi}$,

$$E_C = (1.03\text{ MeV})\left(\frac{83}{209^{1/3} + 1}\right) = 12.33\text{ MeV}$$

Fig. 30-3

30.10.   Refer to Problem 30.9. Compare the Coulomb barrier, E_C, with the threshold energy for the reactions

$$^{16}_{8}\text{O}(p, d)^{15}_{8}\text{O} \qquad ^{93}_{41}\text{Nb}(p, d)^{92}_{41}\text{Nb} \qquad ^{209}_{83}\text{Bi}(p, d)^{208}_{83}\text{Bi}$$

    The Q-value for a reaction is

$$Q = (M_i + m_i - M_f - m_f)c^2$$

and, from (4) of Problem 30.8,

$$K_{th} = -Q\left(\frac{M_i + m_i}{M_i}\right)$$

For $^{16}_{8}O(p, d)^{15}_{8}O$:

$$Q = (15.994915\ u + 1.007825\ u - 15.003070\ u - 2.014102\ u)(931.5\ MeV/u) = -13.44\ MeV$$

$$K_{th} = (13.44\ MeV)\left(\frac{16\ u + 1\ u}{16\ u}\right) = 14.28\ MeV$$

For $^{93}_{41}Nb(p, d)^{92}_{41}Nb$:

$$Q = (92.906382\ u + 1.007825\ u - 91.907211\ u - 2.014102\ u)(931.5\ MeV/u) = -6.62\ MeV$$

$$K_{th} = (6.62\ MeV)\left(\frac{93\ u + 1\ u}{93\ u}\right) = 6.69\ MeV$$

For $^{209}_{83}Bi(p, d)^{208}_{83}Bi$:

$$Q = (208.980394\ u + 1.007825\ u - 207.979731\ u - 2.014102\ u)(931.5\ MeV/u) = -5.23\ MeV$$

$$K_{th} = (5.23\ MeV)\left(\frac{209\ u + 1\ u}{209\ u}\right) = 5.26\ MeV$$

For $^{16}_{8}O(p, d)^{15}_{8}O$, $K_{th} \gg E_C$ and the reaction will occur with a large probability at the threshold energy. For $^{209}_{83}Bi(p, d)^{208}_{83}Bi$, $K_{th} \ll E_C$ and the reaction will hardly ever occur at the threshold energy, because the proton never gets close to the $^{209}_{83}Bi$ nucleus. In the $^{83}_{41}Nb(p, d)^{82}_{41}Nb$ reaction the threshold energy (6.69 MeV) is slightly less than the Coulomb barrier (7.61 MeV), so one might expect no reaction, because the proton just doesn't reach the $^{93}_{41}Nb$ nucleus. But in fact the reaction $^{93}_{41}Nb(p, d)^{92}_{41}Nb$ is seen at the threshold energy. This is an example of Coulomb barrier *tunneling*, where the proton, even though below the Coulomb barrier, does manage to reach the $^{93}_{41}Nb$ nucleus.

30.11. If there are n scattering centers (nuclei) per unit volume, each of area σ, in a thin target of thickness dT, find the ratio R of the area covered by scattering centers to the total area of the target.

In a thin target no nucleus hides another nucleus. Hence,

$$R = \frac{\text{total area of scattering centers}}{\text{area of target}} = \frac{\text{volume of target} \times n \times \sigma}{\text{area of target}} = \frac{(A\ dT) \times n \times \sigma}{A} = n\sigma\ dT$$

30.12. Obtain an expression for the number of particles scattered from a beam of area A containing N_0 particles, after it traverses a thickness T of target material containing n scattering centers per unit volume, each of cross-sectional area σ.

Consider a thin slice, of thickness dT, of the target material. Any time an incident particle encounters one of the scattering centers in this thin slice, the incident particle will be scattered. Therefore, the ratio of the number of scattered particles to the number of particles N incident on the thin slice will be the same as the ratio of the total area of the scattering centers to the area of the beam, so that from Problem 30.11 we have

$$\frac{\text{number of scattered particles}}{\text{number of incident particles}} = \frac{\text{total area of scattering centers}}{\text{area of target}}$$

or

$$\frac{dN_{sc}}{N} = -\frac{dN}{N} = n\sigma\ dT$$

(The minus sign is used because an increase of scattered particles, dN_{sc}, corresponds to a decrease of incident particles, $-dN$.) Integrating this expression we obtain

$$-\int_{N_0}^{N_f}\frac{dN}{N} = \int_0^T n\sigma\ dT \qquad \text{or} \qquad -\ln\frac{N_f}{N_0} = n\sigma T \qquad \text{or} \qquad N_f = N_0 e^{-n\sigma T}$$

with N_0 and N_f the initial and final numbers of particles in the beam. The number of scattered particles is then given by

$$N_{sc} = N_0 - N_f = N_0(1 - e^{-n\sigma T})$$

30.13. For a hypothetical scattering target $10^{-3}\%$ of an incoming neutron beam is scattered. If the target has a density of 1.06×10^4 kg/m³, $A = 200$ and the total neutron cross section per nucleus, σ, is 1.1 barns, find the target thickness.

The number of scattering centers per unit volume is

$$n = \left(\frac{6.02 \times 10^{26} \text{ nuclei/kmol}}{200 \text{ kg/kmol}} \right)(1.06 \times 10^4 \text{ kg/m}^3) = 3.19 \times 10^{28} \text{ nuclei/m}^3$$

and

$$n\sigma = (3.19 \times 10^{28} \text{ m}^{-3})(1.1 \times 10^{-28} \text{ m}^2) = 3.51 \text{ m}^{-1}$$

From Problem 30.12, the number of scattered particles is given by

$$N_{sc} = N_0(1 - e^{-n\sigma T})$$

and with $N_{sc}/N_0 = 10^{-5}$ we have

$$10^{-5} = 1 - e^{-(3.51 \text{ m}^{-1})T} \qquad \text{or} \qquad e^{-(3.51 \text{ m}^{-1})T} = 1 - 10^{-5}$$

For small x, $e^{-x} \approx 1 - x$, and we have

$$(3.51 \text{ m}^{-1})T = 10^{-5} \qquad \text{or} \qquad T = \frac{10^{-5}}{3.51 \text{ m}^{-1}} = 2.85 \times 10^{-6} \text{ m}$$

30.14. When 5.30 MeV α-particles from a $^{210}_{84}$Po source are incident on a $^{9}_{4}$Be target it is found that uncharged but otherwise unknown radiation is produced. Assuming that the unknown radiation is γ-rays, calculate the energy the γ-rays have as they leave the $^{9}_{4}$Be target in the forward direction. [This problem, together with Problems 30.15 and 30.16, illustrates the reasoning that led Chadwick in 1932 to the discovery of the neutron.]

The assumed reaction is $^{9}_{4}$Be$(\alpha, \gamma)^{13}_{6}$C. Taking the $^{9}_{4}$Be nucleus to be at rest and the α-particle kinetic energy to be 5.30 MeV, we have from conservation of mass-energy

$$(M_{Be} + M_\alpha)c^2 + K_\alpha = M_C c^2 + K_C + K_\gamma$$

$$(9.012186 \text{ u} + 4.002603 \text{ u})(931.5 \text{ MeV/u}) + 5.30 \text{ MeV} = (13.003354 \text{ u})(931.5 \text{ MeV/u}) + K_C + K_\gamma$$

$$K_\gamma + K_C = 16.0 \text{ MeV} \qquad (1)$$

When the γ-ray and $^{13}_{6}$C nucleus move in the same direction as the incident α-particle, we have from conservation of momentum

$$p_\alpha = p_\gamma + p_C \qquad \text{or} \qquad p_\alpha c = p_\gamma c + p_C c \qquad (2)$$

For the material particles, in a nonrelativistic treatment,

$$K = \frac{p^2}{2M} = \frac{(pc)^2}{2Mc^2} \qquad \text{or} \qquad pc = \sqrt{2(Mc^2)K}$$

Thus

$$p_\alpha c = \sqrt{2(4 \text{ u} \times 931.5 \text{ MeV/u})(5.30 \text{ MeV})} = 199 \text{ MeV}$$

$$p_C c = \sqrt{2(13 \text{ u} \times 931.5 \text{ MeV/u})K_C} = 156 K_C^{1/2}$$

and for the γ-ray photon, $E_\gamma = K_\gamma = p_\gamma c$. Substituting in (2):

$$199 \text{ MeV} = K_\gamma + 156 K_C^{1/2} \qquad (3)$$

Solving (1) and (3) simultaneously, we obtain

$$K_\gamma = 14.6 \text{ MeV} \qquad K_C = 1.4 \text{ MeV}$$

30.15. In separate experiments the unknown radiation of Problem 30.14 is incident on a proton-rich paraffin target and a $^{14}_7N$ target. Still assuming this radiation to be photons, determine the minimum photon energies to produce the observed 5.7 MeV recoil protons and the 1.4 MeV recoil $^{14}_7N$ nuclei, and compare these energies with the result of Problem 30.14.

The photons will interact with the target nuclei by Compton scattering. The minimum E_γ will correspond to a head-on collision. In analyzing this collision we may use nonrelativistic expressions for the particles since the observed kinetic energies are much less than the rest energies of the target particles. Thus (primes refer to conditions after collision):

$$h\nu_{min} = h\nu' + K' \qquad \text{(energy conservation)}$$

and, since all momenta are along the x-axis,

$$\frac{h\nu}{c} = -\frac{h\nu'}{c} + m_0v' \qquad \text{(momentum conservation)}$$

Multiplying the second equation by c and adding it to the first equation, we obtain, after using $m_0v' = \sqrt{2m_0K'}$,

$$2h\nu_{min} = \sqrt{2m_0c^2K'} + K' = \sqrt{K'}\left(\sqrt{2m_0c^2} + \sqrt{K'}\right)$$

Since $2m_0c^2 \gg K'$, we may neglect the $\sqrt{K'}$ in the parentheses to obtain

$$h\nu_{min} = \sqrt{\frac{K'(m_0c^2)}{2}}$$

For the proton target,

$$h\nu_{min} \approx \sqrt{(5.7\text{ MeV})(938\text{ MeV})/2} = 52\text{ MeV}$$

For the $^{14}_7N$ target,

$$h\nu_{min} \approx \sqrt{(1.4\text{ MeV})(14\text{ u} \times 931.5\text{ MeV/u})/2} = 96\text{ MeV}$$

Both these energies far exceed the $K_\gamma = 14.6$ MeV calculated in Problem 30.14, showing that the assumption that the unknown radiation is γ-rays is inconsistent with the observed data.

30.16. Assuming that the recoil protons and $^{14}_7N$ nuclei of Problem 30.15 were the results of head-on collisions with a massive incident particle, find its mass and initial kinetic energy.

Using subscripts 1 and 2 for the projectile and target particles, respectively, we have

$$\tfrac{1}{2}m_1v_1^2 = \tfrac{1}{2}m_1v_1'^2 + \tfrac{1}{2}m_2v_2'^2 \qquad \text{(nonrelativistic energy conservation)}$$

$$m_1v_1 = m_1v_1' + m_2v_2' \qquad \text{(momentum conservation)}$$

The velocity v_1' is not measured in the experiments. Solving for v_1' from the second equation and substituting into the first equation, we obtain the relation

$$v_2' = \frac{2m_1v_1}{m_1 + m_2}$$

The final kinetic energy of particle 2 is then

$$K_2' = \frac{1}{2}m_2v_2'^2 = \frac{1}{2}m_2\left[\frac{2m_1v_1}{m_1 + m_2}\right]^2 = \frac{4m_1m_2}{(m_1 + m_2)^2}K_1$$

Substituting the target masses and observed energies, we obtain two equations for the quantities m_1 and K_1:

$$\text{proton target:} \qquad 5.7\text{ MeV} = \frac{4m_1(1\text{ u})}{(m_1 + 1\text{ u})^2}K_1$$

$$^{14}_7N\text{ target:} \qquad 1.4\text{ MeV} = \frac{4m_1(14\text{ u})}{(m_1 + 14\text{ u})^2}K_1$$

Solving:

$$m_1 = 0.98 \text{ u} \qquad K_1 = 5.7 \text{ MeV}$$

The value $m_1 \approx 1$ u agrees reasonably well with the mass of a neutron as we now know it. Also, if in Problem 30.14 the reaction is taken as ${}^9_4\text{Be}(\alpha, n){}^{12}_6\text{C}$ instead of ${}^9_4\text{Be}(\alpha, \gamma){}^{13}_6\text{C}$, it will be found that the kinetic energy of the neutron will be approximately equal to the above value of 5.7 MeV (see Problem 30.22).

Supplementary Problems

Atomic masses are tabulated in the Appendix

30.17. Determine the unknown particle in the nuclear reactions (a) ${}^{182}_{74}\text{W}({}^3_2\text{He}, n)X$, (b) ${}^{42}_{20}\text{Ca}({}^6_3\text{Li}, X){}^{45}_{21}\text{Sc}$. *Ans.* (a) ${}^{184}_{76}\text{Os}$; (b) ${}^3_2\text{He}$

30.18. Find the velocity of $[{}^{42}_{21}\text{Sc}]^*$ in the reaction

$$ {}^{41}_{20}\text{Ca} + p \rightarrow \left[{}^{42}_{21}\text{Sc} \right]^* \rightarrow {}^{40}_{20}\text{Ca} + d $$

if the proton energy in the laboratory is 7.2 MeV. *Ans.* 8.9×10^5 m/s

30.19. Calculate the Q-value for the reaction ${}^{42}_{20}\text{Ca}(p, d){}^{41}_{20}\text{Ca}$. *Ans.* -9.25 MeV

30.20. Find the mass excess in u for (a) ${}^4_2\text{He}$ and (b) ${}^{88}_{38}\text{Sr}$. *Ans.* (a) 0.002603 u; (b) -0.094359 u

30.21. For a certain scattering target, $10^{-6}\%$ of an incoming neutron beam is scattered. If the target density is 4.1×10^3 kg/m³, $A = 30$, and the target thickness is 10^{-8} m, find the total neutron cross section. *Ans.* 0.122 barn

30.22. Assuming in Problem 30.14 that the reaction is ${}^9_4\text{Be}(\alpha, n){}^{12}_6\text{C}$, calculate the kinetic energy of the neutron and compare it to the value found in Problem 30.16. *Ans.* 5.7 MeV

30.23. Show that in a two-body elastic collision each particle's speed will be unchanged by the collision when measured in the center-of-mass system.

Chapter 31

Fission and Fusion

31.1 NUCLEAR FISSION

One of the most practical nuclear reactions is the formation of a compound nucleus when a nucleus with $A > 230$ absorbs an incident neutron. Many of these compound nuclei will then split into two medium-mass nuclear fragments and additional neutrons. This type of reaction is called *nuclear fission*.

In a nuclear reactor, the number of fissions per unit time is controlled by the absorption of excess neutrons so that, on the average, one neutron from each fission produces a new fission. The liberated heat is used to make steam to drive turbines and generate electrical power. If the reaction is uncontrolled, so that each fission results in more than one neutron capable of producing further fissions, the number of fissions will increase geometrically, resulting in all the energy of the source being released over a short time interval, producing a nuclear bomb.

A typical fission reaction is

$$^{235}_{92}U + ^{1}_{0}n \rightarrow \left[^{236}_{92}U \right]^{*} \rightarrow ^{A_1}_{Z_1}X + ^{A_2}_{Z_2}Y + \epsilon ^{1}_{0}n$$

with $Z_1 + Z_2 = 92$, $A_1 + A_2 + \epsilon = 236$, and ϵ an integer. The ratio of the masses of the fission fragments, M_1/M_2, is found experimentally to be roughly 3/2. The number ϵ of neutrons released in the fissioning of a particular element will depend upon the final fragments that are produced. For the above reaction the average number of neutrons released in a fission is found experimentally to be about 2.44, the fractional number resulting from an average taken over all reaction products.

The two decay fragments usually have a neutron-proton ratio approximately equal to that of the original nucleus. Therefore, in Fig. 31-1, they lie above the stability curve, in a region where nuclei are neutron-rich and undergo beta decay. Usually it will require a chain of several beta decays, each decay reducing the N/Z ratio, before a stable nucleus is reached (Problem 31.7).

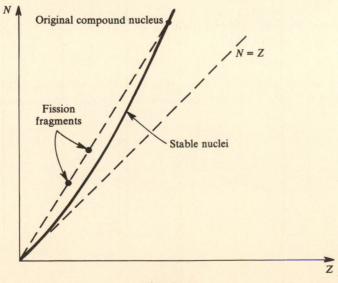

Fig. 31-1

209

A fission reaction liberates about 200 MeV of energy for each fission (Problem 31.8). This is much greater than the few MeV released in a typical exothermic reaction where the final products include only one particle comparable in mass to the original target nucleus. This 200 MeV is distributed as follows:

(a)   170 MeV is kinetic energy of the fission fragments
(b)   5 MeV is the combined kinetic energy of fission neutrons
(c)   15 MeV is β^-- and γ-ray energy
(d)   10 MeV is neutrino energy liberated in the β^- decays of the fission fragments

In many fission reactions the formation of the compound nucleus occurs most readily with thermal neutrons of energy $E \approx 0.04$ eV. From the above it is seen that the neutrons released in a typical fission reaction have large kinetic energies of about 2 MeV. How these fast neutrons are slowed down to facilitate further fissions is demonstrated in Problems 31.2 and 31.17.

31.2   NUCLEAR FUSION

As implied by its name, the *fusion* reaction is one in which two nucleons or relatively light ($A < 20$) nuclei combine to form a heavier nucleus, with a resulting release of energy. An example of a fusion reaction is the formation of a deuteron from a proton and a neutron:

$$\ _1^1\text{H} + \ _0^1n \rightarrow \ _1^2\text{H} \qquad Q = 2.23 \text{ MeV}$$

Another fusion reaction is the formation of an α-particle by the fusion of two deuterons:

$$\ _1^2\text{H} + \ _1^2\text{H} \rightarrow \ _2^4\text{He} \qquad Q = 23.8 \text{ MeV}$$

Although these energies are much smaller than the energy released in a typical fission reaction (≈ 200 MeV), the energy per unit mass is larger because of the smaller masses of the participating particles.

The release of energy in fusion can be understood from Fig. 28-1, which shows that for light nuclei the binding energy per nucleon generally increases with increasing mass number A. Consequently, the heavier nucleus formed from the fusion of two lighter nuclei will have a larger binding energy per nucleon than either of the two original nuclei. But higher binding energy means lower rest mass (Section 28.1), and the lost rest mass appears as released energy.

The reactions which seem most promising for use in the first practical fusion reactor are the *D–D reactions*

$$\ _1^2\text{H}(d, n)\ _2^3\text{He} \qquad Q = 3.27 \text{ MeV}$$

$$\ _1^2\text{H}(d, p)\ _1^3\text{H} \qquad Q = 4.03 \text{ MeV}$$

and the *D–T reaction*

$$\ _1^3\text{H}(d, n)\ _2^4\text{He} \qquad Q = 17.59 \text{ MeV}$$

Reaction series known as the *carbon* or *Bethe cycle* and the *proton-proton* or *Critchfield cycle* are believed to occur in stars. These cycles are illustrated in Problems 31.11 through 31.13 and Problem 31.18.

Solved Problems

31.1. What is the kinetic energy of a 300 K thermal neutron?

The thermal energy of a particle is of the order of kT, where k is Boltzmann's constant. Thus,

$$K_n \approx (8.617 \times 10^{-5} \text{ eV/K})(300 \text{ K}) = 0.026 \text{ eV}$$

31.2. On the average, neutrons lose half their energy per collision with quasi-free protons (see Problem 31.17 for the effects of a *head-on* collision). How many collisions, on the average, are required to reduce a 2 MeV neutron to a thermal energy of 0.04 eV?

If N is the number of collisions, the ratio of the final to initial energy is

$$\frac{K_f}{K_i} = \frac{0.04 \text{ eV}}{2 \times 10^6 \text{ eV}} = (0.5)^N \qquad \text{or} \qquad (0.5)^N = 2 \times 10^{-8}$$

Taking the log of both sides of the equation, we obtain

$$(-0.301)N = -7.70 \qquad \text{or} \qquad N \approx 26$$

The energy of the neutrons produced in a nuclear fission is approximately 2 MeV, and it takes on the average about 26 proton collisions to reduce the energy to thermal levels. Thermal neutrons have large probabilities for producing further fissions.

31.3. Determine the total final kinetic energy in the photofission of $^{235}_{92}U$ by a 6 MeV γ-ray into $^{90}_{36}Kr$, $^{142}_{56}Ba$, and three neutrons.

The fission reaction is

$$^{235}_{92}U + \gamma \rightarrow {}^{90}_{36}Kr + {}^{142}_{56}Ba + 3{}^{1}_{0}n$$

From conservation of mass-energy,

$$M_U c^2 + K_\gamma = (M_{Kr} + M_{Ba} + 3m_n)c^2 + K_f$$

or

$$K_f = [235.043915 \text{ u} - (89.91972 \text{ u} + 141.91635 \text{ u} + 3 \times 1.008665 \text{ u})](931.5 \text{ MeV/u}) + 6 \text{ MeV} = 175.4 \text{ MeV}$$

31.4. About 185 MeV of usable energy is released in the neutron-induced fissioning of a $^{235}_{92}U$ nucleus. If $^{235}_{92}U$ in a reactor is continuously generating 100 MW of power, how long will it take for 1 kg of the uranium to be used up?

The fission rate corresponding to the given power output is

$$\left(10^8 \, \frac{\text{J}}{\text{s}}\right)\left(\frac{10^{-6} \text{ MeV}}{1.6 \times 10^{-19} \text{ J}}\right)\left(\frac{1 \text{ fission}}{185 \text{ MeV}}\right) = 3.38 \times 10^{18} \, \frac{\text{fissions}}{\text{s}}$$

One kilogram of ^{235}U contains

$$\left(\frac{1 \text{ kg}}{235 \text{ kg/kmol}}\right)\left(6.023 \times 10^{26} \, \frac{\text{nuclei}}{\text{kmol}}\right) = 2.56 \times 10^{24} \text{ nuclei}$$

and so it will last

$$t = \frac{2.56 \times 10^{24}}{3.38 \times 10^{18} \text{ s}^{-1}} = 7.58 \times 10^5 \text{ s} = 8.78 \text{ d}$$

31.5. Estimate the temperature required to produce fusion in a deuterium plasma (a neutral mixture of negatively charged electrons and positively charged deuterium nuclei).

Taking the range of nuclear forces to be 2 fm, the Coulomb repulsion energy between two deuterons separated by this distance is

$$E_C = \frac{ke^2}{R} = \frac{1.44 \text{ MeV} \cdot \text{fm}}{2 \text{ fm}} = 0.72 \text{ MeV}$$

The average kinetic energy in a system of particles at a temperature T is of the order of kT, giving

$$E_C = kT$$

$$0.72 \text{ MeV} = (8.617 \times 10^{-11} \text{ MeV/K})T$$

$$T = 8.35 \times 10^9 \text{ K}$$

A more detailed analysis that takes into account barrier penetration shows that fusion will begin at about 10^7 K.

31.6.   What will be the energy released if two deuterium nuclei fuse into an α-particle?

The reaction is

$$^{2}_{1}\text{H} + ^{2}_{1}\text{H} \rightarrow ^{4}_{2}\text{He}$$

Conservation of mass-energy gives

$$2M_H c^2 = M_{He} c^2 + Q$$

$$Q = (2M_H - M_{He})c^2$$

$$= (2 \times 2.014102 \text{ u} - 4.002603 \text{ u})(931.5 \text{ MeV/u}) = 23.80 \text{ MeV}$$

In Problem 31.5 it was found that about 0.7 MeV of energy is required to begin the fusion process, while 23.8 MeV of energy is released after fusion takes place.

31.7.   In a sequential process $^{235}_{92}\text{U}$ plus a neutron forms the compound nucleus $[^{236}_{92}\text{U}]^*$, which then fissions; the fission then produces further decays.   If the initial fission fragments are $^{143}_{56}\text{Ba}$ and $^{90}_{36}\text{Kr}$, illustrate a process leading to final stable nuclei.

The initial process is

$$^{235}_{92}\text{U} + ^{1}_{0}n \rightarrow \left[^{236}_{92}\text{U}\right]^* \rightarrow ^{143}_{56}\text{Ba} + ^{90}_{36}\text{Kr} + 3^{1}_{0}n$$

$^{143}_{56}\text{Ba}$ then starts the series of beta decays

$$^{143}_{56}\text{Ba} \rightarrow ^{143}_{57}\text{La} + e^- + \bar{\nu}$$
$$\hookrightarrow ^{143}_{58}\text{Ce} + e^- + \bar{\nu}$$
$$\hookrightarrow ^{143}_{59}\text{Pr} + e^- + \bar{\nu}$$
$$\hookrightarrow ^{143}_{60}\text{Nd} + e^- + \bar{\nu}$$

the nucleus $^{143}_{60}\text{Nd}$ being stable. $^{90}_{36}\text{Kr}$ starts the beta decays

$$^{90}_{36}\text{Kr} \rightarrow ^{90}_{37}\text{Rb} + e^- + \bar{\nu}$$
$$\hookrightarrow ^{90}_{38}\text{Sr} + e^- + \bar{\nu}$$
$$\hookrightarrow ^{90}_{39}\text{Y} + e^- + \bar{\nu}$$
$$\hookrightarrow ^{90}_{40}\text{Zr} + e^- + \bar{\nu}$$

the nucleus $^{90}_{40}\text{Zr}$ being stable.
The total reaction then looks like

$$^{235}_{92}\text{U} + ^{1}_{0}n \rightarrow \left[^{236}_{92}\text{U}\right]^* \rightarrow ^{143}_{60}\text{Nd} + ^{90}_{40}\text{Zr} + 3^{1}_{0}n + 8e^- + 8\bar{\nu} \qquad (1)$$

31.8. Calculate the energy released in the fission reaction of Problem 31.7.

If atomic rest masses are used in calculating Q from (1) of Problem 31.7, the term $8e^-$ drops out. Thus,

$$Q = [M_U - M_{Nd} - M_{Zr} - (3-1)m_n]c^2$$

$$= [235.043915 \text{ u} - 142.909779 \text{ u} - 89.904700 \text{ u} - 2(1.008665 \text{ u})](931.5 \text{ MeV/u}) = 197.6 \text{ MeV}$$

31.9. Estimate the Coulomb energy of repulsion for the $^{143}_{56}$Ba and $^{90}_{36}$Kr nuclei of Problem 31.7 just after they are formed.

Just after formation, the nuclei are assumed to be spherical and touching. The Coulomb energy is then

$$E_c = k\frac{(Z_1 e)(Z_2 e)}{R_1 + R_2} = \frac{(ke^2)Z_1 Z_2}{r_0(A_1^{1/3} + A_2^{1/3})} = \frac{(1.44 \text{ MeV} \cdot \text{fm})(56)(36)}{(1.4 \text{ fm})(143^{1/3} + 90^{1/3})} = 214 \text{ MeV}$$

This is approximately the energy released in the reaction, as determined in Problem 31.8.

31.10. For the D–T fusion reaction, calculate the rate at which deuterium and tritium are consumed to produce 1 MW. (Assume all energy from the fusion reaction is available.)

In the D–T reaction, $^3_1$H$(d, n)^4_2$He, the energy released in each fusion is $Q = 17.6$ MeV (Problem 31.14). The rate at which the reactions must occur is

$$R = \left(1 \times 10^6 \frac{\text{J}}{\text{s}}\right)\left(\frac{1 \text{ eV}}{1.6 \times 10^{-19} \text{ J}}\right)\left(\frac{1 \text{ reaction}}{17.6 \times 10^6 \text{ eV}}\right) = 3.55 \times 10^{17} \frac{\text{reactions}}{\text{s}}$$

In each reaction one atom of deuterium and one of tritium are used up. Therefore, for deuterium ($A = 2$):

$$-\frac{dm}{dt} = \left(3.55 \times 10^{17} \frac{\text{atoms}}{\text{s}}\right)\left(\frac{1 \text{ kmol}}{6.023 \times 10^{26} \text{ atoms}}\right)\left(\frac{2 \text{ kg}}{1 \text{ kmol}}\right) = 1.18 \times 10^{-9} \frac{\text{kg}}{\text{s}}$$

and for tritium ($A = 3$):

$$-\frac{dm}{dt} = \frac{3}{2}\left(1.18 \times 10^{-9} \frac{\text{kg}}{\text{s}}\right) = 1.77 \times 10^{-9} \frac{\text{kg}}{\text{s}}$$

31.11. Calculate the total energy released in the following carbon (Bethe) cycle:

$$p + {}^{12}_6\text{C} \rightarrow {}^{13}_7\text{N}$$

$$^{13}_7\text{N} \rightarrow {}^{13}_6\text{C} + e^+ + \nu$$

$$p + {}^{13}_6\text{C} \rightarrow {}^{14}_7\text{N}$$

$$p + {}^{14}_7\text{N} \rightarrow {}^{15}_8\text{O}$$

$$^{15}_8\text{O} \rightarrow {}^{15}_7\text{N} + e^+ + \nu$$

$$p + {}^{15}_7\text{N} \rightarrow {}^{12}_6\text{C} + {}^4_2\text{He}$$

Instead of finding the energy released in each reaction, we can add all the reactions together to get

$$(p + {}^{12}\text{C}) + ({}^{13}\text{N}) + (p + {}^{13}\text{C}) + (p + {}^{14}\text{N}) + ({}^{15}\text{O}) + (p + {}^{15}\text{N}) \rightarrow$$

$$({}^{13}\text{N}) + ({}^{13}\text{C} + e^+ + \nu) + ({}^{14}\text{N}) + ({}^{15}\text{O}) + ({}^{15}\text{N} + e^+ + \nu) + ({}^{12}\text{C} + {}^4\text{He})$$

After canceling common terms from both sides we are left with

$$4p \rightarrow {}^4\text{He} + 2e^+ + 2\nu$$

Thus all the reactions are equivalent to the fusion of four protons into a helium nucleus. Applying conservation of mass-energy, we obtain

$$4(M_H - m_e)c^2 = (M_{He} - 2m_e)c^2 + 2m_ec^2 + Q$$

$$Q = (4M_H - M_{He} - 4m_e)c^2$$

$$= [4(1.007825 \text{ u}) - 4.002603 \text{ u} - 4(0.000549 \text{ u})](931.5 \text{ MeV/u}) = 24.69 \text{ MeV}$$

It is seen that the carbon atom acts as a sort of catalyzer, since it is regenerated at the end of the carbon cycle.

31.12. Determine the energy released for each kilogram of hydrogen that is consumed in the cycle of Problem 3.11.

From the equivalent cycle $4p \rightarrow {}^4\text{He} + 2e^+ + 2\nu$ it is seen that 24.69 MeV of energy is released for each four protons that are consumed, so that we have

$$\frac{24.69 \text{ MeV}}{4 \text{ protons}} \times \frac{1 \text{ proton}}{1.673 \times 10^{-27} \text{ kg}} = 3.69 \times 10^{27} \frac{\text{MeV}}{\text{kg}}$$

or 5.90×10^{14} J/kg.

31.13. Refer to Problem 31.12. It is estimated that the carbon cycle releases about 4×10^{26} W of power. Determine the rate at which hydrogen is consumed.

$$\left(4 \times 10^{26} \frac{\text{J}}{\text{s}}\right)\left(\frac{1 \text{ kg hydrogen}}{5.90 \times 10^{14} \text{ J}}\right) = 6.8 \times 10^{11} \frac{\text{kg hydrogen}}{\text{s}}$$

For comparison, the mass of the sun is about 2×10^{30} kg.

Supplementary Problems

Atomic masses are tabulated in the Appendix.

31.14. Calculate the Q-value for the D–T fusion reaction, ${}^3_1\text{H}(d, n){}^4_2\text{He}$.   *Ans.*   17.6 MeV

31.15. Find the Q-values for the D–D reactions (*a*) ${}^2_1\text{H}(d, n){}^3_2\text{He}$, (*b*) ${}^2_1\text{H}(d, p){}^3_1\text{H}$.
Ans.   (*a*) 3.27 MeV; (*b*) 4.03 MeV

31.16. Refer to Problem 31.4. What is the power output of a ${}^{235}_{92}\text{U}$ reactor if it takes 30 days to use up 2 kg of fuel?   *Ans.*   62.5 MW

31.17. Using results of Problem 30.16, find the kinetic energy that a 2 MeV neutron has after it undergoes a head-on collision with a quasi-free proton at rest.   *Ans.*   0.347 eV

31.18. Show the equivalence of the following proton-proton (Critchfield) cycle to the carbon cycle of Problem 31.11:

$$p + p \rightarrow d + e^+ + \nu$$
$$p + d \rightarrow {}^3\text{He}$$
$${}^3_2\text{He} + {}^3_2\text{He} \rightarrow {}^4_2\text{He} + 2p$$

31.19. Calculate the energy released in the fusion process

$${}^4_2\text{He} + {}^4_2\text{He} + {}^4_2\text{He} \rightarrow {}^{12}\text{C}$$

Ans.   7.27 MeV

31.20. Consider a reaction series similar to that of Problem 31.11, but with p and ${}^{14}\text{N}$ as the initial reactants. If the intermediate nuclei formed are ${}^{15}_8\text{O}$, ${}^{15}_7\text{N}$, ${}^{16}_8\text{O}$, ${}^{17}_9\text{F}$, and ${}^{17}_8\text{O}$, give the reactions which result in the regeneration of ${}^{14}_7\text{N}$ and give the overall reaction.   *Ans.* $4p \rightarrow {}^4_2\text{He} + 2e^+ + 2\nu$ (overall)

Chapter 32

Elementary Particles

32.1 ELEMENTARY PARTICLE GENEALOGY

At present more than 30 long-lived *elementary particles* and *antiparticles* have been detected experimentally. An antiparticle has the same mass and spin as its associated particle, but the electromagnetic properties, such as charge and magnetic moment, are opposite in particle and antiparticle. These particles are listed in Table 32-1, which also gives some of their properties. In addition to these particles, about 50 *resonances* have been observed since 1963. In contrast with the relatively stable elementary particles (mean lifetime $T_m \gg 10^{-21}$ s), a resonance is extremely short-lived, with $T_m < 10^{-21}$ s. As indicated in Table 32-1, elementary particles are grouped into four families. The classification arises mainly from the particle's spin, mass, and type of interaction, according to the scheme of Table 32-2.

Table 32-1*

	Particle (Antiparticle)	Rest Mass, MeV	Particle Mean Lifetime, s	Charge Number $\mathcal{Q}$	Spin	Lepton Number $\mathcal{L}_e$	Lepton Number $\mathcal{L}_\mu$	Baryon Number $\mathcal{B}$	Strangeness $\mathcal{S}$
Massless bosons	γ (γ)	0	stable	0	1				
Leptons	ν_e $(\bar{\nu}_e)$	0	stable	0 (0)	1/2	$+1$ (-1)			
	ν_μ $(\bar{\nu}_\mu)$	0	stable	0 (0)	1/2		$+1$ (-1)		
	e^- (e^+)	0.511	stable	-1 $(+1)$	1/2	$+1$ (-1)			
	μ^- (μ^+)	105.7	2.2×10^{-6}	-1 $(+1)$	1/2		$+1$ (-1)		
Mesons	π^+ (π^-)	139.6	2.6×10^{-8}	$+1$ (-1)	0				0 (0)
	π^0 (π^0)	135.0	0.8×10^{-16}	0	0				0 (0)
	π^- (π^+)	139.6	2.6×10^{-8}	-1 $(+1)$	0				0 (0)
	K^+ (K^-)	493.7	1.2×10^{-8}	$+1$ (-1)	0				$+1$ (-1)
	K^0 $(\bar{K}^0)$	497.7	8.8×10^{-11}	0 (0)	0				$+1$ (-1)
	$\bar{K}^0$ (K^0)	497.7	5.2×10^{-8}	0 (0)	0				-1 $(+1)$
	K^- (K^+)	493.7	1.2×10^{-8}	-1 $(+1)$	0				-1 $(+1)$
	η^0 (η^0)	549	2.5×10^{-19}	0	0				0 (0)
Baryons	p $(\bar{p})$	938.3	stable	$+1$ (-1)	1/2			$+1$ (-1)	0 (0)
	n $(\bar{n})$	939.6	932	0 (0)	1/2			$+1$ (-1)	0 (0)
	Λ^0 $(\bar{\Lambda}^0)$	1116	2.5×10^{-10}	0 (0)	1/2			$+1$ (-1)	-1 $(+1)$
	Σ^+ $(\bar{\Sigma}^-)$	1189	8.0×10^{-11}	$+1$ (-1)	1/2			$+1$ (-1)	-1 $(+1)$
	Σ^0 $(\bar{\Sigma}^0)$	1192	10^{-14}	0 (0)	1/2			$+1$ (-1)	-1 $(+1)$
	Σ^- $(\bar{\Sigma}^+)$	1197	1.5×10^{-10}	-1 $(+1)$	1/2			$+1$ (-1)	-1 $(+1)$
	Ξ^0 $(\bar{\Xi}^0)$	1315	3.0×10^{-10}	0 (0)	1/2			$+1$ (-1)	-2 $(+2)$
	Ξ^- $(\bar{\Xi}^+)$	1321	1.7×10^{-10}	-1 $(+1)$	1/2			$+1$ (-1)	-2 $(+2)$
	Ω^- $(\bar{\Omega}^+)$	1672	1.3×10^{-10}	-1 $(+1)$	3/2			$+1$ (-1)	-3 $(+3)$

* Adapted from Thomas G. Trippe, et al., *Rev. Mod. Phys.*, **48**:2, Part II (1976).

Table 32-2

Family	Spin	Mass (m_e = mass of electron)	Type of Interaction
Massless bosons	Integer	0	Electromagnetic, Gravitational
Leptons	Half-integer	$0 \leqslant M < 207m_e$	Weak, Electromagnetic
Mesons	Integer	$273m_e < M < 1075m_e$	Strong, Weak, Electromagnetic, Gravitational
Baryons	Half-integer	$1836m_e < M$	Strong, Weak, Electromagnetic, Gravitational

32.2   PARTICLE INTERACTIONS

Table 32-2 shows that in addition to the familiar gravitational and electromagnetic interactions there are two other types of forces by which particles may interact with each other, namely the *strong* and *weak interactions*.

The force that holds the nucleons together in a nucleus is an example of a strong interaction. Since protons are bound within a nucleus, the strong interaction must be much greater than the electromagnetic interaction, which tends to force the protons apart.   In addition, it is found that the strong interaction does not depend on the charge of the body.   An example of a strong interaction is

$$\pi^- + p \rightarrow \pi^0 + n$$

The existence of weak interactions is necessary to explain how a neutrino interacts with nuclear matter.   Since a neutrino is massless and carries no charge, it cannot undergo gravitational or electromagnetic interactions.   Moreover, since it is not a nucleon, a neutrino does not participate in nuclear or strong interactions.   An example of a weak interaction is

$$n \rightarrow p + e^- + \bar{\nu}$$

Both the interaction strengths and decay times (lifetimes before decay) of elementary particles are characterized by dimensionless *coupling constants*.   At low energies the larger the coupling constant, the stronger the interaction and the shorter the lifetime.   For the electromagnetic interaction the coupling constant is

$$\frac{ke^2}{\hbar c} = \frac{1}{137}$$

and mean lifetimes against electromagnetic decays are about 10^{-16} s.   For strong interactions the coupling constant is $g^2/\hbar c \approx 13$ (g is a constant appearing in Yukawa's theory) and strong decays have lifetimes of about 10^{-23} s.   The weak coupling constant is about 3×10^{-12} and weak decays have mean lifetimes of about 10^{-8} s.   The gravitational coupling constant, $Gm^2/\hbar c$, where m is a nuclear mass, is about 10^{-40}.   All particles listed in Table 32-1 have lifetimes which are large compared to 10^{-23} s, the mean lifetime for strong decays.

32.3   CONSERVATION LAWS

All elementary particle reactions and decays appear to obey certain conservation laws and selection rules.   These include the familiar conservation laws for:
  (*a*)   Mass-energy
  (*b*)   Linear momentum
  (*c*)   Angular momentum (spin)
  (*d*)   Charge

which are found to hold whether the process goes by the strong, weak or electromagnetic interaction.

The last of these conservation laws differs from the others in that not only is charge conserved but it is also quantized in units of e, the magnitude of the electron charge. Conservation of quantized charge can be expressed by assigning a *charge quantum number*, $\mathcal{Q}$ = charge/e, to every particle. In a reaction the initial and final values of the total $\mathcal{Q}$ will then be equal. For example, in antiproton production,

$$p + p \rightarrow p + p + p + \bar{p}$$
$$\mathcal{Q}: \quad +1 + 1 = +1 + 1 + 1 - 1$$

32.4 CONSERVATION OF LEPTONS

Several other conservation laws or selection rules are found to hold for other quantum numbers. A *lepton number* is defined as $\mathcal{L} = +1$ for lepton particles, $\mathcal{L} = -1$ for lepton antiparticles, and $\mathcal{L} = 0$ for all other particles. Lepton numbers for electrons and their associated neutrinos (ν_e), and also lepton numbers for μ-mesons and their associated neutrinos (ν_μ), are separately conserved in all processes. Examples of the conservation of leptons are:

$$\mu^- \rightarrow e^- + \bar{\nu}_e + \nu_\mu$$
$$\mathcal{L}_\mu: \quad +1 = 0 + 0 + 1$$
$$\mathcal{L}_e: \quad 0 = +1 - 1 + 0$$
$$K^0 \rightarrow \pi^+ + e^- + \bar{\nu}_e$$
$$\mathcal{L}_e: \quad 0 = 0 + 1 - 1$$

32.5 CONSERVATION OF BARYONS

Similarly, a *baryon number*, $\mathcal{B}$, can be defined as being $+1$ for baryon particles, -1 for baryon antiparticles, and 0 for all other particles. For any reaction or decay, the total baryon number is conserved. Examples of this conservation law are:

$$n \rightarrow p + e^- + \bar{\nu}_e$$
$$\mathcal{B}: \quad +1 = +1 + 0 + 0$$
$$K^- + p \rightarrow \Lambda^0 + \pi^+ + \pi^-$$
$$\mathcal{B}: \quad 0 + 1 = +1 + 0 + 0$$

32.6 CONSERVATION OF ISOTOPIC SPIN

Table 32-1 shows that the mesons and baryons occur in groups, or *multiplets*, of like mass, the particles in a particular multiplet differing with respect to charge. For example, the three pions (π^+, π^0, π^-) all have a mass of about 140 MeV, and the two nucleons (n, p) have a mass of approximately 940 MeV.

Mesons and baryons interact with each other by means of the strong interaction. Because of the charge independence of the strong interaction, all particles in a multiplet should interact with another particle strongly in the same way; however, the (much weaker) electromagnetic interaction causes small differences. This charge independence has led to the introduction of another quantum number, called the *isotopic spin*, I, defined such that $2I + 1$ gives the number of particles in the particular multiplet. Thus, for pions $I = 1$ and for nucleons $I = \frac{1}{2}$.

Isotopic spin is treated as a vector with magnitude $\sqrt{I(I+1)}$, like angular momentum. However, unlike angular momentum, isotopic spin is a dimensionless quantity. The z-component of the isotopic spin, m_I, is quantized according to

$$m_I = I, I - 1, \ldots, -I \tag{32.1}$$

Each particle in the multiplet corresponds to a value of m_I, with the values arranged in order of decreasing charge. Thus, for the pions, $m_I = +1, 0, -1$ for π^+, π^0, π^-, respectively, and, for the nucleons, $m_I = +\frac{1}{2}, -\frac{1}{2}$ for the proton and neutron, respectively. Antiparticle multiplets have the same isotopic spin as the corresponding particle multiplet, but m_I for an antiparticle is the negative of m_I for the corresponding particle. Table 32-3 gives I and m_I for the mesons and baryons.

Table 32-3

Baryons	I \\ m_I	1	1/2	0	$-1/2$	-1
940 MeV	1/2		p		n	
1110 MeV	0			Λ^0		
1190 MeV	1	Σ^+		Σ^0		Σ^-
1320 MeV	1/2		Ξ^0		Ξ^-	
1670 MeV	0			Ω^-		
Mesons						
138 MeV	1	π^+		π^0		π^-
496 MeV	1/2		K^+		K^0	
549 MeV	0			η^0		

It is found that *in all strong interactions the total isotopic spin (added as a vector) is conserved*. For the case of two particles, $\mathbf{I} = \mathbf{I}_1 + \mathbf{I}_2$, the z-component of the resultant $\mathbf{I}$ is given by

$$m_I = m_{I_1} + m_{I_2} \tag{32.2}$$

and the allowed values of I are given by

$$I = I_1 + I_2, \; I_1 + I_2 - 1, \; I_1 + I_2 - 2, \ldots \tag{32.3}$$

the sequence terminating in $|I_1 - I_2|$ or $|m_I|$, whichever is greater. For example, in the strong reaction

$$\pi_0 + p \rightarrow \pi^+ + n$$

we have for the reactants

$$m_I = 0 + \tfrac{1}{2} = \tfrac{1}{2} \qquad I = 1 + \tfrac{1}{2}, \; 1 + \tfrac{1}{2} - 1 = \tfrac{3}{2}, \tfrac{1}{2}$$

and for the products

$$m_I = 1 - \tfrac{1}{2} = \tfrac{1}{2} \qquad I = 1 + \tfrac{1}{2}, \; 1 + \tfrac{1}{2} - 1 = \tfrac{3}{2}, \tfrac{1}{2}$$

It is further found that *in all strong and electromagnetic processes the total m_I is conserved*. The above example illustrates this conservation law.

It should be noted that the conservation laws for isotopic spin do not apply to weak interactions. For example, in the weak decay

$$K^0 \rightarrow \pi^0 + \pi^0$$
$$m_I: \quad -\tfrac{1}{2} \neq 0 + 0$$
$$I: \quad \tfrac{1}{2} \neq 2, 1, 0$$

so that neither m_I nor I is conserved.

32.7 CONSERVATION OF STRANGENESS

It is found experimentally that the K-mesons—and the Λ, Σ, Ξ, and Ω baryons (this group being referred to as *hyperons*)—are always produced in pairs in strong interactions, a phenomenon called *associated production*. Moreover, the lifetimes of these particles are very much greater than 10^{-23} s, showing that they do not decay by the strong interaction (which they might be expected to do). To explain this "strange" phenomenon a new quantum number, the *strangeness*, $\mathcal{S}$, was introduced in Table 32-1, and particles with $\mathcal{S} \neq 0$ were designated as *strange* particles. It was found that *the total strangeness (added as a scalar) is conserved in strong and electromagnetic reactions (or decays). In weak interactions it is found that* $\Delta\mathcal{S} = 0, \pm 1$. This second condition, though not a conservation law, forbids certain reactions and is called a *selection rule*.

An example of strangeness conservation in a strong process is

$$\pi^+ + p \rightarrow \Sigma^+ + K^+$$
$$\mathcal{S}: \quad 0 \;\; + 0 = -1 \;\; + 1$$

while in the weak decay

$$\Lambda^0 \rightarrow \pi^- + p$$
$$\mathcal{S}: \quad -1 \neq 0 \;\; + 0$$

strangeness is not conserved but the selection rule for strange particles is satisfied ($\Delta\mathcal{S} = +1$).

The strangeness of a particle can be expressed in terms of its charge, baryon number, and z-component of isotopic spin.

$$\mathcal{Q} = m_I + \tfrac{1}{2}(\mathcal{B} + \mathcal{S}) \tag{32.4}$$

With this definition the strangeness of a particle will be an integer, and the strangeness of an antiparticle will have the opposite sign to that of its associated particle.

32.8 CONSERVATION OF PARITY

Another quantity that we mention for completeness is *parity*, which is conserved in a reaction if the mirror image of the reaction (involving the antiparticles) also occurs. It is found that *parity is conserved in strong and electromagnetic interactions but is not conserved in weak interactions.*

32.9 SHORT-LIVED PARTICLES AND THE RESONANCES

Because of their extremely short lifetimes, particles such as the π^0 and η^0 ($T_m < 10^{-16}$ s) and the resonances ($T_m < 10^{-21}$ s) do not leave observable tracks in instruments like bubble chambers. Their existence is inferred by measuring the energies and momenta of the final decay products and working backward through the conservation laws to see if the measured results are consistent with the assumption of the existence of the unobservable intermediate particle. For example, when the K^+ decays, what is observed is a π^+ particle and two γ-rays, so that it might be believed that the decay scheme is

$$K^+ \rightarrow \pi^+ + 2\gamma$$

However, it is found experimentally that in the center-of-mass system the π^+ particle is monoenergetic (see Problem 32.22). This fact rules out a three-particle decay (which would be associated with a *distribution* of π^+ energies — see Section 29.5). The correct decay is

$$K^+ \rightarrow \pi^+ + \pi^0$$
$$\mathrel{\llcorner\!\!\rightarrow} 2\gamma$$

Solved Problems

32.1. A 150 MeV K^+ particle decays into $2\pi^+ + \pi^-$. Range measurements in a photographic emulsion give the kinetic energies of the π^+'s as 68.6 MeV and 80.8 MeV, and that of the π^- as 75.5 MeV. Find the Q for the reaction and the mass of the K^+.

The reaction is $K^+ \to \pi^+ + \pi^+ + \pi^-$, so

$$Q = K_{\pi^+} + K_{\pi^+} + K_{\pi^-} - K_{K^+} = 68.6 \text{ MeV} + 80.8 \text{ MeV} + 75.5 \text{ MeV} - 150 \text{ MeV} = 74.9 \text{ MeV}$$

From $Q = (m_{K^+} - 2m_{\pi^+} - m_{\pi^-})c^2$ we then obtain

$$m_{K^+}c^2 = Q + (2m_{\pi^+} + m_{\pi^-})c^2 = 74.9 \text{ MeV} + 3(139.6 \text{ MeV}) = 493.7 \text{ MeV}$$

32.2. Give the possible values of the isotopic spin and its z-component for the following systems of particles: (a) $\pi^+ + p$, (b) $\pi^- + p$.

(a) For π^+, $I = 1$, $m_I = 1$, and for p, $I = \frac{1}{2}$, $m_I = \frac{1}{2}$; so that the total m_I is $1 + \frac{1}{2} = \frac{3}{2}$. By (32.3), the only possible value of the total isotopic spin is

$$I = 1 + \frac{1}{2} = \frac{3}{2}$$

(b) For π^-, $I = 1$, $m_I = -1$, and for p, $I = \frac{1}{2}$, $m_I = \frac{1}{2}$; so that the total m_I is $-1 + \frac{1}{2} = -\frac{1}{2}$. By (32.3), there are two possible values of the total isotopic spin:

$$I = 1 + \frac{1}{2}, \ 1 + \frac{1}{2} - 1 = \frac{3}{2}, \ \frac{1}{2}$$

32.3. In the following reactions, what particles are possible for the unknown particle X? (All reactions are strong.)

$$(a) \quad \bar{K}^- + p \to K^+ + X$$
$$(b) \quad \pi^- + p \to K^0 + X$$
$$(c) \quad p + p \to \pi^+ + n + \Lambda^0 + X$$

Writing the various conservation laws for each reaction we have:

(a)

charge number:	$-1 + 1 = +1 + \mathcal{Q}$ or	$\mathcal{Q} = -1$
lepton number:	$0 + 0 = 0 + \mathcal{L}$	$\mathcal{L} = 0$
baryon number:	$0 + 1 = 0 + \mathcal{B}$	$\mathcal{B} = 1$
strangeness:	$-1 + 0 = +1 + \mathcal{S}$	$\mathcal{S} = -2$
z-component of isotopic spin:	$-\frac{1}{2} + \frac{1}{2} = +\frac{1}{2} + m_I$	$m_I = -\frac{1}{2}$

These properties fix X as a Ξ^- particle.

(b)

charge number:	$-1 + 1 = 0 + \mathcal{Q}$ or	$\mathcal{Q} = 0$
lepton number:	$0 + 0 = 0 + \mathcal{L}$	$\mathcal{L} = 0$
baryon number:	$0 + 1 = 0 + \mathcal{B}$	$\mathcal{B} = +1$
strangeness:	$0 + 0 = 1 + \mathcal{S}$	$\mathcal{S} = -1$
z-component of isotopic spin:	$-1 + \frac{1}{2} = -\frac{1}{2} + m_I$	$m_I = 0$

These properties fix X as either a Σ^0 or a Λ^0 particle.

(c)

charge number:	$+1+1 = +1+0+0+\mathfrak{Q}$ or	$\mathfrak{Q} = +1$
lepton number:	$0+0 = 0+0+0+\mathfrak{L}$	$\mathfrak{L} = 0$
baryon number:	$+1+1 = 0+1+1+\mathfrak{B}$	$\mathfrak{B} = 0$
strangeness:	$0+0 = 0+0-1+\mathfrak{S}$	$\mathfrak{S} = +1$
z-component of isotopic spin:	$+\frac{1}{2}+\frac{1}{2} = 1-\frac{1}{2}+0+m_I$	$m_I = +\frac{1}{2}$

These properties fix X as a K^+-meson.

32.4. In the following pairs determine which of the reactions is possible.

(a)
$$\pi^- + p \to \Sigma^0 + \eta^0$$
$$\pi^- + p \to \Sigma^0 + K^0$$
(strong interaction)

(b)
$$\Sigma^- \to \pi^- + \eta$$
$$\Sigma^- \to \pi^- + p$$
(weak decay)

(c)
$$p + p \to K^+ + \Sigma^+$$
$$p + p \to K^+ + p + \Lambda^0$$
(strong interaction)

(d)
$$\pi^- + p \to n + \gamma$$
$$\pi^- + p \to \pi^0 + \Lambda^0$$
(strong interaction)

(e)
$$n \to p + e^- + \nu_e$$
$$n \to p + e^- + \bar{\nu}_e$$
(weak decay)

The reactions marked "none" in the last column of Table 32-4 violate none of the applicable conservation laws and so are possible.

Table 32-4

	Reaction Pair	Charge Number ($\mathfrak{Q}$)	Lepton Number ($\mathfrak{L}_e$)	Baryon Number ($\mathfrak{B}$)	Strangeness ($\mathfrak{S}$)	z-Component of Isotopic Spin (m_I)	Violated Conservation Laws
(a)	$\pi^- + p \to \Sigma^0 + \eta^0$	$-1+1=0+0$	$0+0=0+0$	$0+1=1+0$	$0+0 \neq -1+0$	$-1+\frac{1}{2} \neq 0+0$	$\mathfrak{S}, m_I$
	$\pi^- + p \to \Sigma^0 + K^0$	$-1+1=0+0$	$0+0=0+0$	$0+1=1+0$	$0+0=-1+1$	$-1+\frac{1}{2}=0-\frac{1}{2}$	none
(b)	$\Sigma^- \to \pi^- + n$	$-1=-1+0$	$0+0=0+0$	$1=0+1$	(not applicable)	(not applicable)	none
	$\Sigma^- \to \pi^- + p$	$-1 \neq -1+1$	$0+0=0+0$	$1=0+1$			$\mathfrak{Q}$
(c)	$p + p \to K^+ + \Sigma^+$	$1+1=1+1$	$0+0=0+0$	$1+1 \neq 0+1$	$0+0=1-1$	$\frac{1}{2}+\frac{1}{2} \neq \frac{1}{2}+1$	$\mathfrak{B}, m_I$
	$p + p \to K^+ + p + \Lambda^0$	$1+1=1+1+0$	$0+0=0+0$	$1+1=0+1+1$	$0+0=1+0-1$	$\frac{1}{2}+\frac{1}{2}=\frac{1}{2}+\frac{1}{2}+0$	none
(d)	$\pi^- + p \to n + \gamma$	$-1+1=0+0$	$0+0=0+0$	$0+1=1+0$	$0+0=0+0$	$-1+\frac{1}{2}=-\frac{1}{2}+0$	none
	$\pi^- + p \to \pi^0 + \Lambda^0$	$-1+1=0+0$	$0+0=0+0$	$0+1=0+1$	$0+0 \neq 0-1$	$-1+\frac{1}{2} \neq 0+0$	$\mathfrak{S}, m_I$
(e)	$n \to p + e^- + \nu_e$	$0=+1-1+0$	$0 \neq 0+1+1$	$0=0+0+0$	(not applicable)	(not applicable)	$\mathfrak{L}$
	$n \to p + e^- + \bar{\nu}_e$	$0=+1-1+0$	$0=0+1-1$	$0=0+0+0$			none

32.5. Explain why the decay $\Sigma^0 \to \Lambda^0 + \gamma$ is observed but not
$$\Sigma^0 \to p + \pi^- \quad \text{or} \quad \Sigma^0 \to n + \pi^0$$

In the decay

$$\Sigma^0 \rightarrow \Lambda^0 + \gamma$$
$$\mathcal{S}: \quad -1 \; = \; -1 + 0$$

strangeness is conserved, and the lifetime of the Σ^0 ($\approx 10^{-14}$ s) indicates that it decays by the faster electromagnetic process and not by the slower weak interaction (lifetime $\approx 10^{-10}$ s). The decays

$$\Sigma^0 \rightarrow p + \pi^-$$
$$\mathcal{S}: \quad -1 \neq 0 + 0 \qquad (\Delta \mathcal{S} = +1)$$
$$\Sigma^0 \rightarrow n + \pi^0$$
$$\mathcal{S}: \quad -1 \neq 0 + 0 \qquad (\Delta \mathcal{S} = +1)$$

would be weak because the strangeness changes.

32.6. Figure 32-1 shows two sets of bubble chamber tracks (a magnetic field is directed into the paper). Identify the unknown neutral particles (dashed tracks).

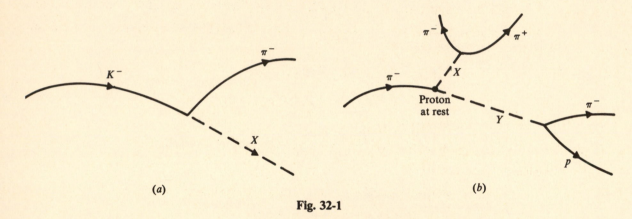

(a)   (b)

Fig. 32-1

(a) The reaction is $K^- \rightarrow \pi^- + X$. The conservation laws that must be satisfied are

> spin: $\qquad\qquad 0 = 0 + \mathcal{S}$
> lepton number: $\quad 0 = 0 + \mathcal{L}$
> baryon number: $\quad 0 = 0 + \mathcal{B}$

Therefore $\mathcal{S} = \mathcal{L} = \mathcal{B} = 0$. From Table 32-1 it is seen that the unknown particle is an uncharged meson, either π^0, K^0, $\overline{K}^0$, or η^0. Since the rest mass of the parent K^- must be greater than the combined rest masses of the daughters ($Q > 0$ for a spontaneous decay), the only possibility is that X is a π^0-meson.

(b) The decay of particle X is $X \rightarrow \pi^+ + \pi^-$. Conservation laws applied to this decay yield

> spin: $\qquad\qquad \mathcal{S} = +1 - 1 = 0$
> lepton number: $\quad \mathcal{L} = \quad 0 + 0 = 0$
> baryon number: $\quad \mathcal{B} = \quad 0 + 0 = 0$

As in (a), a π^0, K^0, $\overline{K}^0$ or η^0 is indicated. This time, $Q > 0$ rules out the π^0-meson. The lifetime of the η^0-meson is so short ($T_m < 10^{-18}$ s) that its path would not be observable on the diagram. Thus the particle X can only be a K^0- or $\overline{K}^0$-meson. The correct choice will be made by determining particle Y.

The decay of particle Y is $Y \rightarrow \pi^- + p$. Application of conservation laws to this decay yields

> spin: $\qquad\qquad \mathcal{S} = 0 + \frac{1}{2} = \frac{1}{2}$
> lepton number: $\quad \mathcal{L} = 0 + 0 = 0$
> baryon number: $\quad \mathcal{B} = 0 + 1 = +1$

From Table 32-1 these properties indicate that particle Y is an uncharged baryon, either n, Λ^0, Σ^0, or Ξ^0. Since the mass of the parent must be greater than the combined mass of the daughters, the neutron is ruled out. Because Y decays weakly, we know that $\Delta S = 0, \pm 1$; and since $\pi^- + p$ has $S = 0$, we know that particle Y has $S = 0, \pm 1$, thereby ruling out Ξ^0. From Problem 32.5 we see that the Σ^0 decay is

$$\Sigma^0 \rightarrow \Lambda^0 + \gamma$$

so the particle Y must be a Λ^0, which has $S = -1$ and $I = m_I = 0$.

To fix X we return to the original strong interaction reaction, which must be either

$$\pi^- + p \rightarrow K^0 + \Lambda^0$$
$$S: \quad 0 \quad +0 = 1 \quad -1$$

or

$$\pi^- + p \rightarrow \overline{K}^0 + \Lambda^0$$
$$S: \quad 0 \quad +0 \neq -1 -1$$

The second reaction doesn't conserve strangeness, so $\overline{K}^0$ can be ruled out, leaving X as a K^0-meson.

32.7. A Σ^0 particle decays at rest to a Λ^0 particle. Determine the energy of the released photon.

The reaction is $\Sigma^0 \rightarrow \Lambda^0 + \gamma$. From conservation of momentum we have, using the relativistic relationship $E^2 = (pc)^2 + (m_0 c^2)^2$,

$$p_\Lambda = p_\gamma = \frac{E_\gamma}{c}$$
$$p_\Lambda^2 c^2 = E_\Lambda^2 - \left(M_\Lambda c^2\right)^2 = E_\gamma^2 \qquad (1)$$

From conservation of energy we have

$$M_\Sigma c^2 = E_\Lambda + E_\gamma \quad \text{or} \quad E_\Lambda^2 = \left(M_\Sigma c^2\right)^2 + E_\gamma^2 - 2M_\Sigma c^2 E_\gamma \qquad (2)$$

Combining (1) and (2), we obtain

$$E_\gamma = \frac{\left(M_\Sigma c^2\right)^2 - \left(M_\Lambda c^2\right)^2}{2M_\Sigma c^2} = \frac{(1192 \text{ MeV})^2 - (1116 \text{ MeV})^2}{2(1192 \text{ MeV})} = 73.6 \text{ MeV}$$

32.8. Determine the energies of the products in the reaction

$$\pi^- + p \rightarrow n + \pi^0$$

if the π^- and p were initially at rest.

From conservation of momentum,

$$p_n = p_{\pi^0} \quad \text{or} \quad (p_n c)^2 = (p_{\pi^0} c)^2 \qquad (1)$$

Using the relativistic relationship between energy and momentum for each of the final products, we have

$$E_n^2 = E_{0n}^2 + (p_n c)^2 \qquad E_{\pi^0}^2 = E_{0\pi^0}^2 + (p_{\pi^0} c)^2$$

Subtracting these and using (1), we obtain

$$E_n^2 - E_{\pi^0}^2 = E_{0n}^2 - E_{0\pi^0}^2 = (939.6 \text{ MeV})^2 - (135.0 \text{ MeV})^2 = 8.646 \times 10^5 \text{ (MeV)}^2$$

or

$$(E_n + E_{\pi^0})(E_n - E_{\pi^0}) = 8.646 \times 10^5 \text{ (MeV)}^2 \qquad (2)$$

From conservation of energy,

$$E_n + E_{\pi^0} = E_{0\pi^-} + E_{0p} = 139.6 \text{ MeV} + 938.3 \text{ MeV} = 1007.9 \text{ MeV} \qquad (3)$$

Substituting this in (2), we obtain

$$(1077.9 \text{ MeV})(E_n - E_{\pi^0}) = 8.646 \times 10^5 \text{ (MeV)}^2$$
$$E_n - E_{\pi^0} = 802.1 \text{ MeV} \qquad (4)$$

Solving (3) and (4) simultaneously, we obtain

$$E_n = 940.0 \text{ MeV} \qquad E_{\pi^0} = 137.9 \text{ MeV}$$

from which

$$K_n = E_n - E_{0n} = 940.0 \text{ MeV} - 939.6 \text{ MeV} = 0.4 \text{ MeV}$$

$$K_{\pi^0} = E_{\pi^0} - E_{0\pi^0} = 137.9 \text{ MeV} - 135 \text{ MeV} = 2.9 \text{ MeV}$$

32.9. Find the threshold energy for the high-energy reaction

$$m_1 + m_2 \rightarrow M_1 + M_2 + \cdots + M_n$$

if the target, m_2, is stationary.

The calculation must be done relativistically. In the laboratory system, where m_2 is at rest,

$$E_{\text{lab}} = (m_1 c^2 + K_1) + m_2 c^2 \qquad (1)$$

In the center-of-mass system the total momentum is zero, and at the threshold energy all the final particles are created at rest. Therefore,

$$E_{\text{cm}} = (M_1 + M_2 + \cdots + M_n)c^2 \qquad (2)$$

For a system of particles the quantity $E^2 - (pc)^2$ is invariant, where E is the sum of the energies of the particles and p is the magnitude of the vector sum of the particle momenta. Therefore, since the total momentum in the laboratory system is just the momentum of the projectile m_1,

$$E_{\text{lab}}^2 - (p_1 c)^2 = E_{\text{cm}}^2$$

$$\left[(m_1 c^2 + K_1) + m_2 c^2 \right]^2 - (p_1 c)^2 = \left[(M_1 + M_2 + \cdots + M_n)c^2 \right]^2 \qquad (3)$$

Also, for the particle m_1,

$$(p_1 c)^2 = E_1^2 - (m_1 c)^2 = (K_1 + m_1 c^2)^2 - (m_1 c^2)^2 \qquad (4)$$

Elimination of $(p_1 c)^2$ between (3) and (4) gives a linear equation for K_1, with solution

$$K_{\text{th}} = K_1 = -\frac{1}{2m_2} \left[(m_1 + m_2 - M_1 - M_2 - \cdots - M_n)c^2 \right](m_1 + m_2 + M_1 + M_2 + \cdots + M_n)$$

$$= -\frac{1}{2m_2} Q(m_1 + m_2 + M_1 + M_2 + \cdots + M_n) \qquad (5)$$

in terms of the (negative) Q-value of the reaction.

Note that, in a low-energy approximation, we could use the classical mass relationship

$$M_1 + M_2 + \cdots + M_n = m_1 + m_2$$

We would then obtain from (5)

$$K_{\text{th}} = \frac{-Q}{m_2} (m_1 + m_2)$$

in agreement with the nonrelativistic result of Problem 30.8.

32.10. Find the threshold energy for the reaction $p + p \rightarrow p + p + \pi^0$.

For the reaction,

$$Q = \left[m_p + m_p - (m_p + m_p + m_\pi) \right]c^2 = -m_\pi c^2 = -135 \text{ MeV}$$

so, by Problem 32.9,

$$K_{\text{th}} = -\frac{Q}{2m_p} (m_p + m_p + m_p + m_p + m_\pi) = -\frac{Q}{2m_p c^2} (4m_p + m_\pi)c^2$$

$$= \frac{135 \text{ MeV}}{2(938 \text{ MeV})} [4(938 \text{ MeV}) + 135 \text{ MeV}] = 280 \text{ MeV}$$

This is the minimum energy that an accelerator must give to a proton to produce a π^0-meson by the above reaction.

32.11. From the reaction $\pi^- + p \rightarrow n + \gamma$ determine the possible values of the spin of a π^--meson.

From conservation of intrinsic angular momentum we have

$$\mathbf{s}_\pi + \mathbf{s}_p = \mathbf{s}_n + \mathbf{s}_\gamma$$

where $|\mathbf{s}| = \sqrt{s(s+1)}\,\hbar$. On the right side we have $s_n = \frac{1}{2}$ and $s_\gamma = 1$, so the total angular momentum is $\frac{3}{2}$ or $\frac{1}{2}$. The proton has $s_p = \frac{1}{2}$, so s_π is 0 or 1.

32.12. Evaluate the quantity $\tau_0 \equiv \hbar / m_\pi c^2$.

$$\tau_0 = \frac{\hbar}{m_\pi c^2} = \frac{6.58 \times 10^{-16}\ \text{eV} \cdot \text{s}}{140 \times 10^6\ \text{eV}} = 4.7 \times 10^{-24}\ \text{s}$$

If one views the strong interaction as the exchange of a π-meson, then inside a nucleus processes such as

$$n \rightarrow p + \pi^-$$

can occur, even though mass-energy conservation forbids the process for free nucleons. According to quantum mechanics, conservation of energy can be violated in the amount $m_\pi c^2$ if the time for the process is of the order given by the Heisenberg uncertainty principle:

$$\Delta t\ \Delta E \approx \hbar \qquad \text{or} \qquad \tau_0 (m_\pi c^2) \approx \hbar \qquad \text{or} \qquad \tau_0 \approx \frac{\hbar}{m_\pi c^2}$$

Therefore, strong interaction processes occur on a time scale of about 10^{-24} s.

32.13. Estimate the mass of a π-meson

If the range a of the π-meson field is about the size of a nucleus and if it is assumed that the π-meson travels at nearly the speed of light, then $a = c\tau_0$, where τ_0 is the time for the π-meson to travel the extent of the nucleus. Setting $a = 1.4$ fm (the approximate nuclear size) and $\tau_0 = \hbar / m_\pi c^2$, the time consistent with the Heisenberg uncertainty principle (Problem 32.12), we obtain

$$1.4\ \text{fm} = \frac{c\hbar}{m_\pi c^2} = \frac{197\ \text{MeV} \cdot \text{fm}}{m_\pi c^2} \qquad \text{or} \qquad m_\pi c^2 = 141\ \text{MeV}$$

This compares well with the observed masses of π-mesons (Table 32-1).

Supplementary Problems

32.14.   Find the threshold energy for the reaction $p + p \rightarrow p + p + p + \bar{p}$.   *Ans.*   5630 MeV

32.15.   Give the possible values of the isotopic spin and its z-component for the system of particles $K^+ + p$.
Ans. $m_I = 1$, $I = 1$

32.16.   Which of the following weak decays is not possible for the Ω^-?

$$(a) \quad \Omega^- \rightarrow \Xi^- + \pi^0 \qquad (b) \quad \Omega^- \rightarrow \Lambda^0 + K^- \qquad (c) \quad \Omega^- \rightarrow \Sigma^- + K^0$$

Ans.   *(c)*

32.17.   The μ^+ decays weakly to a positron plus neutrinos. Which neutrinos are observed in the final state?
Ans. $\nu_e + \bar{\nu}_\mu$

32.18.   Which of the following strong reactions is possible?

$$(a) \quad p + p \rightarrow p + n + K^+ \qquad (b) \quad p + p \rightarrow \Lambda^0 + K^0 + p + \pi^+$$

Ans.   *(b)*

32.19.   Which conservation laws are violated in the reaction $p \rightarrow \pi^0 + e^+ + e^-$?   *Ans.* $\mathcal{B}$ and $\mathcal{Q}$

32.30.   For the decay $\Lambda^0 \rightarrow p + \pi^-$, with the Λ^0 initially at rest, find the π^- kinetic energy.   *Ans.*   32.7 MeV

32.21.   For the decay $\pi^- \rightarrow e^- + \bar{\nu}_e$, with the π^- initially at rest, find the e^- kinetic energy.   *Ans.*   69.3 MeV

32.22.   For the decay $K^+ \rightarrow \pi^+ + \pi^0$ calculate the π^+ kinetic energy in the K^+ rest frame (which is the
center-of-mass frame).   *Ans.*   108 MeV
$\quad\quad\quad\quad\quad\quad\quad\quad\quad\quad \llcorner 2\gamma$

Chapter 33

Molecular Bonding

33.1 IONIC BONDING

Ionic bonding is the type of bonding found in most salts, in which an *alkali metal* (the first column in the periodic table—Li, Na, K, . . .) is bound to a *halogen* (the seventh column in the periodic table —F, Cl, Br, . . .). To explain ionic bonding let us analyze the typical salt KCl. Potassium ($_{19}$K) has one $4s$ electron beyond a closed inert argon core ($1s^2\,2s^2\,2p^6\,3s^2\,3p^6\,4s^1 = {}_{18}$Ar $+ 4s^1$). Because the last electron is weakly bound, having an ionization energy of only 4.34 eV, it is very easy to form a potassium ion K$^+$. Chlorine ($_{17}$Cl) lacks one electron from closing the $3p$ shell ($1s^2\,2s^2\,2p^6\,3s^2\,3p^5 = $ Ar $- 3p^1$), so that it is relatively easy to bind an extra electron to a chlorine atom to form the negative chlorine ion Cl$^-$. Neutral atoms like Cl that have the ability to accept an extra electron are said to have an *electron affinity*. The energy, called the *electron affinity energy*, required to remove the extra electron from a Cl$^-$ ion to form a neutral Cl atom is found to be 3.62 eV. The fact that energy must be added means that the total energy of the chlorine ion is actually lower by 3.62 eV than the total energy of the neutral chlorine atom.

The formation of a KCl molecule can be thought of as occurring in two steps. First, an electron is removed from neutral potassium, forming a K$^+$ ion, and transferred to a neutral chlorine atom to form a Cl$^-$ ion. In this process 4.34 eV must be added to the potassium atom, but 3.62 eV is given back by the Cl atom; hence the net energy required to form the K$^+$ Cl$^-$ pair is 4.34 eV $-$ 3.62 eV, or 0.72 eV. In the next step, the K$^+$ and Cl$^-$ ions can be imagined to be brought together to form the neutral KCl atom. In this process the energy of the system will increase in a negative manner because of the Coulombic attraction between the two oppositely charged ions. The final total energy of the system will be the sum of the positive 0.72 eV required to form the original K$^+$ Cl$^-$ pair and the negative Coulomb energy of the combined ions. If this total energy is negative, the KCl molecule will be stable, since it will require energy to dissociate the molecule into the original K and Cl atoms. Experimentally it is found that the ions in KCl are separated by a distance of $r_0 = 2.79$ Å and that KCl has a dissociation energy of 4.42 eV.

Since the Coulomb force between the two ions is attractive it might at first be expected that there would be no stable configuration. However, at sufficiently small distances the electrons in the ions produce repulsive effects because of the Coulomb force between them and also because of the Pauli exclusion principle. Thus, because there is an attractive force at large separations and a repulsive force at small separations, there exists some intermediate separation at which the K$^+$ and Cl$^-$ ions will be in equilibrium.

33.2 COVALENT BONDING

Molecules such as H$_2$, Cl$_2$, NO are bound by a mechanism known as *covalent bonding*. To explain covalent bonding in detail would require the quantum mechanical solution for a many-electron system under the influence of two nuclei. Though quantum mechanical solutions are beyond the scope of this book, their results can be given to explain covalent bonding.

As an example of covalent bonding consider the simplest covalent molecule, H$_2$. The H$_2$ molecule can be visualized as two positive charges with two electrons moving in their electromagnetic fields. The two electrons can be found with total spin $S = 0$ (singlet state, with spins antialigned) or $S = 1$ (triplet state, with spins aligned). The quantum mechanical solution shows that the singlet state has the lower energy. Moreover, it is found that the most probable location for the two electrons is *between* the hydrogen nuclei. The protons on either side of the electrons are thus attracted to the negative electrons between them, resulting in a force binding the two protons together.

The H_2^+ molecule will bond essentially the same way as the neutral H_2 molecule, but because only one electron is present to attract the protons, the bonding is much weaker. For H_2 the bonding energy is 4.48 eV and the atomic separation 0.74 A, while for H_2^+ the bonding energy is 2.65 eV and the atomic separation is 1.06 Å.

The difference between covalent and ionic bonding is that in covalent bonding the electrons are in effect *shared* by the atoms, while in ionic bonding an electron is effectively *transferred* from one atom to another. In most chemical bonds there is some contribution from each type of bonding.

The differences between the two types of bonding are also manifested in the relative sizes of the electric dipole moments of ionically and covalently bound molecules. An electric dipole can be thought of roughly as two equal but oppositely charged particles separated by a small distance d; the magnitude of the electric dipole moment vector $\mathbf{p}$ is then given by $p = Qd$, where Q is the magnitude of the charge of either particle, with the direction of the dipole moment vector being from the negative to the positive charge. The dipole moment of an ionically bound molecule is relatively large because of the large separation between the two charged ions. For a covalently bound molecule, on the other hand, where the electrons are located *between* the two nuclei, there are effectively two dipole moments formed by the two positively charged nuclei and the negatively charged electrons. Because they point in opposite directions, the two dipoles tend to cancel each other, resulting in essentially no dipole moment for a covalently bound molecule.

33.3 OTHER TYPES OF BONDING

It is possible for two or more atoms, each normally possessing zero dipole moment, to induce dipole moments in each other. The induced dipole can be thought of as arising from the separation of the atom's positive and negative charge because of the close proximity of another atom. The weak attractive force between the induced dipoles is called a *van der Waals* bond. It is the sole force that bonds the inert elements in the liquid and solid state.

In metals, atoms do not share or exchange electrons to bond together. Instead, many electrons (roughly one for each atom) are more or less free to move throughout the metal, so that each electron can interact with many of the fixed atoms. The effects of this interaction, which can be explained only in terms of a quantum mechanical analysis, are responsible for the *metallic bonding* holding the metal together.

Solved Problems

33.1. Obtain an expression for the dissociation energy of an ionically bound, diatomic molecule of a salt in terms of the interatomic spacing, the ionization energy of the alkali, and the electron affinity energy of the halogen. Compare the results predicted by the equation with the experimental values of D given in Table 33-1.

The dissociation of an ionically bound, diatomic molecule can be thought of as occurring in three steps: (1) increasing the separation of the two ions from a distance r_0 to infinity, which requires an energy $-E_C = +ke^2/r_0$ (the negative of the Coulomb energy); (2) the removal of an electron from the negative halogen ion to neutralize it, requiring an energy equal to the halogen's electron affinity energy, F; (3) the placing of this electron on the positive alkali ion to neutralize it, requiring energy $-I$, where I is the ionization energy of the alkali.

The energy D required to dissociate the molecule is thus

$$D = \frac{ke^2}{r_0} + F - I = \frac{14.40 \text{ eV} \cdot \text{Å}}{r_0} + F - I$$

Table 33-1

Alkali	Ionization Energy, I	Halogen	Electron Affinity Energy, F
Li	5.39 eV	F	3.45 eV
Na	5.14 eV	Cl	3.62 eV
K	4.34 eV	Br	3.36 eV

Salt	Bond Length, r_0	Dissociation Energy, D
LiF	1.56 Å	5.95 eV
NaCl	2.51 Å	3.58 eV
KCl	2.79 Å	4.42 eV
KBr	2.94 Å	3.96 eV

Substituting the data from Table 33-1, we obtain

$$\text{LiF:} \qquad D = \frac{14.40 \text{ eV} \cdot \text{Å}}{1.56 \text{ Å}} + 3.45 \text{ eV} - 5.39 \text{ eV} = 7.29 \text{ eV}$$

$$\text{NaCl:} \qquad D = \frac{14.40 \text{ eV} \cdot \text{Å}}{2.51 \text{ Å}} + 3.62 \text{ eV} - 5.14 \text{ eV} = 4.22 \text{ eV}$$

$$\text{KCl:} \qquad D = \frac{14.40 \text{ eV} \cdot \text{Å}}{2.79 \text{ Å}} + 3.62 \text{ eV} - 4.34 \text{ eV} = 4.44 \text{ eV}$$

$$\text{KBr:} \qquad D = \frac{14.40 \text{ eV} \cdot \text{Å}}{2.94 \text{ Å}} + 3.36 \text{ eV} - 4.34 \text{ eV} = 3.92 \text{ eV}$$

It is seen that the predicted values agree fairly well with the observed dissociation energies.

33.2. Estimate the dipole moment for KCl ($r_0 = 2.79$ Å).

If we assume the charge separation to be the atomic separation, the dipole moment is given by

$$p = Qr_0 = (1.6 \times 10^{-19} \text{ C})(2.79 \times 10^{-10} \text{ m}) = 4.46 \times 10^{-29} \text{ C} \cdot \text{m}$$

(Dipole moments are often given in *debyes*; 1 D $= 3.335641 \times 10^{-30}$ C $\cdot$ m.)   The actual dipole moment is found to be 2.64×10^{-29} C $\cdot$ m, showing the centers of charge to be closer than the atomic separation.

33.3. The two protons in a H_2 molecule are separated by 0.74 Å.   How much negative electric charge must be placed at a point midway between the two protons to give the system the observed binding energy of 4.5 eV?

The binding energy, the energy necessary to remove all particles to infinity, will be equal in magnitude to the total Coulomb energy of the charges forming the H_2 molecule.   The Coulomb energy consists of a positive contribution from the two protons,

$$E_+ = \frac{ke^2}{r_0} = \frac{14.40 \text{ eV} \cdot \text{Å}}{0.74 \text{ Å}} = 19.5 \text{ eV}$$

and a negative contribution from the unknown negative charge $\delta(-e)$ located midway between the two protons,

$$E_- = \delta(-e)V_{\text{mid}} = -\delta e\left(\frac{ke}{r_0/2} + \frac{ke}{r_0/2}\right) = -2\delta\frac{ke^2}{r_0/2}$$

$$= -2\delta\frac{14.40 \text{ eV} \cdot \text{Å}}{(0.74 \text{ Å})/2} = -(77.8 \text{ eV})\delta$$

From this we find

$$-\text{BE} = E_+ + E_-$$

$$-4.5 \text{ eV} = 19.5 \text{ eV} - (77.8 \text{ eV})\delta$$

$$\delta = 0.308$$

This number is much smaller than the $\delta = 2$ electrons that actually exist in a H_2 molecule. These two electrons are not located exactly at the midpoint but can be found anywhere around the two protons. However, the most probable place to find the two electrons is, from a quantum mechanical analysis, at the midpoint between the two protons. Thus the electrons spend most of their time around the midpoint, thereby producing the attractive force that results in the covalent bonding of an H_2 molecule.

33.4. An approximate expression for the potential energy of two ions as a function of their separation is

$$PE = -\frac{ke^2}{r} + \frac{b}{r^9}$$

The first term is the usual Coulomb interaction, while the second term is introduced to account for the repulsive effects of the two ions at small distances. Find b as a function of the equilibrium spacing r_0.

The minimum of the expression for the potential energy occurs at the equilibrium separation r_0. The minimum is found as follows:

$$\frac{d(PE)}{dr}\bigg|_{r=r_0} = \frac{ke^2}{r_0^2} - \frac{9b}{r_0^{10}} = 0 \qquad \text{or} \qquad b = \frac{ke^2 r_0^8}{9}$$

The general shape of PE as a function of r is shown in Fig. 33-1.

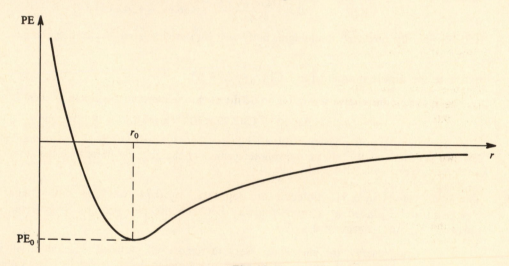

Fig. 33-1

33.5. Calculate the potential energy of KCl at its equilibrium spacing ($r_0 = 2.79$ Å).

From Problem 33.4 the minimum PE is

$$PE_0 = -\frac{ke^2}{r_0} + \frac{b}{r_0^9} = -\frac{ke^2}{r_0} + \frac{ke^2}{9r_0} = -\frac{8ke^2}{9r_0}$$

$$= \frac{-8(14.40 \text{ eV} \cdot \text{Å})}{9(2.79 \text{ Å})} = -4.59 \text{ eV}$$

33.6. An expression for the potential energy of two neutral atoms as a function of their separation r is given by the *Morse potential*,

$$PE = P_0\big[1 - e^{-a(r-r_0)}\big]^2$$

Show that r_0 is the atomic spacing and P_0 the dissociation energy.

The minimum in the potential energy function is found from

$$\frac{d(\text{PE})}{dr} = 2P_0 a e^{-a(r-r_0)}[1 - e^{-a(r-r_0)}] = 0$$

giving $r = r_0$ for the equilibrium separation.   At $r = r_0$, PE $= 0$; as $r \to \infty$, PE $\to P_0$.   Thus an amount of work

$$P_0 - 0 = P_0$$

is needed to bring about an infinite separation of the atoms, and this is the dissociation energy.

Supplementary Problems

33.7.   Determine the minimum separation for Na^+ and Cl^- if they are just to be bound.   Consider the ions as point charges.   (The ionization energy of Na is 5.14 eV, the electron affinity energy for Cl is 3.62 eV, and the dissociation energy of NaCl is 3.58 eV.)   *Ans.*   2.82 Å

33.8.   The dissociation energy of KI is 3.33 eV.   Calculate the bond length (interionic distance) for KI given that the electron affinity energy of I is 3.06 eV and the ionization energy of K is 4.34 eV.   (The measured KI bond length is 3.23 Å.)   *Ans.*   3.12 Å

33.9.   The dipole moment of KI is 3.05×10^{-29} C·m.   Estimate the bond length of KI.   (Experimentally it is found that $r_0 = 3.23$ Å.)   *Ans.*   1.91 Å

33.10.   Suppose the two electrons in H_2 were located at a point midway between the two protons.   What must then be the separation between the protons to account for the observed binding energy of 4.5 eV?   *Ans.*   9.6 Å

33.11.   The two protons in a H_2^+ molecule are separated by 1.06 Å and the binding energy is 2.6 eV.   How much negative electric charge must be placed at a point midway between the two protons to be consistent with these values?   *Ans.*   0.298e

33.12.   Using the data in Problem 33.8, calculate the potential energy of KI at its equilibrium spacing.   *Ans.* -4.10 eV

Chapter 34

Excitations of Diatomic Molecules

In complex molecules one finds that besides the normal electron excitations the molecules can also possess rotational and vibrational motions. We shall restrict our discussion to diatomic (two-atom) molecules, which exhibit all the important features of molecular excitations without undue mathematical complexity.

34.1 MOLECULAR ROTATIONS

The rotational motion of a diatomic molecule can be considered as the planar rotation of a dumbbell about its center of mass (see Fig. 34-1). If the moment of inertia of the system about the center of mass is I, the kinetic energy E_r due to the rotational motion will be

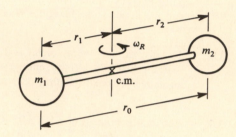

Fig. 34-1

$$E_r = \frac{1}{2} I \omega_r^2 = \frac{1}{2} \frac{(I\omega_r)^2}{I} = \frac{1}{2} \frac{L^2}{I} \qquad (34.1)$$

where ω_r is the rotational angular velocity and $\mathbf{L} = I\omega_r$ is the angular momentum of the system.

As with many-electron atoms, the angular momentum of molecules is quantized and can assume only one of the discrete values

$$|\mathbf{L}| = \sqrt{l(l+1)}\,\hbar \qquad l = 0, 1, 2, \ldots \qquad (34.2)$$

Thus the rotational kinetic energy given by (34.1) will be quantized, taking on only the values

$$E_r = \frac{l(l+1)\hbar^2}{2I} \qquad (34.3)$$

A typical rotational energy-level diagram is given in Fig. 34-2(a).

34.2 MOLECULAR VIBRATIONS

The chemical bonds holding a diatomic molecule together allow vibrational motions to occur in a manner similar to the vibrations of two masses at opposite ends of a spring. For sufficiently small energies the vibrational motions of molecules can be considered to be approximately simple harmonic. A quantum mechanical treatment shows that a harmonic oscillator can assume only discrete vibrational kinetic energies, E_v, with values

$$E_v = (n + \tfrac{1}{2})\hbar\omega_v \qquad n = 0, 1, 2, \ldots \qquad (34.4)$$

where ω_v is the vibrational angular frequency, which depends upon the restoring constant and the reduced mass of the system.

It is seen from (34.4) that the lowest vibrational energy is $E_0 = \frac{1}{2}\hbar\omega_v$. Thus at its lowest energy the molecule is not at rest but is performing some minimum vibration about its equilibrium point. Equation (34.4) predicts that the energy levels of a harmonic oscillator will be equally spaced in intervals of $\hbar\omega_v$. At higher energies, however, the simple harmonic oscillator approximation breaks

232

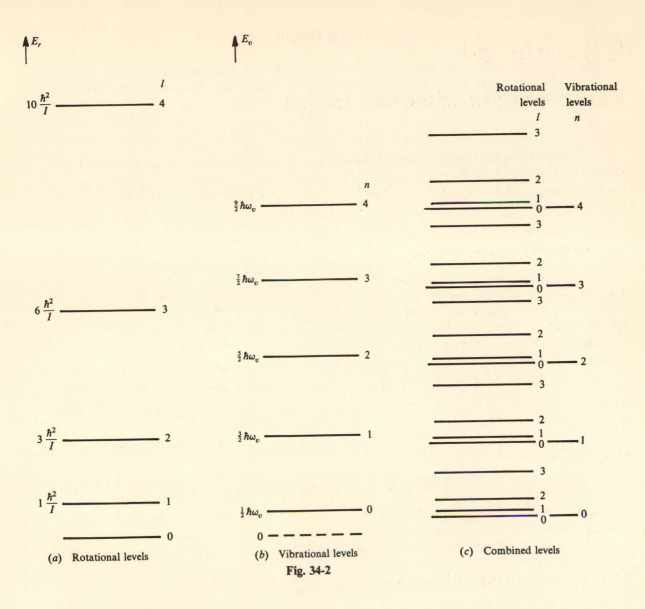

(a) Rotational levels   (b) Vibrational levels   (c) Combined levels

Fig. 34-2

down and it is found that the energy levels actually become more closely spaced than predicted by (34.4). An energy level diagram for a typical harmonic oscillator is shown in Fig. 34-2(b).

34.3 COMBINED EXCITATIONS

It is found that the vibrational energy level spacings are roughly 10–100 times the rotational energy level spacings (see Problem 34.7). In classical terms this means that there are many cycles of vibration during each rotation, so that the two motions can be treated separately. Figure 34-2(c) shows the small rotational energy levels superimposed on the larger vibrational energy levels.

Transitions between the levels can occur either by photon emission in molecular deexcitation or by photon absorption. In most cases at low energies, the transitions are subject to the selection rules

$$\Delta l = \pm 1 \qquad \Delta n = \pm 1$$

One of the uses of vibrational and rotational spectra is to obtain bond distances and bond stiffnesses of molecules (see Problems 34.4 and 34.8).

Solved Problems

34.1. Consider a diatomic molecule to be a dumbbell with masses m_1 and m_2 at the ends of a massless rod of length r_0. Show that the moment of inertia about the center of mass for an axis perpendicular to the dumbbell axis is

$$I = \frac{m_1 m_2}{m_1 + m_2} r_0^2 \equiv \mu r_0^2$$

where μ is the reduced mass (Section 19.3).

The moment of inertia of two point masses m_1 and m_2 about the given axis through the center of mass is

$$I = m_1 r_1^2 + m_2 r_2^2 \qquad (1)$$

(see Fig. 34-1). From the definition of the center of mass we obtain

$$r_1 = \frac{m_2}{m_1 + m_2} r_0 \qquad r_2 = \frac{m_1}{m_1 + m_2} r_0$$

and (1) becomes

$$I = m_1 \left(\frac{m_2 r_0}{m_1 + m_2} \right)^2 + m_2 \left(\frac{m_1 r_0}{m_1 + m_2} \right)^2 = \frac{m_1 m_2}{m_1 + m_2} r_0^2$$

34.2. Determine the interatomic spacing of O_2, given that $\hbar^2/2I = 1.78 \times 10^{-4}$ eV.

For an O_2 molecule the reduced mass is $m/2$, where $m = 16$ u is the mass of each O atom. Therefore, from Problem 34.1,

$$r_0 = \sqrt{\frac{I}{m/2}} = \sqrt{\frac{2Ic^2}{mc^2}} = \sqrt{\frac{\hbar^2 c^2}{(1.78 \times 10^{-4}\,\text{eV})mc^2}}$$

$$= \frac{\hbar c}{\sqrt{(1.78 \times 10^{-4}\,\text{eV})mc^2}} = \frac{1973\,\text{eV}\cdot\text{Å}}{\sqrt{(1.78 \times 10^{-4}\,\text{eV})(16\,\text{u})(931.5 \times 10^{6}\,\text{eV/u})}} = 1.21\,\text{Å}$$

34.3. Show that the rotational frequency spectrum of a diatomic molecule will consist of equally spaced lines separated by an amount $\Delta\nu = h/4\pi^2 I$, where I is the moment of inertia of the molecule.

From (34.3) the possible rotational energy values are

$$E_r = \frac{l(l+1)\hbar^2}{2I}$$

A transition between any two levels will result in the emission (or absorption) of a photon, whose frequency ν is found from

$$E_\gamma = h\nu = E_2 - E_1 = \frac{(h/2\pi)^2}{2I} \left[l_2(l_2 + 1) - l_1(l_1 + 1) \right]$$

$$\nu = \frac{h}{8\pi^2 I} \left[l_2(l_2 + 1) - l_1(l_1 + 1) \right]$$

Because of the selection rule $\Delta l = \pm 1$ we must have $l_2 = l_1 + 1$, so that

$$\nu = \frac{h}{8\pi^2 I} \left[(l_1 + 1)(l_1 + 1 + 1) - l_1(l_1 + 1) \right] = \frac{h}{4\pi^2 I} (l_1 + 1)$$

The frequencies of the emitted photons are determined from $l_1 = 0, 1, \ldots$, so that for adjacent lines the frequencies will be equally spaced by an amount

$$\Delta\nu = \frac{h}{4\pi^2 I}$$

34.4.   The frequency separation between adjacent lines in the rotational spectrum of $^{35}Cl^{19}F$ is measured as 11.2 GHz.   Determine the separation between the atoms.

From Problem 34.3,

$$I = \frac{h}{4\pi^2 \,\Delta\nu} = \frac{6.625 \times 10^{-34} \text{ J} \cdot \text{s}}{4\pi^2 (11.2 \times 10^9 \text{ s}^{-1})} = 1.50 \times 10^{-45} \text{ kg} \cdot \text{m}^2$$

Then, from Problem 34.1,

$$r_0 = \sqrt{\frac{(m_1 + m_2)I}{m_1 m_2}} = \sqrt{\frac{(35 \text{ u} + 19 \text{ u})(1.50 \times 10^{-45} \text{ kg} \cdot \text{m}^2)}{(35 \text{ u})(19 \text{ u})} \times \frac{1 \text{ u}}{1.66 \times 10^{-27} \text{ kg}}} = 2.71 \times 10^{-10} \text{ m}$$

or 2.71 Å.

34.5.   The interatomic spacing of a $^{12}C^{16}O$ molecule is 1.13 Å.   Determine the approximate wavelength separation between adjacent rotational lines arising from electronic transitions in the visible region (5000 Å).

From Problems 34.1 and 34.3,

$$\Delta\nu = \frac{h}{4\pi^2 I} = \frac{h}{4\pi\mu r_0^2}$$

and, for small changes,

$$\Delta\nu = \Delta\left(\frac{c}{\lambda}\right) = -\frac{c}{\lambda^2}\,\Delta\lambda$$

Hence,

$$|\Delta\lambda| = \frac{\lambda^2}{c}\frac{h}{4\pi^2\mu r_0^2} = \frac{\lambda^2 (hc)}{4\pi^2\left(\dfrac{m_1 m_2}{m_1 + m_2} c^2\right)r_0^2}$$

$$= \frac{(5000 \text{ Å})^2 (12.4 \text{ MeV} \cdot \text{Å})}{4\pi^2\left[\left(\dfrac{12 \cdot 16}{12 + 16} \text{ u}\right)(931.5 \text{ MeV/u})\right](1.13 \text{ Å})^2} = 0.962 \text{ Å}$$

34.6.   Determine the rotational energy levels for H_2, whose equilibrium spacing is 0.74 Å.

The rotational energy levels and moment of inertia of H_2 are given by

$$E_r = \frac{\hbar^2}{2I}\,l(l+1) \qquad I = \mu r_0^2 = \frac{m_H}{2}\,r_0^2$$

Therefore

$$E_r = \frac{\hbar^2}{m_H r_0^2}\,l(l+1) = \frac{(\hbar c)^2}{(m_H c^2)r_0^2}\,l(l+1)$$

$$= \frac{(1973 \text{ eV} \cdot \text{Å})^2}{(1.008 \text{ u})(931.5 \times 10^6 \text{ eV/u})(0.74 \text{ Å})^2}\,l(l+1) = (7.57 \times 10^{-3} \text{ eV})l(l+1)$$

The following table gives the first four rotational levels for H_2.

l	0	1	2	3	4
E_r, eV $\times 10^{-2}$	0	1.51	4.54	9.08	15.1

34.7. Estimate the vibrational energy-level spacing for H_2 ($r_0 = 0.74$ Å).

Consider the hydrogen nuclei to electrostatically repel and to be held together by a spring. If the spring is stretched an amount equal to the nuclear spacing, then the spring constant, K, is found from

$$F = \frac{ke^2}{r_0^2} = Kr_0 \qquad \text{or} \qquad K = \frac{ke^2}{r_0^3} = \frac{14.4 \text{ eV} \cdot \text{Å}}{(0.74 \text{ Å})^3} = 35.5 \frac{\text{eV}}{\text{Å}^2}$$

Relating K to the harmonic oscillator angular frequency gives

$$\hbar\omega_v = \hbar\sqrt{\frac{K}{\mu}} = \hbar c\sqrt{\frac{K}{\mu c^2}}$$

where μ, the reduced mass, is $m_H/2$. Therefore

$$\hbar\omega_v = \hbar c\sqrt{\frac{2K}{m_H c^2}} = (1973 \text{ eV} \cdot \text{Å})\sqrt{\frac{2 \times 35.5 \text{ eV/Å}^2}{938 \times 10^6 \text{ eV}}} = 0.543 \text{ eV}$$

which is the order of the vibrational energy-level spacing. From Problem 34.6 we see that the rotational energy-level spacings are a hundredth the vibrational energy-level spacings. For this reason one can regard rotational states as being built upon vibrational states, as shown in Fig. 34-2(c).

34.8. Infrared radiation of wavelength 3.465 μm is strongly absorbed by HCl gas. What is the force constant for a HCl molecule?

If the HCl molecule is treated like a quantized harmonic oscillator, it can assume only energies

$$E_v = \left(n + \tfrac{1}{2}\right)\hbar\omega_v$$

The absorbed infrared radiation will produce an increase in the energy of the harmonic oscillator. Because of the selection rule $\Delta n = \pm 1$, this increase will be

$$\Delta E_v = \left(n + 1 + \tfrac{1}{2}\right)\hbar\omega_v - \left(n + \tfrac{1}{2}\right)\hbar\omega_v = \hbar\omega_v = \frac{h}{2\pi}\omega_v$$

Setting this increase equal to the energy of the photon, we have

$$h\frac{c}{\lambda} = \frac{h}{2\pi}\omega_v$$

$$\omega_v = 2\pi\frac{c}{\lambda} = \frac{2\pi(3 \times 10^8 \text{ m/s})}{3.465 \times 10^{-6} \text{ m}} = 5.44 \times 10^{14} \text{ Hz}$$

The reduced mass of an HCl molecule is

$$\mu = \frac{m_H m_{Cl}}{m_H + m_{Cl}} = \frac{(1 \text{ u})(35 \text{ u})}{1 \text{ u} + 35 \text{ u}}\left(1.661 \times 10^{-27} \frac{\text{kg}}{\text{u}}\right) = 1.61 \times 10^{-27} \text{ kg}$$

The angular frequency of a harmonic oscillator is related to its reduced mass and force constant by $\omega_v = \sqrt{K/\mu}$; hence,

$$K = \mu\omega_v^2 = (1.61 \times 10^{-27} \text{ kg})(5.44 \times 10^{14} \text{ s}^{-1})^2 = 476 \text{ N/m}$$

34.9. Refer to Problem 34.8. Determine the total vibrational energy of one mole of HCl at absolute zero temperature.

At 0 K all the HCl molecules will be at the lowest possible energy, corresponding to $n = 0$ in (34.4). Thus, using Avogadro's number, N_0,

$$E_{\text{total}} = N_0\left(\tfrac{1}{2}\hbar\omega_v\right) = (6.023 \times 10^{23} \text{ mol}^{-1})\tfrac{1}{2}(1.055 \times 10^{-34} \text{ J} \cdot \text{s})(5.44 \times 10^{14} \text{ s}^{-1})$$

$$= 17.3 \text{ kJ/mol}$$

34.10. Molecules of N_2 are excited into the $n = 1$ vibrational levels and deexcite through the emission of photons. What are the energies of the emitted photons? (Consider only the first five rotational levels for each vibrational level.) For N_2, $\hbar^2/2I = 2.5 \times 10^{-4}$ eV, $\hbar\omega_v = 0.29$ eV.

The energy levels are given by

$$E = \hbar\omega_v\left(n + \tfrac{1}{2}\right) + \frac{\hbar^2}{2I}l(l+1)$$

and are shown in Fig. 34-3. The transitions all obey $\Delta n = -1$; they are given by

$$\Delta E = \hbar\omega_v + \frac{\hbar^2}{2I}[l'(l'+1) - l(l+1)]$$

From the rotational selection rule $\Delta l = +1$ we must have $l' = l + 1$, so that

$$\Delta E = \hbar\omega_v + \frac{\hbar^2}{2I}[2(l+1)] \qquad l = 0, 1, 2, \ldots$$

and from the selection rule $\Delta l = -1$ we must have $l' = l - 1$, so that

$$\Delta E = \hbar\omega_v + \frac{\hbar^2}{2I}[-2l] \qquad l = 1, 2, 3, \ldots$$

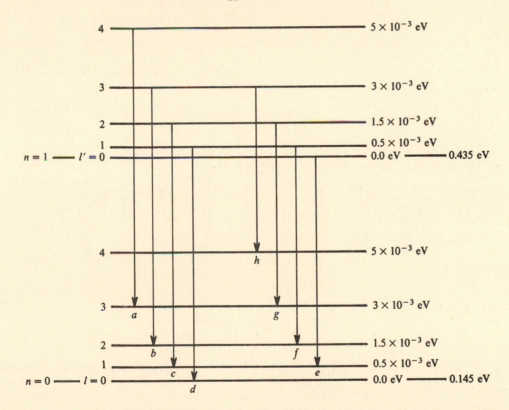

Fig. 34-3

These transitions are summarized in Table 34-1.   It is seen that the energies are separated by an amount $\hbar^2/I = 0.5 \times 10^{-3}$ eV, except that there is no line occurring at $\Delta E = \hbar\omega_v = 0.29$ eV.

Table 34-1

Transition	l	ΔE, eV
a	3	$0.29 + 2.0 \times 10^{-3}$
b	2	$0.29 + 1.5 \times 10^{-3}$
c	1	$0.29 + 1.0 \times 10^{-3}$
d	0	$0.29 + 0.5 \times 10^{-3}$
e	1	$0.29 - 0.5 \times 10^{-3}$
f	2	$0.29 - 1.0 \times 10^{-3}$
g	3	$0.29 - 1.5 \times 10^{-3}$
h	4	$0.29 - 2.0 \times 10^{-3}$

34.11. The absorption spectrum for a certain gas at room temperature is shown in Fig. 34-4. Determine the zero-point vibration energy of a molecule and the moment of inertia of a molecule.

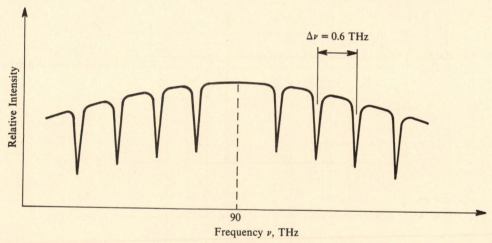

Fig. 34-4

The absorption spectrum shown is a combination vibration-rotation spectrum. Absorption of photons by the molecules of the gas produce the excitations shown in Fig. 34-5. At room temperature ($T \approx 300$ K) the molecules will not have enough energy to occupy vibrational bands higher than $n = 0$. However, they will possess sufficient thermal energy to be excited into the various rotational levels of the $n = 0$ vibrational band.   Absorption of photons then produces excitations into the $n = 1$ band as shown in Fig. 34-5.   The selection rule $\Delta n = \pm 1$ is satisfied.   Because of the selection rule $\Delta l = \pm 1$, the transitions occur in two groups, $l \to l + 1$ and $l \to l - 1$, as shown.   The energy difference between final and initial levels, which is the sum of the vibrational energy difference and the rotational energy difference, will be equal to the energy $h\nu$ of the absorbed photon:

$$h\nu = \frac{h}{2\pi}\,\omega_v + [l_u(l_u + 1) - l_l(l_l + 1)]\frac{1}{2I}\left(\frac{h}{2\pi}\right)^2$$

or

$$\nu = \frac{\omega_v}{2\pi} + [l_u(l_u + 1) - l_l(l_l + 1)]\frac{h}{8\pi^2 I}$$

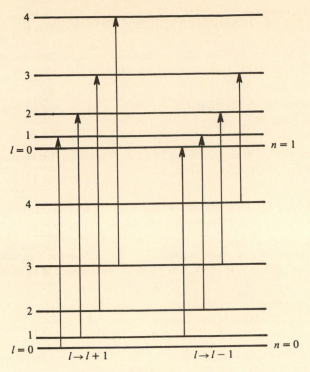

Fig. 34-5

For the two groups of transitions we have

$$l_u = l_l + 1: \qquad \nu_1 = \frac{\omega_v}{2\pi} + (l_l + 1)\frac{h}{4\pi^2 I} \qquad\qquad l_l = 0, 1, 2, \ldots$$

$$l_u = l_l - 1: \qquad \nu_2 = \frac{\omega_v}{2\pi} - l_l\frac{h}{4\pi^2 I} \qquad\qquad l_l = 1, 2, 3, \ldots$$

The frequencies of the absorbed photons are thus seen to increase in steps of $h/4\pi^2 I$ from

$$\frac{\omega_v}{2\pi} - 4\frac{h}{4\pi^2 I} \quad (l_l = 4 \text{ in } \nu_2) \qquad \text{to} \qquad \frac{\omega_v}{2\pi} + 4\frac{h}{4\pi^2 I} \quad (l_l = 3 \text{ in } \nu_1)$$

with the exception of a gap at $\nu_{\text{gap}} = \omega_v/2\pi$. Hence by measuring the frequency difference $\Delta\nu$ between adjacent levels one can find the moment of inertia from

$$\Delta\nu = \frac{h}{4\pi^2 I}$$

$$I = \frac{h}{4\pi^2\,\Delta\nu} = \frac{6.63 \times 10^{-34}\,\text{J}\cdot\text{s}}{4\pi^2(0.6 \times 10^{12}\,\text{s}^{-1})} = 2.80 \times 10^{-47}\,\text{kg}\cdot\text{m}^2$$

Knowing the vibrational frequency ($\nu_{\text{gap}} = 90 \times 10^{12}\,\text{s}^{-1}$), we have for the zero-point vibration energy:

$$E_0 = \tfrac{1}{2}\hbar\omega_v = \tfrac{1}{2}h\nu_{\text{gap}} = \tfrac{1}{2}(4.136 \times 10^{-15}\,\text{eV}\cdot\text{s})(90 \times 10^{12}\,\text{s}^{-1}) = 0.186\,\text{eV}$$

Supplementary Problems

34.12. The frequency separation between adjacent lines in the rotational spectrum of a certain molecule is measured as 40 GHz. What is the moment of inertia of the molecule?   *Ans.* 4.20×10^{-46} kg $\cdot$ m^2

34.13. The interatomic spacing of ^{79}Br^{19}F is 1.76 Å. What is its moment of inertia?
Ans. 7.88×10^{-46} kg $\cdot$ m^2

34.14. Refer to Problem 34.13. What will be the frequency separation between adjacent lines in the rotational spectrum of ^{79}Br^{19}F?   *Ans.* 21.3 GHz

34.15. For $n = 0$ determine the ratio of the wavelengths of the photons emitted in the $l = 2 \rightarrow l = 1$ and $l = 1 \rightarrow l = 0$ transitions in a molecule.   *Ans.* 1/2

34.16. The energy difference of the photons emitted in the $l = 0 \rightarrow l = 1$ and the $l = 1 \rightarrow l = 2$ transitions in ^{11}B^{16}O is measured as 4.46×10^{-4} eV. What is the interatomic spacing of BO?   *Ans.* 1.20 Å

34.17. The wavelength separation in the rotational spectrum of a certain molecule is measured as 3.62 Å in the visible region (5000 Å). What is the moment of inertia of the molecule?   *Ans.* 3.86×10^{-47} kg $\cdot$ m^2

34.18. Calculate the first three rotational energy levels of ^{79}Br^{19}F ($r_0 = 1.76$ Å).
Ans. 0 eV; 8.81×10^{-5} eV; 26.4×10^{-5} eV; 52.8×10^{-5} eV

34.19. Refer to Problem 34.18. What photon wavelengths would produce the rotational transitions $l = 0 \rightarrow l = 1$ and $l = 1 \rightarrow l = 2$ in ^{79}Br^{19}F?   *Ans.* 1.41 cm; 0.705 cm

34.20. What is the vibrational energy-level spacing of ^{79}Br^{19}F ($r_0 = 1.76$ Å)? (Compare the magnitude of these energy levels with the magnitude of the rotational energy levels found in Problem 34.18.)
Ans. 2.68×10^{-2} eV

Chapter 35

Kinetic Theory

In the kinetic theory of gases the classical laws of mechanics, applied to a system containing a large number of particles, are used to derive various thermodynamic relations—in particular, the *ideal gas law*.

35.1 THE IDEAL GAS LAW

In an *ideal gas* the pressure of the gas is assumed to arise from perfectly elastic collisions of molecules with the walls of the container. Problem 35.1 shows how a classical analysis of the collision process leads to an expression for the pressure p exerted on the walls of a container of volume V, containing N molecules each with mass m, of the form

$$pV = \frac{2}{3} N \left[\tfrac{1}{2} m (v^2)_{\text{avg}} \right] \tag{35.1}$$

Here the quantity $(v^2)_{\text{avg}}$ is defined as follows. Let the total number of particles in the container with an x-component of velocity of magnitude v_{xi} be designated by n_i. The average value of the square of the x-velocity components of the N molecules in the container is

$$(v_x^2)_{\text{avg}} \equiv \frac{1}{N} \left[n_1 v_{x1}^2 + n_2 v_{x2}^2 + \cdots + n_i v_{xi}^2 + \cdots \right] = \frac{1}{N} \sum_i n_i v_{xi}^2 \tag{35.2}$$

where $N = \sum_i n_i$. If it is further assumed that there is no difference between the x-, y-, and z-directions, we can also write

$$(v_x^2)_{\text{avg}} = (v_y^2)_{\text{avg}} = (v_z^2)_{\text{avg}} = \frac{1}{3} (v^2)_{\text{avg}} \tag{35.3}$$

where $(v^2)_{\text{avg}} = (v_x^2)_{\text{avg}} + (v_y^2)_{\text{avg}} + (v_z^2)_{\text{avg}}$ is the average value of the square of the speed. The square root of $(v^2)_{\text{avg}}$ is called the *root-mean-square* speed:

$$v_{\text{rms}} \equiv \sqrt{(v^2)_{\text{avg}}} \tag{35.4}$$

The quantity $\frac{1}{2} m (v^2)_{\text{avg}} = K_{\text{avg}}$ is the average kinetic energy per molecule; therefore

$$U = N \left[\tfrac{1}{2} m (v^2)_{\text{avg}} \right] \tag{35.5}$$

represents the total internal kinetic energy of the gas, and (*35.1*) can be written as

$$pV = \frac{2}{3} U \tag{35.6}$$

This result can be carried one step further by using the relationship between the total number of particles, N, and the total number of moles, $\mathfrak{N}$: $N = \mathfrak{N} N_0$, where N_0 is Avogadro's number,

$$N_0 = 6.023 \times 10^{23} \text{ molecules/gm mole}$$

(When expressing molecular weights in kilograms, we shall use $N_0 = 6.023 \times 10^{26}$ kmol^{-1}.) Then (*35.1*) becomes

$$pV = \mathfrak{N} \left[\frac{2}{3} N_0 \, \tfrac{1}{2} m (v^2)_{\text{avg}} \right]$$

If this expression is compared with the ideal gas equation,

$$pV = \mathscr{N}RT \qquad (35.7)$$

it is seen that

$$\tfrac{1}{2}m(v^2)_{\text{avg}} = \frac{3}{2}\frac{R}{N_0}T = \frac{3}{2}kT \qquad (35.8)$$

where the ratio $R/N_0 = 1.38 \times 10^{-23}$ J/K $= 8.617 \times 10^{-5}$ eV/K is called the *Boltzmann constant*, k. Thus the absolute temperature T of an ideal gas is a measure of the average kinetic energy of the molecules composing the gas.

Suppose, however, that we did not know the ideal gas equation, (35.7). Then, in order to proceed further with (35.1), we would have to be able to evaluate

$$(v^2)_{\text{avg}} = \frac{1}{N}\sum_i n_i v_i^2$$

Methods for finding such averages involve distribution functions and will be treated in Chapters 36, 37, and 38.

Solved Problems

35.1.   Assume that a gas is composed of point molecules that make perfectly elastic collisions with the walls of its container. Show then that the pressure exerted on the walls of a rectangular box of volume $V = abc$, containing N identical molecules each of mass m, is given by

$$p = \frac{2}{3}\frac{N}{V}\left[\tfrac{1}{2}m(v^2)_{\text{avg}}\right]$$

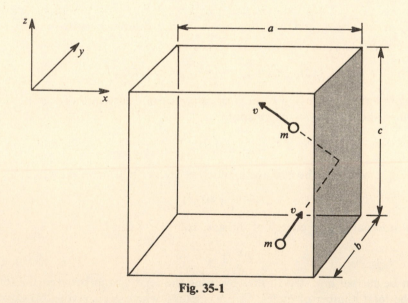

Fig. 35-1

Consider first the collision of one molecule with a wall (Fig. 35-1). In an elastic collision the x-component of the particle's velocity will be changed from $+v_x$ to $-v_x$. The corresponding change in the particle's x-component of momentum is then

$$\Delta p_x = m(-v_x) - m(v_x) = -2mv_x$$

and the collision results in a small force to the left being exerted on the molecule by the wall. The reaction (Newton's third law) is a small force to the right on the wall, i.e. pressure.

Suppose now that instead of one particle there are in the box n_i particles, all with the same x-velocity magnitude, v_{xi}.   In a small time interval dt, the wall of area $A = bc$ will be struck by all those particles within a small volume $A(v_{xi}\, dt)$ to the left of the wall that are moving with x-velocity $+v_{xi}$. Thus, the number of collisions in time dt is

$$\frac{1}{2}\left(\frac{n_i}{V}\right)A(v_{xi}\, dt)$$

the factor $1/2$ occurring because half the particles are moving away from the wall.   The total change in x-momentum, dp_{xi}, as a result of the collisions is

total change in momentum = (change in momentum per particle)

$\times$ (number of particles striking the wall)

$$dp_{xi} = (-2mv_{xi})\left[\frac{1}{2}\left(\frac{n_i}{V}\right)A(v_{xi}\, dt)\right] = -\frac{n_i m A v_{xi}^2\, dt}{V}$$

so that, by Newton's second law, the net force on the particles is

$$F_{xi} = \frac{dp_{xi}}{dt} = -\frac{n_i m A v_{xi}^2}{V}$$

and the force on the wall is $-F_{xi}$.

In the above argument we neglected the effects of any collisions between the particles.   The time interval dt, however, can be made small enough so that the number of collisions will be negligible.   Also we shall soon see that in a real gas in equilibrium, collisions are really not a problem because, on the average, every particle whose velocity is changed by a collision is replaced by another particle which, because of a collision, has acquired the first particle's velocity.

Of course, all the N molecules in a gas will not have the same magnitude of x-component of velocity.   Instead, there is a spread or distribution in the x-velocity component among all the molecules. Therefore, to find the total force exerted on the area A we must add the contributions from all groups of velocities, obtaining

$$F_x = \sum_i (-F_{xi}) = \frac{mA}{V}\sum_i n_i v_{xi}^2 = \frac{mA}{V}\left[N(v_x^2)_{\text{avg}}\right] = \frac{NmA(v^2)_{\text{avg}}}{3V}$$

where we have used (35.2) and (35.3).   The pressure p is then

$$p = \frac{F_x}{A} = \frac{Nm(v^2)_{\text{avg}}}{3V} = \frac{2}{3}\frac{N}{V}\left[\tfrac{1}{2}m(v^2)_{\text{avg}}\right]$$

Note that the expression for p does not depend on the particular wall that we chose to consider.

35.2.   What is the average kinetic energy of a molecule at room temperature (300 K)?

$$K_{\text{avg}} = \frac{3}{2}kT = \frac{3}{2}\left(8.62\times10^{-5}\,\frac{\text{eV}}{\text{K}}\right)(300\text{ K}) = 3.88\times10^{-2}\text{ eV}$$

It is useful to remember that this number is about $1/25$ eV.

35.3.   If the molecule in Problem 35.2 is O_2, calculate the rms velocity of the molecule.

$$K_{\text{avg}} = \tfrac{1}{2}m(v^2)_{\text{avg}} = \tfrac{1}{2}mv_{\text{rms}}^2$$

$$v_{\text{rms}} = \sqrt{\frac{2K_{\text{avg}}}{m}} = \sqrt{\frac{2(3.88\times10^{-2}\text{ eV})(1.6\times10^{-19}\text{ J/eV})}{(32\text{ u})(1.66\times10^{-27}\text{ kg/u})}} = 484\text{ m/s}$$

35.4.   The *law of atmospheres* states that at constant temperature the pressure p at a distance z above the surface of the earth will be related to the pressure p_0 at the surface by

$$p = p_0 e^{-mgz/kT} \qquad\qquad (1)$$

Here, m is an average molecular mass.   Show how this result follows from the ideal gas law.

Consider the small volume of gas, $dV = A\,dz$, shown in Fig. 35-2. Since the volume is in equilibrium, its weight, $\rho g\,dV$, must equal the difference between the pressure forces on the horizontal faces:

$$\rho g\,dV = [p(z) - p(z + dz)]A$$
$$\rho g A\,dz = -A\,dp$$
$$dp = -\rho g\,dz$$

From the ideal gas law, $pV = \mathfrak{N}RT = NkT$, we have for the density:

$$\rho = \frac{Nm}{V} = \frac{mp}{kT}$$

Hence

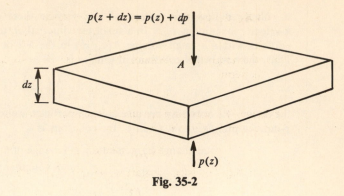

Fig. 35-2

$$dp = -\frac{mp}{kT}\,g\,dz \qquad \text{or} \qquad \frac{dp}{p} = -\frac{mg}{kT}\,dz$$

Integrating and applying the boundary condition $p(0) = p_0$, we obtain (*1*).

35.5. The *mean free path*, L, of a molecule is the average distance it travels between collisions. Estimate the mean free path of a molecule in a gas at standard conditions if the molecular diameter is 4 Å.

Consider a spherical molecule of radius r moving at a constant average speed v, and assume all other molecules in the gas remain fixed. The molecule will collide with any other molecule whose center is within a distance $2r = d$ of its center. Thus, after a time t, it will have collided with all molecules in a zigzag cylindrical volume of cross-sectional area πd^2 and length ut. The number of collisions is then

$$C = n\pi d^2 ut$$

where n is the number of molecules per unit volume. The total distance traveled, divided by the number of collisions, gives the mean distance between collisions:

$$L = \frac{ut}{C} = \frac{1}{n\pi d^2}$$

At standard conditions one kilomole of an ideal gas occupies 22.4 m³, so that

$$n = \frac{6.023 \times 10^{26} \text{ molecules/kmol}}{22.4 \text{ m}^3/\text{kmol}} = 2.69 \times 10^{25} \text{ m}^{-3}$$

and

$$L = \frac{1}{(2.69 \times 10^{25} \text{ m}^{-3})\pi(4 \times 10^{-10} \text{ m})^2} = 7.40 \times 10^{-8} \text{ m} = 740 \text{ Å}$$

The above calculation of L was rough. An exact treatment, involving the Maxwell-Boltzmann distribution (Chapter 37), gives

$$L = \frac{0.707}{n\pi d^2}$$

35.6. A particle suspended in a fluid is acted on by random irregular forces. Show that the average value of the square of the particle's absolute displacement in a time interval t is proportional to t. Assume the time between random forces is much less than the time t, and that each time a force acts on the particle it moves a distance equal to its mean free path L in a random direction. (The situation described here is the classical *random walk*.)

After N interactions we can write

$$\mathbf{X}_N = \mathbf{X}_{N-1} + \mathbf{L}$$

with X_N the position vector after N interactions. Taking the dot product of X_N with itself, we obtain

$$\mathbf{X}_N \cdot \mathbf{X}_N = X_N^2 = X_{N-1}^2 + L^2 + 2\mathbf{L} \cdot \mathbf{X}_{N-1}$$

If we average this expression for X_N^2 over all possible directions of $\mathbf{L}$, the term

$$2\mathbf{L} \cdot \mathbf{X}_{N-1} = 2L X_{N-1} \cos\theta$$

averages to zero because all directions of $\mathbf{L}$ relative to $\mathbf{X}_{N-1}$ are equally probable. Thus

$$\left(X_N^2\right)_{\text{avg}} = \left(X_{N-1}^2\right)_{\text{avg}} + L^2$$

which, together with the starting value $(X_0^2)_{\text{avg}} = 0$, implies that

$$\left(X_N^2\right)_{\text{avg}} = NL^2$$

It is seen that $(X_N^2)_{\text{avg}}$ is directly proportional to the number of collisions, N, which in turn is directly proportional to the time t:

$$\left(X_N^2\right)_{\text{avg}} = NL^2 = \alpha t$$

The quantity $(X^2)_{\text{avg}}$ is experimentally measurable from the observation of a single suspended particle over a large number of time intervals of length t (Fig. 35-3); or, equivalently, from the observation of r suspended particles over a single time interval of length t.

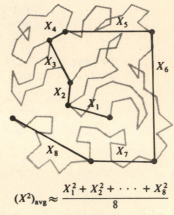

$$(X^2)_{\text{avg}} \approx \frac{X_1^2 + X_2^2 + \cdots + X_8^2}{8}$$

Fig. 35-3

35.7.   Suppose the molecule of Problem 35.5 has an average speed of 454 m/s and moves randomly a distance equal to its mean free path every time it undergoes a collision. What would be its root-mean-square displacement after 10 seconds?

From the definition of mean free path, the number of collisions, N, will be equal to the distance vt traveled in a time t, divided by the mean free path, L: $N = vt/L$. Substituting this into the result of Problem 35.6, we obtain

$$X_{\text{rms}} = N^{1/2}L = \sqrt{vtL} = \sqrt{(454 \text{ m/s})(10 \text{ s})(7.40 \times 10^{-8} \text{ m})} = 1.83 \times 10^{-2} \text{ m} = 1.83 \text{ cm}$$

35.8.   *Brownian motion* occurs when a particle suspended in a fluid undergoes motion as it is randomly bombarded by the molecules composing the fluid. Starting from Newton's laws, show that the mean-square displacement of the particle after a time t will be of the form

$$\left(X^2\right)_{\text{avg}} = \frac{6kT}{\mu} t$$

where μ is a measure of the viscous force exerted on the particle by the fluid.

Consider first one-dimensional motion. In addition to other forces, the particle will experience a viscous drag force proportional to the particle's velocity: $F_{\text{vis}} = -\mu(dx/dt)$. The quantity μ is to be thought of as a given number, because it can be directly measured (for example, from observations of the particle's terminal velocity under the influence of a known external force).

Suppose that in addition to the viscous drag force, the particle also experiences a random, fluctuating external force, F_r. From Newton's second law we then obtain (neglecting the weight of the particle)

$$F_r - \mu \frac{dx}{dt} = m \frac{d^2x}{dt^2}$$

To evaluate $(x^2)_{\text{avg}}$ we proceed as follows: First we multiply the above equation by x and average, to obtain

$$(xF_r)_{\text{avg}} = m\left(x \frac{d^2x}{dt^2}\right)_{\text{avg}} + \mu\left(x \frac{dx}{dt}\right)_{\text{avg}} \qquad (1)$$

Now F_r, being completely random, is uncorrelated with the particle's position x. For a given x, therefore, all values of F_r, both positive and negative, are equally likely, so that the average value of x times F_r will be zero; $(xF_r)_{avg} = 0$. Moreover, it follows from the identity

$$x \frac{d^2x}{dt^2} = \frac{d}{dt}\left(x \frac{dx}{dt}\right) - \left(\frac{dx}{dt}\right)^2$$

that

$$\left(x \frac{d^2x}{dt^2}\right)_{avg} = \left[\frac{d}{dt}\left(x \frac{dx}{dt}\right)\right]_{avg} - \left[\left(\frac{dx}{dt}\right)^2\right]_{avg} = \frac{d}{dt}\left(x \frac{dx}{dt}\right)_{avg} - \left[\left(\frac{dx}{dt}\right)^2\right]_{avg}$$

since the averaging is not over time, but over the random distribution of F_r. Substituting these results into (1), we obtain

$$0 = m \frac{d}{dt}\left[\left(x \frac{dx}{dt}\right)_{avg}\right] - m\left[\left(\frac{dx}{dt}\right)^2\right]_{avg} + \mu\left(x \frac{dx}{dt}\right)_{avg} \tag{2}$$

The middle term of this expression can be written as

$$m\left[\left(\frac{dx}{dt}\right)^2\right]_{avg} = m(v_x^2)_{avg} = m\frac{1}{3}(v^2)_{avg} = \frac{2}{3}\left[\frac{1}{2}m(v^2)_{avg}\right] = kT$$

using (35.8); hence (2) becomes

$$kT = m \frac{d}{dt}\left[\left(x \frac{dx}{dt}\right)_{avg}\right] + \mu\left(x \frac{dx}{dt}\right)_{avg} \tag{3}$$

If we define

$$f \equiv \left(x \frac{dx}{dt}\right)_{avg}$$

(3) then yields the following differential equation for f:

$$m \frac{df}{dt} + \mu f = kT \tag{4}$$

By direct substitution the solution of (4) can be verified to be

$$f = \frac{kT}{\mu} + Ae^{-\frac{\mu}{m}t} \tag{5}$$

For the small masses used in Brownian motion experiments the quantity μ/m is of the order of 10^4 s^{-1}, so the exponential term in (5) quickly damps out. If we assume this term to be essentially zero we have

$$f = \left(x \frac{dx}{dt}\right)_{avg} = \frac{kT}{\mu}$$

This expression can be rewritten in the following manner:

$$\left(x \frac{dx}{dt}\right)_{avg} = \left[\frac{1}{2}\frac{d(x^2)}{dt}\right]_{avg} = \frac{1}{2}\frac{d}{dt}\left[(x^2)_{avg}\right] = \frac{kT}{\mu}$$

so that

$$(x^2)_{avg} = \frac{2kT}{\mu}t$$

where it has been assumed that $(x^2)_{avg} = 0$ at $t = 0$. Finally, one can go over to three dimensions by using

$$(x^2)_{avg} = (y^2)_{avg} = (z^2)_{avg} = \frac{1}{3}(X^2)_{avg}$$

to obtain

$$(X^2)_{avg} = \frac{6kT}{\mu}t$$

This is the same expression as obtained in Problem 35.6, with $\alpha = 6kT/\mu$.

Since μ is known, by measuring $(X^2)_{avg}$ for a fixed time interval t and temperature T it is possible to determine Boltzmann's constant k. The study of Brownian motion was first used by J. Perrin to obtain a value for Avogadro's number N_0, where $N_0 = R/k$.

35.9. Two identical containers, A and B, are filled with the same gas at the same pressure. However, the rms speed, v_B, of the molecules in B is greater than the rms speed, v_A, of the molecules in A. A stopcock between the containers is now opened for a short time. Describe from kinetic theory what happens.

At first one might think that with the pressures equal, opening the stopcock changes nothing. This, however, is not the case. The number of collisions per unit time with the wall, $\dot{N}$, which is proportional to the number of particles per unit time to move through the stopcock once opened, was shown in Problem 35.1 to be proportional to the product nv, where n is the number of particles per unit volume. Therefore,

$$\frac{\dot{N}_A}{\dot{N}_B} = \frac{n_A v_A}{n_B v_B}$$

Initially, $p_A = p_B$, and (35.1) gives $n_A v_A^2 = n_B v_B^2$; hence,

$$\frac{\dot{N}_A}{\dot{N}_B} = \frac{v_B}{v_A} > 1$$

and more particles initially move from A to B than from B to A. Furthermore, the energy flow through the stopcock is given by

$$\dot{E} \propto \dot{N} \times (\text{average energy per particle}) = \dot{N}\left(\tfrac{1}{2}mv^2\right)$$

so that, again using (35.1),

$$\frac{\dot{E}_B}{\dot{E}_A} = \frac{\dot{N}_B}{\dot{N}_A} \frac{\left(\tfrac{1}{2}mv^2\right)_B}{\left(\tfrac{1}{2}mv^2\right)_A} = \frac{n_B v_B}{n_A v_A}\frac{n_A}{n_B} = \frac{v_B}{v_A} > 1$$

and energy initially flows from container B to container A.

Consequently, the system is not in equilibrium with respect to either mass or energy under the initial conditions. In order to have an equilibrium situation, i.e. for $\dot{N}_A = \dot{N}_B$ and $\dot{E}_A = \dot{E}_B$, the two rms velocities must be equal, $v_A = v_B$. Since the temperature is proportional to the square of the rms velocity, it is seen that for an equilibrium situation the temperatures of both gases must be equal.

35.10. Refer to Problem 35.9. Suppose instead of having a stopcock, the two vessels are separated by a wall which, although restrained from moving an appreciable distance, can vibrate without friction between the two vessels. Discuss what will happen under the initial conditions given in Problem 35.9.

Initially,

$$\frac{\dot{E}_B}{\dot{E}_A} = \frac{v_B}{v_A} > 1$$

which shows that the wall is being struck more violently on side B than on side A. As a result, the wall vibrates back and forth in such a manner that the molecules on side B lose energy, while the molecules on side A gain energy. Equilibrium is reached when $\dot{E}_B = \dot{E}_A$, and the mechanical motion of the wall is transmitting as much energy from B to A as from A to B. This mechanical transmission of energy by the wall corresponds to "heat flow" between the two containers. Equilibrium is reached when there is no net flow of heat, and this occurs when a single temperature (rms velocity) is attained throughout A and B.

Supplementary Problems

35.11. Containers of O_2, CO_2, F_2 and He are all at the same temperature. Find the ratios of their rms velocities to the rms velocity of H_2 at the same temperature. *Ans.* 0.354; 0.267; 0.324; 0.707

35.12. From kinetic theory show that $(v_{rms})^2 = 3p/\rho$, where ρ is the density of the ideal gas.

35.13. Let w be any physical quantity that has an average value. From the identity

$$(w - w_{avg})^2 = w^2 - 2ww_{avg} + (w_{avg})^2$$

show that

$$w_{rms} \geq w_{avg}$$

35.14. A man opens a bank account with initial balance zero. Each morning for a year he tosses a coin: if the result is heads, he deposits \$1; if tails, he withdraws \$1 (or borrows \$1 from the bank). (*a*) What are absolute limits for his final balance? (*b*) What is his expected final balance? (*c*) What are likely (rms) limits for his final balance? *Ans.* (*a*) ±\$365; (*b*) 0; (*c*) ±\$19.10

35.15. Find the height above the earth's surface where the pressure will have dropped to one-half the value at the earth's surface. Assume constant temperature of 273 K and take the average mass of an air molecule as 48.5×10^{-27} kg. *Ans.* 5.49 km

35.16. How many collisions per second are made by the molecule of Problem 35.5 if its average speed is 454 m/s? *Ans.* 6.14×10^9

Chapter 36

Distribution Functions

It was found in Chapter 35 that in an ideal gas of N molecules, each of mass m, the pressure p is related to the volume V by

$$pV = \frac{2}{3} N \left[\tfrac{1}{2} m (v^2)_{\text{avg}} \right]$$

where the average value of the square of the velocity was given by

$$(v^2)_{\text{avg}} = \frac{1}{N} \sum_i v_i^2 n_i \qquad (36.1)$$

In the present chapter we shall describe the methods used for calculating the average value of a physical quantity. Many times we shall, for concreteness, be referring to the specific quantity v^2, but the methods will be general.

36.1 DISCRETE DISTRIBUTION FUNCTIONS

To determine the average value of a physical quantity in a system we must first identify the underlying variables on which the quantity depends. As an example, suppose we wish to determine the average value of the kinetic energy, or, equivalently, the average value of the square of the speed. For the variable on which the energy depends we can take the speed v of a particle in the system. Let the equally spaced points $v_1, v_2, v_3, \ldots$ divide the total range of v into intervals of length Δv (Fig. 36-1). Out of the N particles in the system there will, under equilibrium conditions, be a definite number possessing speeds lying within each interval. Let n_i be the number of particles having speeds between v_i and $v_i + \Delta v = v_{i+1}$. To represent this pictorially we could plot n_i versus v. For reasons that will soon be apparent, however, we instead divide n_i by the range Δv and plot the ratio $f_i \equiv n_i / \Delta v$ to obtain the histogram shown in Fig. 36-1. Since the ordinate f_i is the number of particles per unit speed, the number of particles n_i is then equal to the *area* of the ith rectangle:

$$n_i = f_i \, \Delta v \qquad (36.2)$$

Since the total number N of particles must equal the sum of all the n_i, we have the *normalization condition* that the sum of the areas of all the rectangles, the total area under the histogram, gives the total number of particles:

$$N = \sum_i f_i \, \Delta v \qquad (36.3)$$

More generally, the number of particles having speeds between v_α and v_β is just the area under the histogram between those two values.

The histogram of f_i's is called the *distribution function* for the particle speeds. (In statistical work the numbers $f_i \, \Delta v$ are called *frequencies*, and one speaks of a *frequency distribution*.) In terms of the distribution function, (36.1) can be written

$$(v^2)_{\text{avg}} = \frac{1}{N} \sum_i v_i^2 f_i \, \Delta v \qquad (36.4)$$

In a similar manner the average value of any quantity that depends upon the speed can be written in terms of the speed distribution function f_i.

It should be noted that summations like that in (36.4) do *not* in general extend from 1 to N. The summation depends upon the number of intervals chosen, not the number of particles.

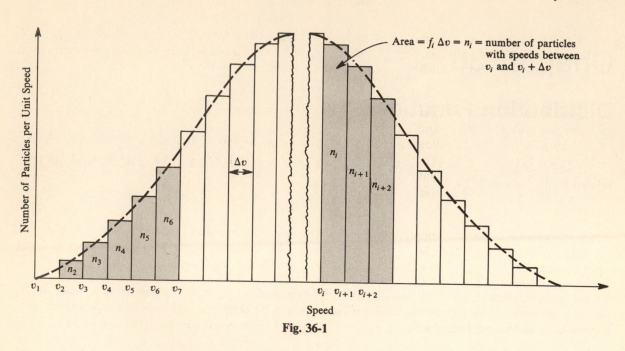

Fig. 36-1

36.2 CONTINUOUS DISTRIBUTION FUNCTIONS

In many cases of interest the total number of particles, N, is very large, and there are many collisions per unit time, so that the speeds can be treated as if they varied continuously. For this situation one allows the speed range Δv to approach zero; $\Delta v \to dv$. One then has a picture from Fig. 36-1 of many more rectangles, each much thinner, with the larger number of rectangles still fitting under the dashed curve shown. The number of particles, dn, with speeds between v and $v + dv$ equals the infinitesimal area of a rectangle and is given from the extension of (36.2) as

$$dn = f(v)\, dv$$

Here $f(v)$ is the continuous distribution function shown by the dashed curve in Fig. 36-1.

In this limit all discrete summations become definite integrals. For example, the normalization condition (36.3) goes over to

$$N = \int_{v_{\min}}^{v_{\max}} f(v)\, dv \qquad (36.5)$$

and the expression (36.4) for the average value of the square of the speed becomes

$$(v^2)_{\text{avg}} = \frac{1}{N} \int_{v_{\min}}^{v_{\max}} v^2 f(v)\, dv \qquad (36.6)$$

36.3 FUNDAMENTAL DISTRIBUTION FUNCTIONS AND DENSITY OF STATES

It is found that both under the laws of classical physics and those of quantum physics there are fundamental distribution functions from which all other distribution functions of interest can be obtained. A fundamental distribution function describes how the particles in a system are, on the average, distributed among the available *states* of the system. Putting it another way, a fundamental distribution function describes, for each state, the *probability* that a particle will be found in that state.

The state of a particle is defined by the smallest collection of variables that is needed to describe completely the particle's motion. As an example, classically, for weakly interacting particles, the state of a particle is completely specified by giving its three velocity components (v_x, v_y, v_z), or equivalently, since $\mathbf{p} = \mathbf{m v}$, its three momentum components (p_x, p_y, p_z).

The fundamental distribution function for systems obeying classical laws is called the *Maxwell-Boltzmann distribution function*, F_{MB}. (We shall denote fundamental distribution functions with capital letters.) The Maxwell-Boltzmann distribution function is defined such that

$$dn_v = F_{MB}\, dv_x\, dv_y\, dv_z \qquad (36.7)$$

is the average number of particles having velocity components in the intervals between v_x and $v_x + dv_x$, v_y and $v_y + dv_y$, v_z and $v_z + dv_z$.

It shall be seen in the following chapters that the classical and quantum mechanical fundamental distribution functions have one thing in common—they all depend only upon the energy of the state and not upon the quantities that define the state. Because of this property it is possible, for example, to write the particle distribution in energy in the form

$$dn_E = Fg(E)\, dE \qquad (36.8)$$

In this expression the fundamental distribution function F is the statistical factor that describes the probability of finding a particle in a given state at energy E, and $g(E)$ is called the *density of states* (with respect to energy). The quantity $g(E)\, dE$ gives the number of states, dS_E, that have energies in the interval between E and $E + dE$:

$$dS_E = g(E)\, dE \qquad (36.9)$$

For weakly interacting systems the total energy E of a particle equals its kinetic energy, and so

$$E = \frac{m}{2}\left(v_x^2 + v_y^2 + v_z^2\right) \qquad (36.10)$$

Consequently, there are many different states (v_x, v_y, v_z) that will have the same energy.

More generally, if a dynamical quantity q depends only upon the energy, the number of particles with values between q and $q + dq$ can always be written in the form

$$dn_q = Fg(q)\, dq \qquad (36.11)$$

containing a statistical factor F, which maintains the same form for a given statistics (e.g. $F = F_{MB}$ for classical statistics), and a density-of-states factor $g(q)$, which depends upon the particular physical property under consideration. The quantity q could, for example, be the speed, energy [as in (36.8)], or the magnitude of the momentum.

Solved Problems

36.1. At a certain instant it is found that in one mole of a gas 4782 molecules have speeds between 495 m/s and 505 m/s. What is the value of the speed distribution function at the speed 500 m/s?

The speed distribution function, f_v, is defined by

$$dn_v = f_v\, dv \qquad \text{or} \qquad f_v = \frac{dn_v}{dv} \approx \frac{\Delta n_v}{\Delta v}$$

From the data, $\Delta n_v = 4782$ molecules and $\Delta v = 505$ m/s $- 495$ m/s $= 10$ m/s, so that

$$f_v = f_{500\,\text{m/s}} = \frac{4782 \text{ molecules}}{10 \text{ m/s}} = 478.2 \ \frac{\text{molecules}}{\text{m/s}}$$

36.2. The speeds of the particles in a certain system at a particular time are as given in Table 36-1.

Table 36-1

Speed interval, m/s	0–5	5–10	10–15	15–20	20–25	25–30	30–35	35–40
Number of particles	1	0	0	1	3	0	2	1
Speed	40–45	45–50	50–55	55–60	60–65	65–70	70–75	75–80
Number	5	1	6	4	2	8	0	2
Speed	80–85	85–90	90–95	95–100	over 100			
Number	3	0	0	1	0			

Plot the speed distribution function, taking speed intervals of 20 m/s, starting from 0 m/s.

For finite velocity intervals the speed distribution function is defined by

$$f_v = \frac{\Delta n_v}{\Delta v}$$

where Δn_v is the number of particles having speeds in the interval between v and $v + \Delta v$. From the data we find the values for f_v given in Table 36-2 and graphed in Fig. 36-2. The two dashed lines in Fig. 36-2 form a continuous approximation to the histogram.

Table 36-2

Δv, m/s	0–20	20–40	40–60	60–80	80–100	over 100
Δn_v, particles	2	6	16	12	4	0
f_v, particles/(m/s)	0.1	0.3	0.8	0.6	0.2	0.0

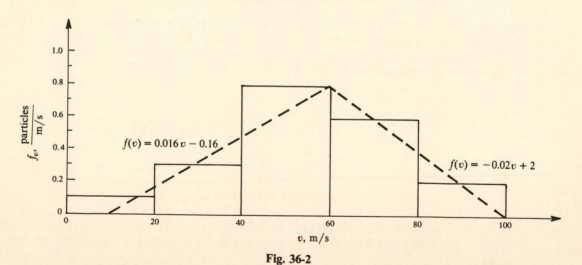

Fig. 36-2

36.3. Refer to Fig. 36-2. Find the number of particles in the system (a) from the histogram, (b) from the triangular approximation.

(a) The area under the histogram is ($\Delta v = 20$ m/s = constant)

$$N = \sum_i f_{vi} \Delta v = \left(0.1\ \frac{\text{particles}}{\text{m/s}} + 0.3\ \frac{\text{particles}}{\text{m/s}} + 0.8\ \frac{\text{particles}}{\text{m/s}}\right.$$
$$\left. + 0.6\ \frac{\text{particles}}{\text{m/s}} + 0.2\ \frac{\text{particles}}{\text{m/s}}\right)(20\ \text{m/s})$$

$$= 40 \text{ particles}$$

(b) The area of the triangle is

$$N = \tfrac{1}{2}[(100 - 10)\ \text{m/s}]\left(0.8\ \frac{\text{particles}}{\text{m/s}}\right) = 36\ \text{particles}$$

36.4.   Refer to Fig. 36-2.   Find the average speed of the particles (a) from the histogram, (b) from the triangular approximation.

(a) Using the value $N = 40$ particles found in Problem 36.3(a),

$$v_{\text{avg}} = \frac{1}{N} \sum_i v_i f_i\, \Delta v$$

$$= \frac{1}{40\ \text{particles}} \left[(10\ \text{m/s})\left(0.1\ \frac{\text{particles}}{\text{m/s}}\right) + (30\ \text{m/s})\left(0.3\ \frac{\text{particles}}{\text{m/s}}\right) \right.$$

$$+ (50\ \text{m/s})\left(0.8\ \frac{\text{particles}}{\text{m/s}}\right) + (70\ \text{m/s})\left(0.6\ \frac{\text{particles}}{\text{m/s}}\right)$$

$$\left. + (90\ \text{m/s})\left(0.2\ \frac{\text{particles}}{\text{m/s}}\right) \right](20\ \text{m/s})$$

$$= 55\ \text{m/s}$$

(b) Using the value $N = 36$ particles found in Problem 36.3(b),

$$v_{\text{avg}} = \frac{1}{N} \int_0^\infty v f(v)\, dv = \frac{1}{36} \int_{10}^{60} v(0.016v - 0.16)\, dv + \frac{1}{36} \int_{60}^{100} v(-0.02v + 2)\, dv$$

$$= \frac{1}{36} \left[0.016\,\frac{v^3}{3} - 0.16\,\frac{v^2}{2} \right]_{10}^{60} + \frac{1}{36} \left[-0.02\,\frac{v^3}{3} + 2\,\frac{v^2}{2} \right]_{60}^{100}$$

$$= 56.7\ \text{m/s}$$

36.5.   Refer to Fig. 36-2.   Find the rms speed of the particles (a) from the histogram, (b) from the triangular approximation.

(a) By definition, $v_{\text{rms}} = \sqrt{(v^2)_{\text{avg}}}$. Using $N = 40$ particles, from Problem 36.3(a),

$$(v^2)_{\text{avg}} = \frac{1}{N} \sum_i v_i^2 f_i\, \Delta v$$

$$= \frac{1}{40\ \text{particles}} \left[(10\ \text{m/s})^2\left(0.1\ \frac{\text{particles}}{\text{m/s}}\right) + (30\ \text{m/s})^2\left(0.3\ \frac{\text{particles}}{\text{m/s}}\right) \right.$$

$$+ (50\ \text{m/s})^2\left(0.8\ \frac{\text{particles}}{\text{m/s}}\right) + (70\ \text{m/s})^2\left(0.6\ \frac{\text{particles}}{\text{m/s}}\right)$$

$$\left. + (90\ \text{m/s})^2\left(0.2\ \frac{\text{particles}}{\text{m/s}}\right) \right](20\ \text{m/s})$$

$$= 3420\ (\text{m/s})^2$$

whence $v_{\text{rms}} = 58.5$ m/s.   Note that v_{rms} exceeds v_{avg} as calculated in Problem 36.4(a).   This is as required by Problem 35.13.

(b) Using $N = 36$ particles, from Problem 36.3(b),

$$(v^2)_{\text{avg}} = \frac{1}{N} \int_0^\infty v^2 f(v)\, dv = \frac{1}{36} \int_{10}^{60} v^2(0.016v - 0.16)\, dv + \frac{1}{36} \int_{60}^{100} v^2(-0.02v + 2)\, dv$$

$$= \frac{1}{36} \left[0.016\,\frac{v^4}{4} - 0.16\,\frac{v^3}{3} \right]_{10}^{60} + \left[-0.02\,\frac{v^4}{4} + 2\,\frac{v^3}{3} \right]_{60}^{100}$$

$$= 3550\ (\text{m/s})^2$$

whence $v_{\text{rms}} = 59.6$ m/s.

36.6.　Find the number of classical states, dS_v, in the speed interval between v and $v + dv$, and thereby show that the density of states, $g(v)$, has the form $g(v) \propto v^2$.

In determining the relationship between states and speed it is convenient to use a coordinate system where the three axes are v_x, v_y, and v_z. A coordinate system like this, where the components of the velocity are plotted along the three axes, is referred to as a "velocity space." A particular value of a particle's velocity, (v_x, v_y, v_z), then appears as a point in this velocity space. Since each set (v_x, v_y, v_z) defines a state of the particle, it is seen that each point in velocity space corresponds to a particular state of the particle. Consider now an infinitesimal "volume" element $dV_v = dv_x\, dv_y\, dv_z$ in the velocity space, as shown in Fig. 36-3. Although the exact number of points inside the volume element cannot be determined (it depends upon the definition of "point"), the number of points will be proportional to the size of the volume element. Since to every point (v_x, v_y, v_z) there corresponds a state of the particle, the number of states, dS_v, determined by the volume element will also be proportional to the size of the volume element, so that we may write

$$dS_v = C\, dV_v = C\, dv_x\, dv_y\, dv_z$$

In the velocity space diagram, $v_x^2 + v_y^2 + v_z^2 = v^2 = \text{constant}$ is a "sphere" of "radius" v, so that all points lying on this sphere correspond to particles that have the same speed v. If we now consider particles with speeds between v and $v + dv$, the corresponding points in the associated velocity space will be within the spherical shell defined by the spheres of radii v and $v + dv$, whose volume is

$$dV = 4\pi v^2\, dv$$

The number of states, dS_v, corresponding to speeds lying within this volume element will be proportional to the size of the volume element, so that

$$dS_v = C\, dV = C4\pi v^2\, dv$$

The density of states is defined by $dS_v = g(v)\, dv$; hence

$$g(v) = C4\pi v^2$$

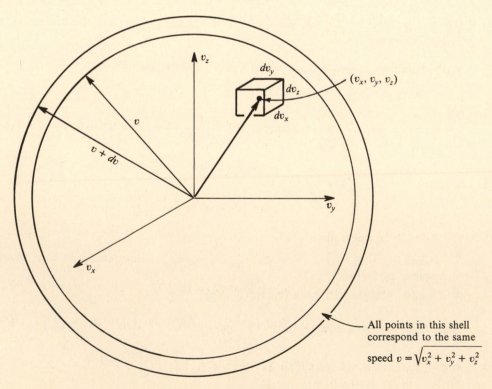

Fig. 36-3

36.7. Find the number of classical states, $dS_E = g(E) \, dE$, in the energy interval between E and $E + dE$, and thereby show that the density of states, $g(E)$, has the form $g(E) \propto E^{1/2}$.

Since $E = \frac{1}{2} mv^2$, a speed interval dv will be related to an energy interval dE by

$$dE = mv \, dv \qquad \text{or} \qquad dv = \frac{dE}{mv} = \frac{dE}{\sqrt{2mE}}$$

Further, the number of states in the energy interval dE must be equal to the number of states in the corresponding speed interval dv, so that, from Problem 36.6,

$$dS_E = dS_v = C4\pi v^2 \, dv = C4\pi \left(\frac{2E}{m} \right)\left(\frac{dE}{\sqrt{2mE}} \right) = \frac{4\pi C}{m^{3/2}} \sqrt{2} \, E^{1/2} \, dE = DE^{1/2} \, dE$$

Therefore the density of states in the energy interval between E and $E + dE$ is $g(E) = DE^{1/2}$.

Supplementary Problems

36.8. If each particle in the system described in Problem 36.2 has a mass of 0.002 kg, find (a) the average energy per particle and (b) the total energy of the system.   Use the histogram distribution.
Ans. (a) 3.42 J; (b) 136.8 J

36.9. Repeat Problem 36.8 using the triangular approximation in Fig. 36-2.   *Ans.* (a) 3.55 J; (b) 127.8 J

36.10. Using the triangular approximation in Fig. 36-2, find the number of particles that have speeds between 20 m/s and 80 m/s.   Compare this answer with the answer obtained from Table 36-2.
Ans. 31.2 particles; 34 particles

36.11. The speed distribution function for a certain system of particles is

$$f(v) = v(500 - v) \text{ particles}/(\text{m}/\text{s})$$

where v can vary between 0 m/s and 500 m/s. Each particle is of mass 2×10^{-12} kg.   Calculate (a) the average particle speed, (b) the average energy per particle, (c) the total energy of the system.
Ans. (a) 250 m/s; (b) 7.51×10^{-8} J; (c) 1.56 J

36.12. For speeds $v = 0$ m/s to $v = 10^3$ m/s the speed distribution function for a group of particles is

$$f(v) = (5 \times 10^{20}) \sin \frac{\pi v}{10^3} \quad \frac{\text{particles}}{\text{m}/\text{s}}$$

and $f(v) = 0$ for speeds above 10^3 m/s.   Find the number of particles in the system.
Ans. 3.18×10^{23}

36.13. What is the average speed of the particles of Problem 36.12?   *Ans.* 500 m/s

36.14. What is the rms speed of the particles of Problem 36.12?   *Ans.* 545 m/s

36.15. If the number of classical states in the momentum interval between p and $p + dp$ is $dS_p = g(p) \, dp$, find the form of the density of states, $g(p)$.   *Ans.* $g(p) = Kp^2$ ($K = $ constant)

Chapter 37

Classical Statistics: The Maxwell-Boltzmann Distribution

For systems of particles that obey classical laws the average number of particles, dn_v, whose velocity components lie between v_x and $v_x + dv_x$, v_y and $v_y + dv_y$, v_z and $v_z + dv_z$ is given by

$$dn_v = F_{MB}\, dv_x\, dv_y\, dv_z \tag{37.1}$$

The Maxwell-Boltzmann distribution function, F_{MB}, which is proportional to the probability that a particle will be found in the state (v_x, v_y, v_z), has the form

$$F_{MB} = N\left(\frac{m}{2\pi kT}\right)^{3/2} e^{-m(v_x^2 + v_y^2 + v_z^2)/2kT} = N\left(\frac{m}{2\pi kT}\right)^{3/2} e^{-E/kT} \tag{37.2}$$

(a derivation is given in Problems 37.13 through 37.17). Here, N is the total number of particles, m is the mass of each particle, k is the Boltzmann constant, and T is the absolute temperature.

As discussed in Section 36.3, if a property q depends only on the energy, the number of particles with values of the property between q and $q + dq$ is given by

$$dn_q = F_{MB}\, g(q)\, dq \tag{37.3}$$

where $g(q)$ is the density of states. For $q = v$, we have (see Problems 36.6 and 37.3)

$$g(v) = C4\pi v^2 = 4\pi v^2 \tag{37.4}$$

and for $q = E$, we have (see Problems 36.7 and 37.2)

$$g(E) = DE^{1/2} = \frac{2^{5/2}\pi}{m^{3/2}} E^{1/2} \tag{37.5}$$

The integrals

$$I_n(\alpha) \equiv \int_0^\infty u^n e^{-\alpha u^2}\, du \qquad (\alpha > 0)$$

often arise in the applications of the Maxwell-Boltzmann distribution. The values for small integers n are given in Table 37-1.

Table 37-1

n	0	1	2	3	4	5
$I_n(\alpha)$	$\frac{1}{2}\sqrt{\frac{\pi}{\alpha}}$	$\frac{1}{2\alpha}$	$\frac{1}{4}\sqrt{\frac{\pi}{\alpha^3}}$	$\frac{1}{2\alpha^2}$	$\frac{3}{8}\sqrt{\frac{\pi}{\alpha^5}}$	$\frac{1}{\alpha^3}$

Solved Problems

37.1. Show that the Maxwell-Boltzmann distribution function (*37.2*) is normalized for a system of N particles.

The total number of particles in the system is given by

$$\int_{-\infty}^{\infty}\int_{-\infty}^{\infty}\int_{-\infty}^{\infty} F_{MB}\, dv_x\, dv_y\, dv_z = N\left(\frac{m}{2\pi kT}\right)^{3/2}\int_{-\infty}^{\infty}\int_{-\infty}^{\infty}\int_{-\infty}^{\infty} e^{-m(v_x^2+v_y^2+v_z^2)/2kT}\, dv_x\, dv_y\, dv_z$$

The triple integral on the right is the product of three integrals of the form

$$\int_{-\infty}^{\infty} e^{-mq^2/2kT}\, dq = 2\int_{0}^{\infty} e^{-mq^2/2kT}\, dq = 2I_0\left(\frac{m}{2kT}\right)$$

Then, from Table 37-1, the total number of particles is

$$N\left(\frac{m}{2\pi kT}\right)^{3/2}\left(\frac{2\pi kT}{m}\right)^{3/2} = N$$

37.2. Evaluate the constant D in the expression for the density of states over energy, $g(E) = DE^{1/2}$ (see Problem 36.7).

We have

$$dn_E = F_{MB}\, g(E)\, dE = N\left(\frac{m}{2\pi kT}\right)^{3/2} e^{-E/kT} DE^{1/2}\, dE$$

Since the total number of particles is N,

$$N = N\left(\frac{m}{2\pi kT}\right)^{3/2} D\int_{0}^{\infty} E^{1/2} e^{-E/kT}\, dE = N\left(\frac{m}{2\pi kT}\right)^{3/2} D\left[2I_2\left(\frac{1}{kT}\right)\right] = N\frac{m^{3/2}D}{2^{5/2}\pi}$$

where we have changed the variable of integration from E to $u = \sqrt{E}$, and used Table 37-1. Thus

$$D = \frac{2^{5/2}\pi}{m^{3/2}}$$

The energy distribution function, f_E, is defined by $dn_E = f_E\, dE$. From the above calculation,

$$f_E = F_{MB}\, g(E) = \frac{2N}{\pi^{1/2}(kT)^{3/2}} E^{1/2} e^{-E/kT}$$

37.3. Evaluate the constant C in the expression for the density of states over speed, $g(v) = C4\pi v^2$ (see Problem 36.6).

We have for the number of particles in the speed interval dv:

$$dn_v = F_{MB}\, g(v)\, dv$$

This must be equal to the number of particles, $dn_E = F_{MB}\, g(E)\, dE$, in the corresponding energy interval dE. From $E = \frac{1}{2}mv^2$,

$$dE = mv\, dv = \sqrt{2mE}\, dv$$

Hence,

$$F_{MB}\, g(v)\, dv = F_{MB}\, g(E)\sqrt{2mE}\, dv \qquad \text{or} \qquad g(v) = g(E)\sqrt{2mE}$$

Since $g(v) = C4\pi v^2 = C8\pi E/m$ and $g(E) = DE^{1/2}$ (Problem 37.2),

$$C\frac{8\pi E}{m} = D2^{1/2}m^{1/2}E$$

$$C = \frac{m^{3/2}}{2^{5/2}\pi} D = 1$$

The speed distribution function, f_v, is defined by $dn_v = f_v\, dv$. From the above calculation,

$$f_v = F_{MB}\,(4\pi v^2) = 4\pi N\left(\frac{m}{2\pi kT}\right)^{3/2} v^2 e^{-mv^2/2kT}$$

37.4.   Evaluate the rms velocity for a Maxwell-Boltzmann distribution.

The average value of the square of the speed is

$$(v^2)_{\text{avg}} = \frac{1}{N} \int v^2 \, dn_v = \frac{1}{N} \int_0^\infty v^2 f_v \, dv$$

Substituting f_v from Problem 37.3, we obtain

$$(v^2)_{\text{avg}} = 4\pi \left(\frac{m}{2\pi kT} \right)^{3/2} I_4 \left(\frac{m}{2kT} \right) = \frac{3kT}{m}$$

and so

$$v_{\text{rms}} = \sqrt{(v^2)_{\text{avg}}} = \sqrt{\frac{3kT}{m}}$$

in agreement with kinetic theory [see (35.8)].

37.5.   Find the ratio of the *most probable speed*, v_p, to the rms velocity, v_{rms}, for the Maxwell-Boltzmann distribution.

The most probable speed in the Maxwell-Boltzmann distribution is that for which $f_v = 4\pi v^2 F_{MB}$ is maximum.   Differentiating the expression found in Problem 37.3,

$$\frac{df_v}{dv} = 0 = 4\pi N \left(\frac{m}{2\pi kT} \right)^{3/2} e^{-mv^2/2kT} v \left[2 - v^2 \left(\frac{m}{kT} \right) \right]$$

The roots $v = 0$ and $v = \infty$ yield minima; hence $v_p = \sqrt{2kT/m}$, and using the result of Problem 37.4,

$$\frac{v_p}{v_{\text{rms}}} = \frac{\sqrt{2kT/m}}{\sqrt{3kT/m}} = \sqrt{\frac{2}{3}} = 0.816$$

37.6.   In many cases of interest the total energy E of a particle can be written in the form

$$E = \sum_{i=1}^n c_i q_i^2$$

where the c_i are constants and the q_i are position and/or momentum (velocity) coordinates. For the Maxwell-Boltzmann distribution function in the form

$$F_{MB} = A e^{-E/kT} = A e^{-(\Sigma c_i q_i^2)/kT}$$

evaluate the constant A.

The normalization condition is

$$N = A \int_{-\infty}^\infty \int_{-\infty}^\infty \cdots \int_{-\infty}^\infty e^{-\left(\sum_{i=1}^n c_i q_i^2 \right)/kT} \, dq_1 \, dq_2 \cdots dq_n$$

$$= A \left[\int_{-\infty}^\infty e^{-c_1 q_1^2/kT} \, dq_1 \right] \left[\int_{-\infty}^\infty e^{-c_2 q_2^2/kT} \, dq_2 \right] \cdots \left[\int_{-\infty}^\infty e^{-c_n q_n^2/kT} \, dq_n \right]$$

$$= A \left[2I_0 \left(\frac{c_1}{kT} \right) \right] \left[2I_0 \left(\frac{c_2}{kT} \right) \right] \cdots \left[2I_0 \left(\frac{c_n}{kT} \right) \right]$$

$$= A \frac{(\pi kT)^{n/2}}{(c_1 c_2 \cdots c_n)^{1/2}}$$

from which

$$A = N \frac{(c_1 c_2 \cdots c_n)^{1/2}}{(\pi kT)^{n/2}}$$

Note that for three velocity coordinates, with $c_1 = c_2 = c_3 = m/2$, this result agrees with the normalization constant in (37.2).

37.7. Refer to Problem 37.6. Show that the average energy per particle is equal to $\frac{1}{2}kT$ times n, the number of squared position and momentum terms in the energy expression. This result, that each quadratic term in the energy adds an amount $\frac{1}{2}kT$ to the average energy per particle, is known as the Equipartition Theorem.

In Problem 37.6 we showed that

$$\int_{-\infty}^{\infty}\int_{-\infty}^{\infty} \cdots \int_{-\infty}^{\infty} e^{-E/kT}\, dq_1\, dq_2 \cdots dq_n = \frac{(\pi kT)^{n/2}}{(c_1 c_2 \cdots c_n)^{1/2}}$$

Differentiate both sides with respect to kT to obtain

$$\frac{1}{(kT)^2}\int_{-\infty}^{\infty}\int_{-\infty}^{\infty} \cdots \int_{-\infty}^{\infty} E e^{-E/kT}\, dq_1\, dq_2 \cdots dq_n = \frac{\pi^{n/2}(n/2)(kT)^{n/2-1}}{(c_1 c_2 \cdots c_n)^{1/2}} = \frac{N}{A}\frac{n/2}{kT}$$

or, transposing terms,

$$\frac{1}{N}\int_{-\infty}^{\infty} \cdots \int_{-\infty}^{\infty} E(Ae^{-E/kT})\, dq_1 \cdots dq_n = n\left(\frac{1}{2}kT\right)$$

which is the Equipartition Theorem, since the left side is the average energy of a particle.

37.8. A model for a diatomic molecule which is free to rotate, but not to vibrate, is a dumbbell with the moment of inertia about the axis passing through the masses approximately equal to zero. Find the molar heat capacity of a diatomic gas, using the Equipartition Theorem.

The energy of one diatomic molecule is equal to

$$E = \frac{1}{2m}\left(p_x^2 + p_y^2 + p_z^2\right) + \frac{1}{2}I\omega_1^2 + \frac{1}{2}I\omega_2^2$$

and this expression has five quadratic terms. From the Equipartition Theorem (Problem 37.7) the energy of one mole of gas is equal to

$$E_{tot} = N_0 n\left(\frac{kT}{2}\right) = \frac{5}{2}N_0 kT$$

where N_0 is Avogadro's number. The molar specific heat, C_V, is equal to

$$C_V = \frac{dE_{tot}}{dT} = \frac{5}{2}N_0 k = \frac{5}{2}R$$

with R the ideal gas constant.

37.9. If a crystal lattice is pictured as a system of regularly spaced atoms connected by springs, find the molar specific heat, using the Equipartition Theorem.

The energy of a vibrating atom in the lattice is given by

$$E = \frac{1}{2m}\left(p_x^2 + p_y^2 + p_z^2\right) + \frac{1}{2}\left(\kappa_x x^2 + \kappa_y y^2 + \kappa_z z^2\right)$$

Since there are six terms of the form cq^2, by the Equipartition Theorem (Problem 37.7) one mole has an energy

$$E_{tot} = N_0(6)\left(\frac{kT}{2}\right) = 3N_0 kT$$

where N_0 is Avogadro's number. The molar specific heat is given by

$$C_V = \frac{dE_{tot}}{dT} = 3N_0 k = 3R$$

which is the Dulong-Petit law. This result is found experimentally to be correct at high temperatures, but at low temperatures quantum effects cause a change. These quantum effects will be discussed in Chapter 38.

37.10. The RLC, high-Q, tuned circuit shown in Fig. 37-1 (R being small) is at a temperature T. Using the Equipartition Theorem, estimate the rms value of the induced voltage in the inductor due to thermal fluctuations.

In the inductance L the energy stored is

$$E = \int_0^I V_L i\; dt = \int_0^I Li\; di = \tfrac{1}{2} LI^2$$

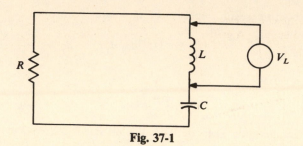

Fig. 37-1

and by the Equipartition Theorem (Problem 37.7)

$$E_{\text{avg}} = \tfrac{1}{2} L(I^2)_{\text{avg}} = \tfrac{1}{2} kT \tag{1}$$

because there is one quadratic term in the energy expression. For a high-Q tuned circuit the noise spectrum is suppressed except at the resonant frequency

$$\omega_0 = \frac{1}{\sqrt{LC}}$$

so the current and accompanying voltages are sinusoidal, with angular frequency ω_0. Then the magnitude of the voltage across the inductor, V_L, is related to the magnitude of the current, I, through it by

$$V_L = \omega_0 L I$$

from which

$$(V_L^2)_{\text{avg}} = (\omega_0 L)^2 (I^2)_{\text{avg}} \tag{2}$$

Equations (1) and (2) together give

$$(V_L^2)_{\text{avg}} = \omega_0^2 L kT \qquad \text{or} \qquad V_{L\text{ rms}} = \omega_0 \sqrt{LkT} = \sqrt{\frac{kT}{C}}$$

37.11. Find the fraction N_V/N of the total number of particles N having speeds above a given speed V in a system of particles described by the Maxwell-Boltzmann distribution.

The desired fraction is given by

$$\frac{N_V}{N} = \frac{1}{N} \int_V^\infty dn_v = \frac{1}{N} \int_V^\infty f_v\; dv = \frac{1}{N} \int_V^\infty 4\pi v^2 F_{MB}\; dv = 4\pi \left(\frac{m}{2\pi kT} \right)^{3/2} \int_V^\infty v^2 e^{-mv^2/2kT}\; dv$$

Changing the integration variable to

$$u^2 = \frac{mv^2}{2kT} \qquad du = \sqrt{\frac{m}{2kT}}\; dv$$

we obtain

$$\frac{N_V}{N} = \frac{4}{\sqrt{\pi}} \int_U^\infty u^2 e^{-u^2}\; du \qquad \left(U = \sqrt{\frac{m}{2kT}}\; V \right)$$

Integrating by parts and using Table 37-1 with $n = 0$,

$$\frac{N_V}{N} = \frac{4}{\sqrt{\pi}}\left\{\left[-\frac{1}{2}ue^{-u^2}\right]_U^\infty + \frac{1}{2}\int_U^\infty e^{-u^2}\,du\right\} = \frac{4}{\sqrt{\pi}}\left\{\frac{1}{2}Ue^{-U^2} + \frac{1}{2}\int_0^\infty e^{-u^2}\,du - \frac{1}{2}\int_0^U e^{-u^2}\,du\right\}$$

$$= \frac{4}{\sqrt{\pi}}\left\{\frac{1}{2}Ue^{-U^2} + \frac{1}{2}\frac{\sqrt{\pi}}{2} - \frac{1}{2}\int_0^U e^{-u^2}\,du\right\} = \frac{2}{\sqrt{\pi}}Ue^{-U^2} + 1 - \frac{2}{\sqrt{\pi}}\int_0^U e^{-u^2}\,du$$

$$= 1 + \frac{2}{\sqrt{\pi}}Ue^{-U^2} - \text{erf}\,U$$

The second term can be evaluated directly, and the third term, which is the *error function* of U, is found from standard tables.   A graph of N_V/N versus U is shown in Fig. 37-2.

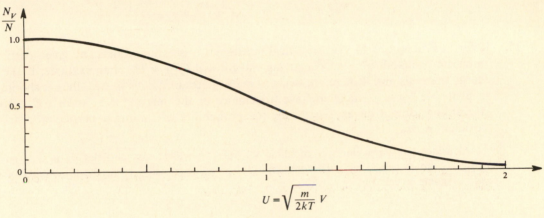

$$U = \sqrt{\frac{m}{2kT}}\,V$$

Fig. 37-2

37.12.   Under a Maxwell-Boltzmann distribution, what percentage of particles have speeds above the most probable speed, v_p?

From Problem 37.5 the most probable speed is

$$V = v_p = \sqrt{\frac{2kT}{m}}$$

which corresponds to $U = 1$ in Problem 37.11.   From Fig. 37-2 we find that when $U = 1$, $N_V/N = 0.57$. Thus 57% of the particles have speeds above v_p.

37.13.   *Stirling's formula* states that for large n

$$\ln n! \approx n \ln n - n$$

i.e. the ratio of the two sides tends to 1.   Evaluate the percentage error entailed in Stirling's formula for $n = 60$.

For $n = 60$ we have the exact result

$$\ln n! = \ln(8.321 \times 10^{81}) = 188.63$$

From Stirling's formula we have the approximate result

$$\ln n! \approx n \ln n - n = 60[(\ln 60) - 1] = 60(4.0943 - 1) = 185.66$$

The percent error, δ, in the approximation is

$$\delta = \frac{188.63 - 185.66}{188.63} \times 100 = 1.57\%$$

For $n = 60$ we see the error is only 1.57%.   For n the order of 10^{23} (i.e. Avogadro's number) the error is insignificant.

37.14. In how many ways can N distinguishable particles be put into r "cells" so that n_1 particles will be in the first cell, n_2 in the second cell, ..., n_r in the rth cell ($\sum n_i = N$)?

Let the number of ways be X, and think of any one of these X combinations as an arrangement of the N particles along a straight line, a group of n_1 followed by a group of n_2 ... followed by a group of n_r. If the elements of the first group were permuted among themselves, we would obtain $n_1!$ different linear arrangements of the N particles. Each of these in turn would give rise to $n_2!$ arrangements upon permutation of the second group; and so on. Thus, a total of

$$C = n_1! \, n_2! \cdots n_r!$$

arrangements would arise from the chosen combination. The X combinations would therefore produce XC arrangements, all distinct, and these must exhaust the $N!$ possible arrangements of the N particles. Hence,

$$XC = N! \quad \text{or} \quad X = \frac{N!}{C} = \frac{N!}{n_1! \, n_2! \cdots n_r!}$$

37.15. Refer to Problem 37.14. In statistical mechanics it is assumed that any given particle has an "intrinsic probability" g_i of occupying the ith cell ($\sum g_i = 1$). For example, if the cells are energy intervals and if there are twice as many states in ΔE_i as in ΔE_j, then it should be twice as probable for a particle to have an energy in the interval ΔE_i as in ΔE_j. Under this assumption, obtain an expression for the probability of finding a particular distribution of particles ($n_1, n_2, \ldots, n_r$).

Consider one particular way of filling the cells, in which, say, particles $\alpha_1, \beta_1, \ldots$ go to cell 1; $\alpha_2, \beta_2, \ldots$ go to cell 2; etc. The probability of α_1 going to cell 1 is g_1; the probability of α_1 *and* β_1 *and* ... going to cell 1 is

$$g_1 \times g_1 \times \cdots \times g_1 = g_1^{n_1}$$

and the probability of the entire filling is

$$g_1^{n_1} g_2^{n_2} \cdots g_r^{n_r}$$

Each way of realizing the distribution has this probability, and, by Problem 37.14, there are

$$\frac{N!}{n_1! \, n_2! \cdots n_r!}$$

such ways. Hence the probability we seek is

$$P_\mathbf{n} = \frac{N!}{n_1! \, n_2! \cdots n_r!} \, g_1^{n_1} g_2^{n_2} \cdots g_r^{n_r} \tag{1}$$

37.16. Refer to Problem 37.15. Statistical mechanics assumes that among all distributions of particles, the one of highest probability corresponds to equilibrium of the system. Find this most probable distribution subject to the conditions that the number of particles and total energy of the system are constant.

The problem is to maximize

$$P_\mathbf{n} = \frac{N!}{n_1! \, n_2! \cdots} \, g_1^{n_1} g_2^{n_2} \cdots \tag{1}$$

(the number r of cells is not important, provided it stays fixed) subject to two constraints:

fixed number of particles: $\qquad \sum n_i = N \tag{2}$

fixed total energy: $\qquad \sum E_i n_i = E_{\text{tot}} \tag{3}$

where N and E_{tot} are given. The energy constraint assumes that each particle in the ith cell has the same energy E_i. It will be convenient to replace $P_\mathbf{n}$ by the monotonically increasing function of $P_\mathbf{n}$

$$\ln \frac{P_\mathbf{n}}{N!} = \sum n_i \ln g_i - \sum \ln n_i!$$

(remember: N is fixed so far as the maximization is concerned).   Further we assume that N is very large ($\approx 10^{23}$) and, more important, that all the n_i are large enough to allow the use of Stirling's formula (Problem 37.13).   Thus we write

$$\ln \frac{P_n}{N!} = \sum n_i \ln g_i - \sum n_i \ln n_i + \sum n_i \qquad (4)$$

dropping the last sum since, by (2), it is constant.

Treating the n_i as continuous variables, we maximize the function (4), subject to constraints (2) and (3), by the *method of Lagrange multipliers*.   That is, we form the function

$$F(n_1, n_2, \ldots ; \lambda_1, \lambda_2) = \sum n_i \ln g_i - \sum n_i \ln n_i - \lambda_1\Big(\sum n_i - N\Big) - \lambda_2\Big(\sum E_i n_i - E_{\text{tot}}\Big)$$

and find its maximum, subject to no constraint on the variables $n_1, n_2, \ldots ; \lambda_1, \lambda_2$.   Taking the partial derivatives of F, we obtain the conditions

$$\frac{\partial F}{\partial n_i} = \ln g_i - \ln n_i - 1 - \lambda_1 - \lambda_2 E_i = 0 \qquad (i = 1, 2, \ldots)$$

plus (2) and (3) from the λ-derivatives.   Solving for the maximizing n_i:

$$n_i = g_i e^{-1-\lambda_1} e^{-\lambda_2 E_i} = A g_i e^{-\beta E_i} \qquad (5)$$

where $A = e^{-1-\lambda_1}$ and $\beta = \lambda_2$ are unknown constants.

37.17. Refer to Problem 37.16.   Show that the Maxwell-Boltzmann distribution follows from (5) under the following assumptions: (1) the "cells" are infinitesimal volumes $dv_x\, dv_y\, dv_z$ in velocity space (see Problem 36.6) and it is equiprobable for a particle to be found in each cell (i.e. all the g_i are equal); (2) the average of the particle energies has the value given by kinetic theory, $E_{\text{avg}} = \frac{3}{2} kT$.

Under the assumption of equiprobability, the intrinsic probability that a particle will be found in the infinitesimal cell $dv_x\, dv_y\, dv_z$ is just

$$g\, dv_x\, dv_y\, dv_z$$

where g is a constant.   Each particle in the cell has the same energy,

$$E = \frac{1}{2} mv_x^2 + \frac{1}{2} mv_y^2 + \frac{1}{2} mv_z^2$$

Thus (5) of Problem 37.16 gives for the number of particles in the cell

$$dn_v = A e^{-\beta E}\, dv_x\, dv_y\, dv_z \qquad (1)$$

where we have absorbed the constant g into A.   Normalizing as in Problem 37.6—except that, there, β was already known—we obtain

$$A = N \frac{(m/2)^{3/2}}{\pi^{3/2}} \beta^{3/2} \qquad (2)$$

Finally, proceeding as in Problem 37.7—where, again, β was already known—we find

$$E_{\text{avg}} = \frac{3}{2\beta} \qquad (3)$$

Equating this to $3kT/2$ gives $\beta = 1/kT$ and, from (2),

$$A = N\Big(\frac{m}{2\pi kT}\Big)^{3/2}$$

With these values of β and A, (1) is the Maxwell-Boltzmann distribution.

37.18. Find the most probable distribution of $N = 5$ distinguishable particles among $r = 3$ cells if the intrinsic probabilities for the cells are

$$g_1 = g_2 = g_3 = \frac{1}{3}$$

By (*1*) of Problem 37.15, the probability of a distribution (n_1, n_2, n_3) is

$$P_n = \frac{5!}{n_1!\, n_2!\, n_3!} \left(\frac{1}{3}\right)^5$$

since $n_1 + n_2 + n_3 = 5$. Therefore, the most probable distribution is that for which the number of ways of realizing it,

$$X = \frac{5!}{n_1!\, n_2!\, n_3!}$$

is greatest. Because the interchanging of the particles in any two cells, e.g.

$$n_1 = 3, n_2 = 2, n_3 = 0 \rightarrow n_1 = 2, n_2 = 3, n_3 = 0$$

results in the same value of X, we calculate X in Table 37-2 only for the case $n_1 \geqslant n_2 \geqslant n_3$.

Table 37-2

n_1	n_2	n_3	X
5	0	0	1
4	1	0	5
3	2	0	10
3	1	1	20
2	2	1	30

From the table it is seen that there will be three most probable distributions, each with $X = 30$:

$$(2, 2, 1), \quad (2, 1, 2), \quad \text{and} \quad (1, 2, 2)$$

37.19. For Problem 37.18, if the energy of a particle in cell 1 is zero, in cell 2 is ϵ, and in cell 3 is 2ϵ, find the most probable distribution if the total energy is fixed at $E_{tot} = 3\epsilon$.

The constraint

$$0n_1 + \epsilon n_2 + 2\epsilon n_3 = 3\epsilon \qquad \text{or} \qquad n_2 + 2n_3 = 3$$

eliminates all distributions except (3, 1, 1) and (2, 3, 0). From Table 37-2 it is seen that the former is the most probable.

For equal intrinsic probabilities, the Maxwell-Boltzmann distribution has the form

$$n_i = Ae^{-\beta E_i}$$

The distribution (3, 1, 1), which puts most of the particles in the lowest energy state, is much closer to the Maxwell-Boltzmann distribution than is (2, 3, 0).

Supplementary Problems

37.20. Find the ratio of the root-mean-square speed, v_{rms}, to the average speed, v_{avg}, for a gas described by the Maxwell-Boltzmann distribution.

Ans. $\sqrt{3\pi/8} = 1.08$

37.21. For the circuit of Problem 37.10, $L = 1$ mH, $C = 1$ μF, $R = 0.1$ Ω and $T = 300$ K. Find the rms voltage produced in the inductance. *Ans.* 6.44×10^{-8} V

37.22. Six distinguishable particles are to be distributed into 3 cells. Find the number of different combinations of particles that can produce (*a*) the distribution (6, 0, 0), (*b*) the distribution (4, 1, 1).
Ans. (*a*) 1; (*b*) 30

37.23.   In Problem 37.22 suppose that the three cells have respective particle energies of 0, ϵ, and 2ϵ.   What is the most probable distribution with total energy 6ϵ if \the intrinsic probabilities for the cells are equal?
Ans.   2 particles in each cell

37.24.   Refer to Problem 37.19.   What is the most probable distribution for a total energy 4ϵ?
Ans.   (2, 2, 1)

37.25.   Consider 50 distinguishable particles distributed into eight cells as follows:

Cell	1	2	3	4	5	6	7	8
Number of particles	6	8	9	0	7	2	10	8

How many combinations can be made of the 50 particles which yield this distribution?
Ans. 1.96×10^{36}

37.26.   For a two-state system with energies ϵ and 2ϵ find for the most probable distribution the average energy, if $g_1 = g_2 = \frac{1}{2}$.

Ans. $E_{avg} = \dfrac{2\epsilon e^{-2\beta\epsilon} + \epsilon e^{-\beta\epsilon}}{e^{-2\beta\epsilon} + e^{-\beta\epsilon}}$

37.27.   Express the distribution (5), Problem 37.16, in terms of the *partition function*,

$$Z = \sum_i g_i e^{-\beta E_i}$$

and the total number of particles, N.

Ans. $n_i = \dfrac{N}{Z} g_i e^{-\beta E_i}$

Chapter 38

Quantum Statistics

Quantum statistics describes systems composed of large numbers of particles obeying the laws of quantum mechanics. Each system will have a large number of discrete states, each described by a complete set of quantum numbers, with each particle in the system occupying one of these states. One of the ways that quantum statistics differs from classical statistics is that built into the derivation of the quantum distribution functions is the fact that particles are *indistinguishable* from each other. In other words, it is impossible in principle to label a particle with, say, a number which will distinguish the particle for all time.

38.1 FERMI-DIRAC STATISTICS

The statistics of particles which possess half-integral spin, called *fermions*, was developed by E. Fermi and P. A. M. Dirac. An example of a fermion system is a large number of weakly interacting electrons (spin $= \frac{1}{2}$), i.e. an electron "gas".

The *Fermi-Dirac distribution function*, F_{FD}, describes the most probable distribution of identical, indistinguishable particles which obey the Pauli exclusion principle. In a system of fermions in equilibrium at the absolute temperature T, the expected number of particles in a particular state i with energy E_i is given by (see Problem 38.35)

$$F_{FD} = \frac{1}{e^\alpha e^{E_i/kT} + 1} = \frac{1}{e^{(E_i - E_f)/kT} + 1} \qquad (38.1)$$

where k is the Boltzmann constant, and where $E_f = -kT\alpha$, the *Fermi energy*, is a characteristic of the system being described. Both α and E_f are functions of T. As $T \to 0$ K, $E_f \to E_{f0}$, a positive constant. The exact form of E_f (or of α) is determined from the normalization condition that the number of particles in the system is a fixed number.

It should be noted that, like F_{MB}, the Fermi-Dirac distribution function F_{FD} *does not* give the number of particles with energy E_i, but rather the number of particles occupying the state i which has an energy E_i. As described in Section 36.3 (for the continuous case), in order to find the number of particles n_i with energy E_i we need to multiply the Fermi-Dirac distribution function by the number of states g_i having the energy E_i:

$$n_i = g_i F_{FD}$$

If the energy levels are so close together that they can be treated as being continuous, the number of particles dn_E having energies in the interval between E and $E + dE$ is

$$dn_E = F_{FD}\, g(E)\, dE$$

where $dS_E = g(E)\, dE$ gives the number of quantum states with energies between E and $E + dE$. As discussed in Section 36.3, the quantity $g(E)$ is called the *density of states*.

38.2 BOSE-EINSTEIN STATISTICS

A. Einstein and S. N. Bose developed the statistics that would apply for a system composed of a large number of weakly interacting, identical and indistinguishable particles, each having an integral spin. These particles, called *bosons*, do not obey the Pauli exclusion principle. Examples of boson systems are photons, H_2 molecules, and liquid helium.

266

The *Bose-Einstein distribution function*, F_{BE}, gives the average number of bosons in a system in equilibrium at the temperature T that will be found in a particular state i with energy E_i (see Problem 38.37):

$$F_{BE} = \frac{1}{e^\alpha e^{E_i/kT} - 1} \tag{38.2}$$

The value of the quantity α depends upon the particular boson system that is being described. For systems of bosons whose numbers are not conserved (for example, photons), $\alpha = 0$ (Problem 38.38). As with the Maxwell-Boltzmann and Fermi-Dirac statistics, the energy distribution of the bosons is given in terms of a density-of-states function by

$$dn_E = F_{BE}\, g(E)\, dE$$

38.3   HIGH-TEMPERATURE LIMIT

When the total number of particles is conserved, the constant α for either quantum statistics increases monotonically with temperature (Problems 38.25 and 38.28). At sufficiently high temperatures (or at sufficiently high energies),

$$e^\alpha e^{E/kT} \gg 1$$

and the two quantum distribution functions reduce to

$$F = Ae^{-E/kT}$$

which is the classical Maxwell-Boltzmann distribution function.

38.4   TWO USEFUL INTEGRALS

(a) With $p > -1$, and in the low-temperature limit $kT \ll E_f$ (i.e. $\alpha \ll -1$),

$$\int_0^\infty E^p F_{FD}\, dE = \int_0^\infty \frac{E^p}{e^{(E-E_f)/kT} + 1}\, dE$$

$$= \frac{1}{p+1}\left[E_f^{p+1} + \sum_{n=1}^\infty 2(kT)^{2n}\left(1 - \frac{1}{2^{2n-1}}\right)\zeta(2n)\, \frac{d^{2n}}{dE_f^{2n}}\left(E_f^{p+1}\right)\right] \tag{38.3}$$

where $\zeta(x)$ is the *Riemann zeta function*,

$$\zeta(x) = \frac{1}{1^x} + \frac{1}{2^x} + \frac{1}{3^x} + \cdots$$

which is extensively tabulated. Some values are:

$$\zeta(2) = \frac{\pi^2}{6} = 1.645 \qquad \zeta(4) = \frac{\pi^4}{90} = 1.082 \qquad \zeta(6) = \frac{\pi^6}{945} = 1.017$$

When p is an integer, the series in (38.3) terminates.

(b) With $p > 0$, $\alpha \geqslant 0$, and $\epsilon = \pm 1$, we have

$$\int_0^\infty \frac{q^p}{e^\alpha e^q - \epsilon}\, dq = \Gamma(p+1)\sum_{n=1}^\infty \frac{\epsilon^{n+1}e^{-n\alpha}}{n^{p+1}} \tag{38.4}$$

where the *gamma function*, $\Gamma(x)$, obeys $\Gamma(x+1) = x\,\Gamma(x)$. Particular values are:

$$\Gamma(1) = 1 \qquad \Gamma\left(\frac{1}{2}\right) = \sqrt{\pi} \qquad \Gamma(n+1) = n! \quad (n \text{ an integer})$$

In the special case $\alpha = 0$, $\epsilon = +1$, (38.4) becomes

$$\int_0^\infty \frac{q^p}{e^q - 1}\, dq = \Gamma(p+1)\,\zeta(p+1) \tag{38.5}$$

Solved Problems

BLACKBODY RADIATION

The walls of a cavity maintained at a temperature T continuously emit and absorb electromagnetic radiation (photons), and in equilibrium the amounts of energy emitted and absorbed by the walls are equal. The radiation inside the cavity can be analyzed by opening a small hole in one of the walls of the cavity; the escaping photons constitute what is called *blackbody radiation*. Quantum physics was born when Max Planck discovered around 1900 the correct expression for the experimentally observed spectral distribution of blackbody radiation, i.e. the fraction of the total radiated energy with frequency between ν and $\nu + d\nu$.

38.1. Consider, for simplicity, a cubic cavity of side l whose edges define a set of axes and whose walls are maintained at a temperature T. Maxwell's equations for electromagnetic waves show that the rectangular components of the wave vector

$$\mathbf{k} = \frac{2\pi}{\lambda^2}\boldsymbol{\lambda}$$

must satisfy the boundary conditions

$$\frac{k_x l}{\pi} = n_x \qquad \frac{k_y l}{\pi} = n_y \qquad \frac{k_z l}{\pi} = n_z$$

where n_x, n_y, and n_z are positive integers. (These boundary conditions ensure an integral number of half waves in each edge of the cube.) Each triplet (n_x, n_y, n_z) represents, in classical terms, an electromagnetic *mode of oscillation* for the cavity; we shall consider these modes as photon *states*. Find the number of modes in the frequency interval between ν and $\nu + d\nu$, given that there are two independent polarization directions for each mode.

Each allowed frequency ν corresponds to a certain mode (n_x, n_y, n_z) and can be written as

$$\nu = \frac{c}{\lambda} = \frac{ck}{2\pi} = \frac{c}{2l}\sqrt{n_x^2 + n_y^2 + n_z^2} = \frac{cN}{2l}$$

where $N = \sqrt{n_x^2 + n_y^2 + n_z^2}$. Instead of trying to find the number of integer states (n_x, n_y, n_z) corresponding to a given N, we use a continuous approximation and find the number of points in a spherical shell of radius N and thickness dN in the first octant of (n_x, n_y, n_z)-space. This is simply the "volume"

$$dM = \frac{1}{8}(4\pi N^2\, dN) = \frac{\pi}{2}N^2\, dN$$

But the number of states between N and $N + dN$ must equal the number of states in the corresponding frequency interval, ν to $\nu + d\nu$. Using $\nu = cN/2l$,

$$dN = \frac{2l}{c}\, d\nu$$

from which

$$dM = \frac{\pi}{2}\left(\frac{2l\nu}{c}\right)^2\left(\frac{2l}{c}\, d\nu\right) = \frac{4\pi l^3}{c^3}\nu^2\, d\nu = \frac{4\pi V}{c^3}\nu^2\, d\nu$$

where V is the volume of the cavity. Because there are two possible independent polarization directions for each mode, we must multiply dM by two to obtain the number of photon states dS:

$$dS = 2\, dM = \frac{8\pi V}{c^3}\nu^2\, d\nu = g(\nu)\, d\nu$$

Thus the density of photon states in a frequency interval $d\nu$ is

$$g(\nu) = \frac{8\pi V}{c^3}\nu^2$$

38.2. Viewing the cavity of Problem 38.1 as a container of photons, which are spin-1 bosons, determine the spectral distribution (the amount of energy per frequency interval) of the blackbody radiation coming from a small hole in the container. Because photons are continuously being emitted and absorbed by the walls of the cavity, their number is not conserved.

Using the density of states $g(\nu)$ determined in Problem 38.1 and the Bose-Einstein distribution function for photons, (38.2) with $\alpha = 0$, we find the number of photons, dn_ν, with frequencies between ν and $\nu + d\nu$ to be

$$dn_\nu = F_{BE}\, g(\nu)\, d\nu = \frac{1}{e^{E/kT} - 1}\, \frac{8\pi V}{c^3}\, \nu^2\, d\nu$$

Each photon has an energy $E = h\nu$, so the amount of energy, dE_ν, carried by the dn_ν photons is

$$dE_\nu = h\nu\, dn_\nu = \frac{8\pi V h}{c^3}\, \frac{\nu^3}{e^{h\nu/kT} - 1}\, d\nu = F(\nu)\, d\nu$$

The spectral distribution, $F(\nu)$, is plotted in Fig. 38-1. It should be mentioned that Planck arrived at the function $F(\nu)$ by a different route.

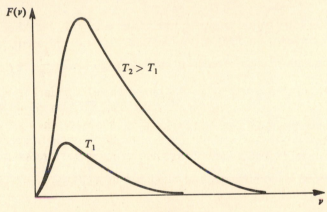

Fig. 38-1

38.3. The *Stefan-Boltzmann law* states that the total electromagnetic energy inside a cavity whose walls are maintained at a temperature T is proportional to T^4. Show how the law follows from the result of Problem 38.2, and evaluate the proportionality factor.

The total energy in the cavity of Problem 38.2 is given by

$$E = \int dE_\nu = \frac{8\pi V h}{c^3} \int_0^\infty \frac{\nu^3\, d\nu}{e^{h\nu/kT} - 1} = \frac{8\pi V k^4 T^4}{c^3 h^3} \int_0^\infty \frac{q^3\, dq}{e^q - 1} = \text{constant} \times T^4$$

From (38.5), the integral has the value

$$\Gamma(4)\zeta(4) = 3!\, \frac{\pi^4}{90} = 6.49$$

and so

$$E = \frac{8\pi k^4 (6.49)}{(hc)^3}\, VT^4 = \frac{8\pi (8.617 \times 10^{-5}\, \text{eV/K})^4 (6.49)}{(12.4 \times 10^{-7}\, \text{eV} \cdot \text{m})^3}\, VT^4$$

$$= (4.71\, \text{keV/K}^4 \cdot \text{m}^3) VT^4$$

38.4. From Fig. 38-1 it is seen that the peak of the spectral distribution curve shifts upward in frequency as the temperature increases. The *Wien displacement law* states that

$$\lambda_{max} T = \text{constant}$$

where λ_{max} is the wavelength at which the maximum value of $F(\nu)$ occurs. Derive the Wien displacement law.

Setting

$$\frac{dF(\nu)}{d\nu} = \frac{d}{d\nu}\left[\frac{8\pi Vh}{c^3}\frac{\nu^3}{e^{h\nu/kT}-1}\right] = \frac{8\pi Vh}{c^3}\frac{\nu^2\left[e^{h\nu/kT}\left(3-\frac{h\nu}{kT}\right)-3\right]}{\left(e^{h\nu/kT}-1\right)^2} = 0$$

we obtain, for a maximum,

$$e^{h\nu_{max}/kT}\left(3-\frac{h\nu_{max}}{kT}\right)-3 \equiv e^y(3-y)-3 = 0$$

This transcendental equation for $y = h\nu_{max}/kT$, which must be solved by approximation methods, will have some solution

$$y = \frac{h\nu_{max}}{kT} = \text{constant}$$

or, since $\nu_{max} = c/\lambda_{max}$,

$$\lambda_{max}T = \text{constant}$$

38.5. Determine the ratio of the energy emitted from a 2000 K blackbody in wavelength bands of width 100 Å centered on 5000 Å (visible) and 50,000 Å (infrared).

The spectral distribution in terms of wavelength can be obtained by setting (working with magnitudes only)

$$\nu = \frac{c}{\lambda} \qquad d\nu = \frac{c}{\lambda^2}\,d\lambda$$

in the expression for dE_ν in Problem 38.2, to obtain

$$dE_\lambda = \frac{8\pi hcV}{\lambda^5}\frac{d\lambda}{e^{hc/kT\lambda}-1} = F(\lambda)\,d\lambda$$

The bandwidth $\Delta\lambda = 100$ Å is sufficiently small that we may treat $F(\lambda)$ as constant over this interval, to obtain

$$\frac{\Delta E_{50,000}}{\Delta E_{5000}} = \frac{(5000\text{ Å})^5}{(50,000\text{ Å})^5}\frac{e^{(12,400\text{ eV}\cdot\text{Å})/(8.62\times10^{-5}\text{ eV/K})(2000\text{ K})(5000\text{ Å})}-1}{e^{(12,400\text{ eV}\cdot\text{Å})/(8.62\times10^{-5}\text{ eV/K})(2000\text{ K})(50,000\text{ Å})}-1} = 5.50$$

This result shows that only a small amount of the overall energy is radiated as visible light.

FREE ELECTRON THEORY OF METALS

In the *free electron theory* of metals it is assumed that the weakly bound valence electrons of the atoms composing the metal are not bound to particular atoms but move throughout the entire solid. It is further assumed that each electron experiences no net force from either the other valence electrons or from the bound electrons and nuclei. These assumptions are equivalent to stating that each electron moves throughout the metal in a constant electrostatic potential. At the boundaries of the metal the potential will rise rapidly because of the net electrostatic force acting on an electron at the boundary. Thus, in this model, the electrons in a metal are treated like a gas composed of noninteracting spin-$\frac{1}{2}$ fermions confined to a three-dimensional box.

38.6. For a one-dimensional infinite square well of length l the allowed energies for noninteracting particles of mass m were found in Problem 17.3 to be $E_n = n^2E_0$, where n is a positive integer and $E_0 = h^2/8ml^2$. The generalization to a three-dimensional infinite well of side l is

$$E = \left(n_x^2 + n_y^2 + n_z^2\right)E_0 \qquad E_0 = \frac{h^2}{8ml^2}$$

where n_x, n_y, n_z are positive integers. It is seen that a number of different states (n_x, n_y, n_z) may have the same energy, a situation called *degeneracy*. For the first 6 energy levels, find the order of degeneracy, i.e. the number of states having the given energy.

See Table 38-1. For higher levels, the equal-energy states need not be permutations of the same three integers; see Problem 38.43.

Table 38-1

Energy	Equal-Energy States (n_x, n_y, n_z)	Order of Degeneracy
$3E_0$	(1, 1, 1)	1
$6E_0$	(2, 1, 1) (1, 2, 1) (1, 1, 2)	3
$9E_0$	(2, 2, 1) (2, 1, 2) (1, 2, 2)	3
$11E_0$	(3, 1, 1) (1, 3, 1) (1, 1, 3)	3
$12E_0$	(2, 2, 2)	1
$14E_0$	(1, 2, 3) (1, 3, 2) (2, 1, 3) (2, 3, 1) (3, 1, 2) (3, 2, 1)	6

38.7. For electrons in a metal or gas molecules in a container, the value of l in Problem 38.6 is so large that the energy levels can be regarded as forming a continuous spectrum. For this case determine the number of states (n_x, n_y, n_z) with energies in the interval between E and $E + dE$.

The problem here is very similar to that of the blackbody, Problem 38.1. Writing $N = \sqrt{n_x^2 + n_y^2 + n_z^2}$, we obtain for the number of states, dS, with energies between E and $E + dE$:

$$dS = \frac{1}{8}\left(4\pi N^2\, dN\right) = \frac{\pi}{2} N^2\, dN$$

From Problem 38.6,

$$N^2 = n_x^2 + n_y^2 + n_z^2 = E\left(\frac{8ml^2}{h^2}\right)$$

so that

$$N = \left(\frac{8ml^2}{h^2}\right)^{1/2} E^{1/2} \qquad dN = \left(\frac{8ml^2}{h^2}\right)^{1/2} \frac{dE}{2E^{1/2}}$$

and

$$dS = \frac{\pi}{2} E\left(\frac{8ml^2}{h^2}\right)\left(\frac{8ml^2}{h^2}\right)^{1/2}\left(\frac{dE}{2E^{1/2}}\right) = \frac{2\pi l^3 (2m)^{3/2}}{h^3} E^{1/2}\, dE$$

Thus the density of states in the energy interval dE is, with $V = l^3$,

$$g(E) = \frac{2\pi V (2m)^{3/2}}{h^3} E^{1/2}$$

38.8. Draw a graph of the Fermi-Dirac distribution function versus energy for $T \approx 0$ K.

Near $T = 0$ K, $E_f \approx E_{f0} > 0$, and so

$$F_{FD} \approx \frac{1}{e^{(E - E_{f0})/kT} + 1}$$

If $E < E_{f0}$ and $T \to 0$ K, $e^{(E - E_{f0})/kT} \to 0$ and $F_{FD} = 1$. If $E > E_{f0}$ and $T \to 0$ K, $e^{(E - E_{f0})/kT} \to \infty$ and $F_{FD} = 0$. Therefore the graph of F_{FD} against E has the form given by the solid curve in Fig. 38-2; if $T > 0$ K it is found that the graph takes on the form given by the dashed curve.

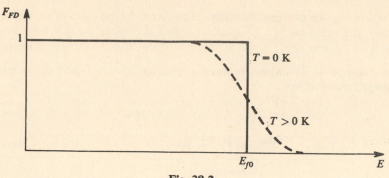

Fig. 38-2

38.9. Obtain an expression for E_{f0}, the Fermi energy at $T = 0$ K, for an electron gas in a metal.

In Problem 38.7 we found that the number of states (n_x, n_y, n_z) in the energy interval between E and $E + dE$ was given by

$$dS = g(E)\,dE = \frac{2\pi V(2m)^{3/2}}{h^3} E^{1/2}\,dE$$

Since for each set of quantum numbers (n_x, n_y, n_z) there are two possible electron spin orientations, we must multiply $g(E)$ by a factor of 2 to get the actual density of states for the electron gas. Thus, the total number N of fermions in the system is given by

$$N = 2\int_0^\infty F_{FD}\,g(E)\,dE = \frac{4\pi V(2m)^{3/2}}{h^3}\int_0^\infty \frac{E^{1/2}\,dE}{e^{(E-E_f)/kT}+1}$$

At $T = 0$ K the Fermi-Dirac distribution function is (see Problem 38.8)

$$F_{FD} = 1 \quad \text{for} \quad E < E_{f0}$$
$$F_{FD} = 0 \quad \text{for} \quad E > E_{f0}$$

so the limits on the integral can be changed, to give

$$N = \frac{4\pi V(2m)^{3/2}}{h^3}\int_0^{E_{f0}} E^{1/2}\,dE = \frac{4\pi V(2m)^{3/2}}{h^3}\left(\frac{2}{3}E_{f0}^{3/2}\right)$$

or

$$E_{f0} = \frac{h^2}{8m}\left(\frac{3N}{\pi V}\right)^{2/3}$$

As the temperature increases, it is found (see Problem 38.13) that the Fermi energy remains about equal to E_{f0}.

38.10. Metallic potassium has a density of 0.86×10^3 kg/m^3 and an atomic weight of 39. Find the Fermi energy for the electrons in the metal if each potassium atom donates one electron to the electron gas.

First we calculate the number of electrons per unit volume, N/V:

$$\frac{N}{V} = \frac{(6.02 \times 10^{26}\text{ atoms/kmol})(0.86 \times 10^3\text{ kg/m}^3)}{39\text{ kg/kmol}} = 1.33 \times 10^{28}\ \frac{\text{atoms}}{\text{m}^3} = 1.33 \times 10^{28}\ \frac{\text{electrons}}{\text{m}^3}$$

Then, from Problem 38.9,

$$E_{f0} = \frac{h^2}{8m}\left(\frac{3N}{\pi V}\right)^{2/3} = \frac{(hc)^2}{8mc^2}\left(\frac{3N}{\pi V}\right)^{2/3} = \frac{(12.4 \times 10^{-7}\text{ eV}\cdot\text{m})^2}{8(0.511 \times 10^6\text{ eV})}\left(\frac{3 \times 1.33 \times 10^{28}\text{ m}^{-3}}{\pi}\right)^{2/3} = 2.05\text{ eV}$$

At 300 K, $kT = 0.026$ eV, so it is seen that $E_f \approx E_{f0}$ is much greater than kT at room temperature.

38.11. Find the average kinetic energy per particle for a Fermi gas of N particles at $T = 0$ K.

Using the density of states $2g(E)$ found in Problem 38.9, we have

$$E_{avg} = \frac{1}{N} \int E \, dn_E = \frac{2}{N} \int_0^\infty E F_{FD} \, g(E) \, dE = \frac{4\pi V (2m)^{3/2}}{Nh^3} \int_0^\infty \frac{E^{3/2} \, dE}{e^{(E - E_f)/kT} + 1}$$

$$\rightarrow \frac{4\pi V (2m)^{3/2}}{Nh^3} \int_0^{E_{f0}} E^{3/2} \, dE = \frac{4\pi V (2m)^{3/2}}{Nh^3} \left(\frac{2}{5} E_{f0}^{5/2} \right)$$

Substituting the value of N found in Problem 38.9, we find that

$$E_{avg} = \frac{3}{5} E_{f0}$$

Thus, even at 0 K, the electrons, on the average, have a sizable kinetic energy. This occurs because the Pauli exclusion principle will not allow all the electrons to occupy the lowest energy levels, so that electrons will be found with all energies up to E_{f0}.

38.12. Using the normalization condition that the total number of particles is a fixed number, obtain, for low temperatures, an expression for the Fermi energy E_f of an electron gas in terms of E_{f0}, the Fermi energy at $T = 0$ K.

The number of particles is given by

$$N = \int_0^\infty F_{FD} g(E) \, dE = C \int_0^\infty E^{1/2} F_{FD} \, dE$$

where, from Problem 38.9, the density of states is $g(E) = CE^{1/2} \, dE$ with $C = 4\pi V (2m)^{3/2}/h^3$. Equation (38.3) gives, for $p = \frac{1}{2}$ and keeping only the first term of the series,

$$N = C \int_0^\infty E^{1/2} F_{FD} \, dE \approx \frac{2C}{3} \left[E_f^{3/2} + 2(kT)^2 \left(1 - \frac{1}{2} \right) \varsigma(2) \left(\frac{3}{4} E_f^{-1/2} \right) \right]$$

$$= \frac{2C}{3} E_f^{3/2} \left[1 + \frac{\pi^2}{8 E_f^2} (kT)^2 \right] \qquad (1)$$

Letting $T \rightarrow 0$ K in (1), we obtain

$$N = \frac{2C}{3} E_{f0}^{3/2}$$

or

$$E_{f0} = \left(\frac{3N}{2C} \right)^{2/3} = \left(\frac{3Nh^3}{8\pi V (2m)^{3/2}} \right)^{2/3} = \frac{h^2}{8m} \left(\frac{3N}{\pi V} \right)^{2/3}$$

which agrees with the result of Problem 38.9. Since kT/E_f is small, we see from (1) that E_f doesn't change rapidly with temperature. Therefore we can set $E_f = E_{f0}$ in the second term on the right of (1), to obtain, after also substituting $N = 2CE_{f0}^{3/2}/3$,

$$\frac{2C}{3} E_{f0}^{3/2} = \frac{2C}{3} E_f^{3/2} \left[1 + \frac{\pi^2}{8} \left(\frac{kT}{E_{f0}} \right)^2 \right]$$

from which

$$E_f = E_{f0} \left[1 + \frac{\pi^2}{8} \left(\frac{kT}{E_{f0}} \right)^2 \right]^{-2/3}$$

Finally, recalling that for small x

$$(1 + x)^{-2/3} = 1 - \frac{2}{3} x$$

we can write

$$E_f = E_{f0} \left[1 - \frac{\pi^2}{12} \left(\frac{kT}{E_{f0}} \right)^2 \right]$$

38.13. Silver has a Fermi energy of 5.5 eV at $T = 0$ K. Using the results of Problem 38.12, estimate the size of the first-order correction at $T = 300$ K.

The first-order correction is

$$\frac{-\pi^2}{12} \frac{(kT)^2}{E_{f0}} = \frac{-\pi^2}{12} \frac{\left[(8.617 \times 10^{-5} \text{ eV/K})(300 \text{ K})\right]^2}{5.5 \text{ eV}} = -10^{-4} \text{ eV}$$

an almost insignificant change.

38.14. At 0 K, silver has a Fermi energy of 5.5 eV and a work function of 4.6 eV. What is the average electrostatic potential energy seen by the free electrons in silver?

The work function, ϕ, is the minimum energy required to remove an electron from a material. The electrons so removed will be those with maximum kinetic energy, which at 0 K is the Fermi energy, E_{f0}. In turn, for a particle in a box, E_{f0} is the energy above the average electrostatic potential energy, $-E_0$. The relationship between the various energies is shown in Fig. 38-3, from which it is seen that

$$E_0 = E_{f0} + \phi = 5.5 \text{ eV} + 4.6 \text{ eV} = 10.1 \text{ eV}$$

Since $E_f \approx E_{f0}$ for moderate temperatures (Problem 38.13), this result explains why the work function of a metal is essentially temperature independent.

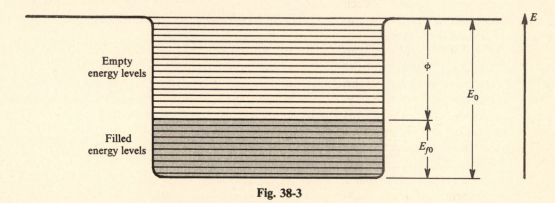

Fig. 38-3

SPECIFIC HEATS OF CRYSTALLINE SOLIDS

The molar specific heat at constant volume, C_V, of a solid is defined as the change in the energy content of one mole of the solid per unit change in temperature as the volume of the solid is held constant:

$$C_V = \frac{1}{\mathfrak{N}} \left(\frac{\partial E_T}{\partial T} \right)_V$$

where E_T is the total energy possessed by $\mathfrak{N}$ moles of the solid. From classical reasoning it is expected that the specific heat of a crystalline solid will remain constant with temperature, and will be given by the Dulong-Petit law (see Problem 37.9), $C_V = 3R$, where R is the ideal gas constant. Experimentally, however, it is found that C_V varies with temperature, as shown in Fig. 38-4.

A successful explanation of the observed behavior of C_V was given by P. Debye in 1912, who improved an earlier theory developed by Einstein in 1906. In the Debye theory a crystalline solid is viewed as being composed of regularly spaced atoms in a three-dimensional lattice. If an atom is displaced from its equilibrium position, it will experience a restoring force due to the surrounding atoms, the atoms behaving like a collection of coupled oscillators. Any disturbance will be transmitted to the surrounding atoms, resulting in a wave propagating through the solid. It is found that

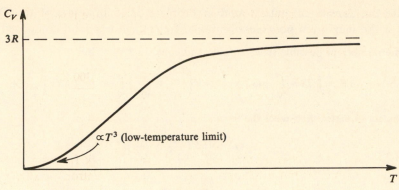

Fig. 38-4

the amount of energy transferred from one atom to its neighbor is quantized in amounts of $h\nu$, with ν the classical frequency at which the atom vibrates about its equilibrium position. Each quantum $h\nu$ of acoustical energy is called a *phonon*, in analogy to the photons of electromagnetic radiation.

Waves propagating in the lattice can be either transverse or longitudinal, with velocities v_t and v_l respectively. A transverse wave has two vibrational degrees of freedom, while a longitudinal wave has only one degree of freedom. Each vibrational degree of freedom (mode of vibration) of the crystal corresponds to a *state* of the system, and the phonons are distributed among these states according to the Bose-Einstein distribution.

In a crystal it is found that there is a maximum frequency of vibration, called the *Debye frequency*, ν_d. The maximum frequency exists because a system of N molecules possesses only $3N$ modes of vibration (each molecule having three independent vibrational degrees of freedom). The specific heat of a crystalline solid is obtained by first finding the Debye frequency and the density of states for the crystal (Problems 38.15 and 38.16), and then using this information to find the vibrational kinetic energy and molar specific heat (Problems 38.17 and 38.18).

In addition to the transfer of energy by atomic vibrations, if the crystalline solid is a conductor, there will also be energy transferred by the free conduction electrons. The total specific heat will then be determined by the sum of the electronic specific heat and the lattice specific heat. However, as shown in Problems 38.23 and 38.24, the electronic contribution to energy transfer becomes important only at very low temperatures.

38.15. Find the number of vibrational states with frequencies in the interval between ν and $\nu + d\nu$.

The sound waves traveling in a solid are exactly analogous to the blackbody oscillations of the photons in the cavity of Problem 38.1. By the same reasoning as in Problem 38.1, the number dM of modes of vibration with frequencies in the interval between ν and $\nu + d\nu$ is

$$dM = \frac{4\pi V}{c^3}\nu^2\,d\nu$$

for each type of vibration, with c the velocity of the wave propagating in the medium. Taking into account the two degrees of vibrational freedom for a transverse wave and the single degree of vibrational freedom for a longitudinal wave, the number of vibrational states, dS, with frequencies in the interval between ν and $\nu + d\nu$ is

$$dS = g(\nu)\,d\nu = 4\pi V\left(\frac{2}{v_t^3} + \frac{1}{v_l^3}\right)\nu^2\,d\nu$$

Thus the density of states in the frequency interval $d\nu$ is

$$g(\nu) = 4\pi V\left(\frac{2}{v_t^3} + \frac{1}{v_l^3}\right)\nu^2$$

38.16. Rewrite the density of states found in Problem 38.15 in terms of the Debye frequency v_d, which is the maximum possible vibrational frequency in a solid composed of N molecules.

The number of possible states is finite and equal to $3N$. Thus:

$$3N = \int dS = \int_0^{v_d} g(v)\, dv = 4\pi V \left(\frac{2}{v_t^3} + \frac{1}{v_l^3} \right) \int_0^{v_d} v^2\, dv = 4\pi V \left(\frac{2}{v_t^3} + \frac{1}{v_l^3} \right) \frac{v_d^3}{3}$$

The density of states then takes the form

$$g(v) = 4\pi V \left(\frac{2}{v_t^3} + \frac{1}{v_l^3} \right) v^2 = \frac{9N}{v_d^3}\, v^2$$

38.17. Assume that acoustical energy is transferred through a crystal lattice in quantized amounts of hv by quasiparticles called *phonons*, which are bosons whose total number, like that of photons, is not fixed. Obtain an expression for the total kinetic energy (vibrational energy) of the crystalline solid.

The number of phonons dn_v with frequencies in the range between v and $v + dv$ is

$$dn_v = F_{BE}\, g(v)\, dv = \frac{1}{e^{E/kT} - 1}\, \frac{9N}{v_d^3}\, v^2\, dv$$

where $g(v)$ was found in Problem 38.16. Since each phonon has an energy $E = hv$, we can write for the energy dE possessed by these dn_v phonons

$$dE = hv\, dn_v = \frac{9Nh}{v_d^3}\, \frac{v^3}{e^{hv/kT} - 1}\, dv$$

The total energy E_T of the solid will be the sum of all the phonon energies:

$$E_T = \int dE = \frac{9Nh}{v_d^3} \int_0^{v_d} \frac{v^3}{e^{hv/kT} - 1}\, dv = 9NkT \left(\frac{T}{T_d} \right)^3 \int_0^{q_d} \frac{q^3\, dq}{e^q - 1}$$

where $q_d = hv_d/kT$ and $T_d = hv_d/k$ (called the *Debye temperature*). The integral must be evaluated numerically.

38.18. From the result of Problem 38.17 obtain an expression for C_V, the molar specific heat at constant volume, in the limit $T \ll T_d$.

From Problem 38.17 the total energy is

$$E_T = 9NkT \left(\frac{T}{T_d} \right)^3 \int_0^{q_d} \frac{q^3\, dq}{e^q - 1}$$

where $q_d = hv_d/kT = T_d/T$. For $T \ll T_d$ we have $q_d \to \infty$, and the value of the integral approaches $\pi^4/15$ (Problem 38.3). Thus,

$$E_T = 9NkT \left(\frac{T}{T_d} \right)^3 \frac{\pi^4}{15}$$

Applying the definition of C_V, we then obtain

$$C_V = \frac{1}{\mathfrak{N}} \left(\frac{\partial E_T}{\partial T} \right)_V = \frac{36kN}{\mathfrak{N}}\, \frac{\pi^4}{15} \left(\frac{T}{T_d} \right)^3 = \frac{12\pi^4 R}{5} \left(\frac{T}{T_d} \right)^3$$

where we have used the result $kN/\mathfrak{N} = R$. Experimentally it is found that at low temperatures C_V does vary as T^3.

38.19. The classical Dulong-Petit law (see Problem 37.9) states that $C_V = 3R$. Show how this law follows from the result of Problem 38.17 in the limit $T \gg T_d$.

From Problem 38.17 we have

$$E_T = 9NkT \left(\frac{T}{T_D} \right)^3 \int_0^{q_d} \frac{q^3 \, dq}{e^q - 1}$$

For $T \gg T_d$, $q_d = T_d/T \to 0$, so that we may write

$$e^q \approx 1 + q$$

over the entire range of integration.   Thus,

$$E_T \approx 9NkT \left(\frac{T}{T_d} \right)^3 \int_0^{T_d/T} q^2 \, dq = 3NkT$$

From the definition of C_V,

$$C_V = \frac{1}{\mathfrak{N}} \left(\frac{\partial E_T}{\partial T} \right)_V = 3 \frac{N}{\mathfrak{N}} k = 3R$$

It is found experimentally that this result holds only at high temperatures.

38.20.   From the result of Problem 38.17 obtain an expression for C_V at an arbitrary temperature.

From Problem 38.16, ν_d is fixed when V is fixed.   Thus,

$$C_V = \frac{1}{\mathfrak{N}} \left(\frac{\partial E_T}{\partial T} \right)_V = \frac{N}{\mathfrak{N}} \frac{\partial}{\partial T} \left[\frac{9Nh}{\nu_d^3} \int_0^{\nu_d} \frac{\nu^3}{e^{h\nu/kT} - 1} \, d\nu \right]$$

$$= \frac{9Nkh^2}{\mathfrak{N}\nu_d^3 (kT)^2} \int_0^{\nu_d} \frac{\nu^4 e^{h\nu/kT}}{(e^{h\nu/kT} - 1)^2} \, d\nu = \frac{9Rh^2}{\nu_d^3 (kT)^2} \int_0^{\nu_d} \frac{\nu^4 e^{h\nu/kT}}{(e^{h\nu/kT} - 1)^2} \, d\nu$$

with $kN/\mathfrak{N} = R$.   In terms of the Debye temperature $T_d = h\nu_d/k$ and the variable $q = h\nu/kT$, this expression becomes

$$C_V = 9R \left(\frac{T}{T_d} \right)^3 \int_0^{T_d/T} \frac{q^4 e^q}{(e^q - 1)^2} \, dq$$

When this expression for C_V is plotted against T one obtains very close agreement with the experimental curve of Fig. 38-4.

38.21.   Find the total energy of an electron gas at low temperature.

The total energy is given by

$$E_T = \int_0^\infty E F_{FD} \, g(E) \, dE$$

where, from Problem 38.9, $g(E) = CE^{1/2}$.   Then, using (38.3) and keeping only the first term of the series,

$$E_T = C \int_0^\infty E^{3/2} F_{FD} \, dE = \frac{2C}{5} \left[E_f^{5/2} + 2(kT)^2 \left(\frac{1}{2} \right) \left(\frac{\pi^2}{6} \right) \left(\frac{5}{2} \cdot \frac{3}{2} E_f^{1/2} \right) \right]$$

$$= \frac{2C}{5} E_f^{5/2} \left[1 + \frac{5\pi^2}{8} \left(\frac{kT}{E_f} \right)^2 \right]$$

Using the results of Problem 38.12 expressing C and E_f in terms of E_{f0}, and recalling that E_f doesn't change rapidly with temperature, we can express E_T as follows:

$$E_T = \frac{2}{5} \left(\frac{3}{2} N E_{f0}^{-3/2} \right) E_{f0}^{5/2} \left[1 - \frac{\pi^2}{12} \left(\frac{kT}{E_{f0}} \right)^2 \right]^{5/2} \left[1 + \frac{5\pi^2}{8} \left(\frac{kT}{E_{f0}} \right)^2 \right]$$

$$\approx \frac{3}{5} N E_{f0} \left[1 + \frac{5\pi^2}{12} \left(\frac{kT}{E_{f0}} \right)^2 \right]$$

where, consistent with earlier approximations, we have kept only terms of the first order in $(kT/E_{f0})^2$.

38.22. For the electron gas of Problem 38.21 find to first order the molar electronic specific heat at constant volume, C_{Ve}.

$$C_{Ve} = \frac{1}{\mathfrak{N}} \left(\frac{\partial E_T}{\partial T} \right)_V = \frac{1}{\mathfrak{N}} \frac{\partial}{\partial T} \left\{ \frac{3}{5} N E_{f0} \left[1 + \frac{5\pi^2}{12} \left(\frac{kT}{E_{f0}} \right)^2 \right] \right\}$$

$$= \frac{3}{5} \frac{N}{\mathfrak{N}} E_{f0} \left(\frac{5}{6} \frac{\pi^2 k^2 T}{E_{f0}^2} \right) = \frac{N\pi^2 k^2}{2\mathfrak{N} E_{f0}} T = \frac{R\pi^2 k}{2 E_{f0}} T$$

with $R = Nk/\mathfrak{N}$.

38.23. Estimate the electronic molar specific heat, C_{Ve}, for silver at room temperature ($T = 300$ K). Silver has a Fermi energy $E_{f0} = 5.5$ eV.

From Problem 38.22,

$$C_{Ve} = \frac{R\pi^2 kT}{2 E_{f0}} = \frac{(8.31 \text{ J/mol} \cdot \text{K})\pi^2 (8.62 \times 10^{-5} \text{ eV/K})(300 \text{ K})}{2(5.5 \text{ eV})} = 0.19 \text{ J/mol} \cdot \text{K}$$

This value is per mole of atoms. However, since silver has valence 1, a mole of atoms corresponds to a mole of electrons. Note that at $T = 300$ K the Debye theory predicts a specific heat due to lattice vibrations of about $3R \approx 25$ J/mol·K (the Dulong-Petit limit), so at this temperature the electron specific heat can be neglected.

38.24. Refer to Problem 38.23. For silver find the temperature at which the electronic molar specific heat, C_{Ve}, and the lattice molar specific heat, C_V, are equal. The Debye temperature for silver is 210 K. The equality occurs at low temperature, so the Debye-theory result can be taken as

$$C_V = \frac{12\pi^4 R}{5} \left(\frac{T}{T_d} \right)^3$$

(see Problem 38.18).

Equating C_V and C_{Ve},

$$\frac{12\pi^4 R}{5} \left(\frac{T}{T_d} \right)^3 = \frac{R\pi^2 kT}{2 E_{f0}}$$

$$T = \left(\frac{5k T_d^3}{24\pi^2 E_{f0}} \right)^{1/2} = \left[\frac{5(8.62 \times 10^{-5} \text{ eV/K})(210 \text{ K})^3}{24\pi^2 (5.5 \text{ eV})} \right]^{1/2} = 1.75 \text{ K}$$

This result shows that the electronic heat conduction will be noticeable only at very low temperatures.

THE QUANTUM MECHANICAL IDEAL GAS

The assumption made for an ideal gas is that the gas molecules are noninteracting. However, since an ideal gas will be composed of either fermions or bosons, it should be analyzed from a quantum statistical viewpoint.

As expected, it is found that at high temperatures the quantum mechanical results go over into those obtained from a classical treatment. At temperatures approaching absolute zero, however, a quantum mechanical analysis predicts marked differences from what is expected classically. For example, it is predicted for a gas of bosons that at a low, but nonzero, temperature, all the particles will be found in the lowest energy state, a phenomenon called *Bose-Einstein condensation* (Problem 38.26). Also, even at absolute zero, a fermion gas will have a finite, nonzero pressure (Problem 38.33). Some gases that obey Bose-Einstein statistics are H_2 and helium, while a gas that follows Fermi-Dirac statistics is atomic hydrogen.

38.25. Consider a gas composed of a fixed number, N, of bosons in a container of volume V. Show that α is a strictly increasing function of the temperature T.

The normalization condition is

$$N = \int dn_E = \int_0^\infty F_{BE}g(E)\,dE = \frac{2\pi V(2m)^{3/2}}{h^3} \int_0^\infty \frac{E^{1/2}\,dE}{e^\alpha e^{E/kT} - 1}$$

where the density of states, $g(E)$, was obtained in Problem 38.7. Rewriting this expression in terms of the variable $q = E/kT$, we obtain

$$N = \frac{2\pi V(2mkT)^{3/2}}{h^3} \int_0^\infty \frac{q^{1/2}\,dq}{e^\alpha e^q - 1} \tag{1}$$

and this equation implicitly defines α as a function of T. As T increases, the expression multiplying the integral increases. Therefore, since N is fixed, the integral decreases, which implies that α increases. Thus, α is a strictly increasing function of T.

Notice that, because N is finite, the integral must always converge, and so α must always be nonnegative.

38.26. Refer to Problem 38.25. For a given particle density, N/V, find the lowest possible temperature of the boson gas consistent with Bose-Einstein statistics.

Since α steadily decreases as the temperature decreases, yet cannot go negative, the minimum temperature, T_0, is that for which $\alpha = 0$. From (1) of Problem 38.25,

$$N = \frac{2\pi V(2mkT_0)^{3/2}}{h^3} \int_0^\infty \frac{q^{1/2}\,dq}{e^q - 1}$$

By (38.5) the integral has the value

$$\Gamma\left(\frac{3}{2}\right)\zeta\left(\frac{3}{2}\right) = \frac{1}{2}\sqrt{\pi}\ \zeta\left(\frac{3}{2}\right) = \frac{1}{2}\sqrt{\pi}\ (2.61)$$

Substituting this value and solving for T_0, we find

$$T_0 = \frac{(hc)^2(2.61)^{-2/3}}{2\pi(mc^2)k}\left(\frac{N}{V}\right)^{2/3} = \frac{(12.4 \times 10^{-7}\ \text{eV}\cdot\text{m})^2(2.61)^{-2/3}}{2\pi(mc^2)(8.62 \times 10^{-5}\ \text{eV/K})}\left(\frac{N}{V}\right)^{2/3}$$

$$= (1.50 \times 10^{-9}\ \text{eV}\cdot\text{K}\cdot\text{m}^2)\frac{(N/V)^{2/3}}{mc^2}$$

At $T \approx T_0$ most of the noninteracting bosons will be in the lowest ($E = 0$) state and the system can be said to be in a condensed state. This phenomenon is referred to as *Bose-Einstein condensation*.

38.27. Evaluate the total energy of the Bose gas of Problem 38.25.

Using the change of variable $q = E/kT$ and (38.4),

$$E_T = \int E\,dn_E = \int_0^\infty EF_{BE}\,g(E)\,dE = \frac{2\pi V(2m)^{3/2}}{h^3}\int_0^\infty \frac{E^{3/2}\,dE}{e^\alpha e^{E/kT} - 1}$$

$$= \frac{2\pi V(2m)^{3/2}(kT)^{5/2}}{h^3}\int_0^\infty \frac{q^{3/2}\,dq}{e^\alpha e^q - 1} = \frac{3V(2\pi mkT)^{3/2}kT}{2h^3}\sum_{n=1}^\infty \frac{e^{-n\alpha}}{n^{5/2}}$$

$$= \frac{3V(2\pi mkT)^{3/2}kT}{2h^3}\sum_{n=1}^\infty \frac{Z^n}{n^{5/2}}$$

where $Z = e^{-\alpha}$.

38.28. Consider a gas composed of N spin-$\frac{1}{2}$ fermions in a container of volume V. Show that α is a strictly increasing function of temperature.

From Problem 38.9, the normalization condition is

$$N = \int dn_E = 2 \int_0^\infty F_{FD}\, g(E)\, dE = \frac{4\pi V(2m)^{3/2}}{h^3} \int_0^\infty \frac{E^{1/2}\, dE}{e^\alpha e^{E/kT} + 1} = \frac{4\pi V(2mkT)^{3/2}}{h^3} \int_0^\infty \frac{q^{1/2}\, dq}{e^\alpha e^q + 1}$$

with $q = E/kT$. The result now follows by the argument of Problem 38.25.

For the Fermi-Dirac distribution, α may assume negative values. It is seen that α also increases monotonically with the mass of the fermions in the system. This fact accounts for the result that at room temperature ($T \approx 300$ K) an electron "gas" exhibits essentially the same quantum behavior as at very low temperatures (see Problem 38.13), while a gas composed of molecules, which are roughly 2000 times more massive than an electron, shows nearly classical behavior at room temperature.

38.29. Evaluate the total energy of the fermion system of Problem 38.28, assuming that $\alpha \geqslant 0$.

The total energy, E_T, is found by use of (38.4) to be

$$E_T = \int E\, dn_E = 2 \int_0^\infty E F_{FD}\, g(E)\, dE$$

$$= \frac{4\pi V(2m)^{3/2}}{h^3} \int_0^\infty \frac{E^{3/2}\, dE}{e^\alpha e^{E/kT} + 1} = \frac{4\pi V(2mkT)^{3/2}kT}{h^3} \int_0^\infty \frac{q^{3/2}\, dq}{e^\alpha e^q + 1}$$

$$= \frac{4\pi V(2mkT)^{3/2}kT}{h^3} \left(\frac{3}{2}\right)\left(\frac{1}{2}\right)\sqrt{\pi} \sum_{n=1}^\infty (-1)^{n+1} \frac{e^{-n\alpha}}{n^{5/2}}$$

$$= \frac{3V(2\pi mkT)^{3/2}kT}{h^3} \sum_{n=1}^\infty (-1)^{n+1} \frac{Z^n}{n^{5/2}}$$

where $Z = e^{-\alpha}$.

38.30. Find the average kinetic energy per particle in the high-temperature limit $T \gg 0$ K for the boson and fermion gases of Problems 38.25 and 38.28.

Using (38.4) to evaluate the integrals for N for the two gases, and using the expressions for E_T found in Problems 38.27 and 38.29, we obtain

$$E_{\text{avg}} = \frac{E_T}{N} = \frac{3}{2} kT \frac{\displaystyle\sum_{n=1}^\infty \epsilon^{n+1} \frac{Z^n}{n^{5/2}}}{\displaystyle\sum_{n=1}^\infty \epsilon^{n+1} \frac{Z^n}{n^{3/2}}}$$

where $Z = e^{-\alpha}$, $\epsilon = +1$ for bosons, and $\epsilon = -1$ for fermions. In the high-temperature limit, $\alpha \to \infty$ (Problems 38.25 and 38.28) and so $Z \to 0$. Thus, keeping only the first two terms in each summation, we obtain, after canceling the common factor $\epsilon^2 Z$,

$$E_{\text{avg}} \approx \frac{3}{2} kT \left(\frac{1 + \epsilon Z/2^{5/2}}{1 + \epsilon Z/2^{3/2}}\right) \approx \frac{3}{2} kT \left(1 + \epsilon \frac{Z}{2^{5/2}} - \epsilon \frac{Z}{2^{3/2}}\right)$$

$$= \frac{3}{2} kT \left(1 - \epsilon \frac{Z}{2^{5/2}}\right) = \frac{3}{2} kT \left(1 - \epsilon \frac{e^{-\alpha}}{2^{5/2}}\right)$$

It is seen that at high temperature E_{avg} for both gases is asymptotically equal to the classical value, $\frac{3}{2} kT$. The first-order correction to the classical result has the same magnitude for both gases, but the average energy per particle is higher than the classical result for fermions and lower than the classical result for bosons.

38.31. From the first law of thermodynamics, $dE_T = dQ - p\,dV$, for a reversible process, it is seen that the pressure p may be written as

$$p = -\left(\frac{\partial E_T}{\partial V}\right)_Q$$

Obtain an expression for the pressure of a quantum mechanical gas in terms of E_T.

We recall from Problem 38.6 that the energy levels for noninteracting particles in a 3-dimensional cubical box of side l are given by

$$E_i = \frac{h^2}{8ml^2}\left(n_x^2 + n_y^2 + n_z^2\right) = \frac{h^2}{8mV^{2/3}}\left(n_x^2 + n_y^2 + n_z^2\right) \tag{1}$$

Because of degeneracy there will be a number of states (n_x, n_y, n_z) corresponding to a given E_i (see Problem 38.6). Let n_{ij} denote the number of particles in the jth state belonging to level E_i. Then the total energy of the system is

$$E_T = \sum_i \sum_j E_i n_{ij}$$

from which

$$dE_T = \sum_i \sum_j E_i\,dn_{ij} + \sum_i \sum_j n_{ij}\,dE_i$$

Since E_i depends on the dimensions of the system, the second term, $\sum_i \sum_j n_{ij}\,dE_i$, corresponds to changing E_T by changing the dimensions or volume of the system by doing work on the system. Thus the first term must be equal to the heat absorbed by the system:

$$dQ = \sum_i \sum_j E_i\,dn_{ij}$$

In other words, a flow of heat into the system corresponds to a change in the occupation numbers n_{ij}. If $Q = \text{const.}$, so that $dQ = 0$, we see that

$$p = -\left(\frac{\partial E_T}{\partial V}\right)_Q = -\sum_i \sum_j n_{ij}\frac{\partial E_i}{\partial V}$$

But, from (1),

$$\frac{\partial E_i}{\partial V} = -\frac{2}{3}\frac{h^2}{8mV^{5/3}}\left(n_x^2 + n_y^2 + n_z^2\right) = -\frac{2}{3V}E_i$$

and so

$$p = \frac{2}{3V}\sum_i \sum_j n_{ij}E_i = \frac{2}{3}\frac{E_T}{V}$$

This relation is formally identical to the classical result (35.6).

38.32. Refer to Problems 38.30 and 38.31. Obtain the first-order correction at high temperatures to the classical ideal gas law, $pV = NkT$, when the quantum mechanical properties of the gas are taken into consideration.

From Problem 38.30,

$$E_T = NE_{\text{avg}} = \frac{3}{2}NkT\left(1 - \epsilon\frac{e^{-\alpha}}{2^{5/2}}\right)$$

where

$$\epsilon = \begin{cases} +1 & \text{for a boson gas} \\ -1 & \text{for a fermion gas} \end{cases}$$

Substituting this into the result of Problem 38.31, we obtain

$$pV = NkT\left(1 - \epsilon\frac{e^{-\alpha}}{2^{5/2}}\right)$$

38.33. Refer to Problems 38.21 and 38.31. Obtain an expression for the pressure of an ideal spin-$\frac{1}{2}$ fermion gas at very low temperatures.

From Problem 38.21 we have for a spin-$\frac{1}{2}$ fermion system at low temperatures

$$E_T = \frac{3}{5} N E_{f0} \left[1 + \frac{5\pi^2}{12} \left(\frac{kT}{E_{f0}} \right)^2 \right]$$

Substituting this into the result of Problem 38.31, we obtain

$$p = \frac{2}{5} \frac{N E_{f0}}{V} \left[1 + \frac{5\pi^2}{12} \left(\frac{kT}{E_{f0}} \right)^2 \right]$$

This result shows that as $T \to 0$ K, the pressure of a fermion gas approaches a finite value. This zero-point pressure occurs because even at 0 K the fermions have, as a result of the Pauli exclusion principle, a finite energy, as discussed in Problem 38.11.

DERIVATION OF THE QUANTUM DISTRIBUTION FUNCTIONS

38.34. For a given energy level E_i in a system of fermions there will be a certain number g_i of states that will have this energy (i.e. E_i will have a degeneracy of order g_i). The maximum number of fermions that can occupy this level will therefore be g_i, since, according to the Pauli exclusion principle, no more than one particle can be found in each state. Find the number of different ways that N identical, indistinguishable fermions can be distributed among energy levels $E_1, E_2, \ldots, E_i, \ldots$, such that the ith energy level will have $n_i \leqslant g_i$ filled states $(n_1 + n_2 + \cdots = N)$.

There is *one* way of assigning n_1 particles to E_1, n_2 particles to $E_2, \ldots$; since the particles are mutually indistinguishable. Next, out of the g_1 states in E_1, we pick the n_1 states that are to be filled by the n_1 particles; this can be done in

$$\binom{g_1}{n_1} = \frac{g_1!}{n_1! \, (g_1 - n_1)!}$$

different ways.   For each of these there are

$$\binom{g_2}{n_2} = \frac{g_2!}{n_2! \, (g_2 - n_2)!}$$

different ways of filling E_2; and so on.   The total number of ways is thus

$$X = 1 \times \binom{g_1}{n_1}\binom{g_2}{n_2} \cdots = \frac{g_1!}{n_1! \, (g_1 - n_1)!} \, \frac{g_2!}{n_2! \, (g_2 - n_2)!} \, \cdots \qquad (1)$$

38.35. Find the most probable distribution for Problem 38.34, subject to the conditions that the number of particles and total energy are held fixed.

The most probable distribution is that which can be realized in the greatest number of ways.   Thus, we must maximize the function $X(n_1, n_2, \ldots)$, as given by (1) of Problem 38.34, subject to the constraints

$$\sum_i n_i = N = \text{constant} \qquad \sum_i E_i n_i = E_T = \text{constant}$$

(The additional constraints $n_i \leqslant g_i$ can be ignored, because we know in advance that the maximum will satisfy them.)   Going over to a continuous problem for the function

$$\ln \frac{X}{g_1! \, g_2! \cdots}$$

using Stirling's formula, and applying the method of Lagrange multipliers, all exactly as in Problem 37.16, we obtain the conditions

$$n_i = \frac{g_i}{e^{\alpha} e^{\beta E_i} + 1}$$

Since n_i is the number of fermions with the energy E_i, and g_i is the number of states with energy E_i, n_i/g_i is the average number of fermions in a state at energy E_i; that is, n_i/g_i, is the Fermi-Dirac distribution function, F_{FD}:

$$F_{FD} = \frac{1}{e^{\alpha} e^{\beta E_i} + 1}$$

38.36. Integer-spin particles (bosons) do not obey the Pauli exclusion principle, so any number of bosons may be found in a given state. Suppose at the energy E_i there are g_i states. Find the number of different ways that N identical, indistinguishable bosons can be distributed among energy levels $E_1, E_2, \ldots, E_i, \ldots$, with a fixed number n_i of particles in the ith level.

Consider first one of the levels, E_i, with g_i states and n_i particles. This level can be pictured as n_i particles in a row divided arbitrarily into the g_i states by $g_i - 1$ lines (see the example in Fig. 38-5). The number of different ways the n_i bosons can be placed in the g_i states, without any limit to the number of particles in a state and temporarily regarding the particles as distinguishable, is equal to the number of different ways the $n_i + g_i - 1$ particles and dividing lines can be permuted, $(n_i + g_i - 1)!$, divided by (since in our picture the interchange of two dividing lines corresponds to the same state) the number of ways the $g_i - 1$ lines can be permuted, $(g_i - 1)!$:

$$\frac{(n_i + g_i - 1)!}{(g_i - 1)!}$$

Because, however, the particles are actually indistinguishable from each other, so that the interchange of two particles results in the same distribution, the above expression must be divided by the $n_i!$ different ways the particles can be permuted, to get the actual number of different ways the ith level can be formed:

$$\frac{(n_i + g_i - 1)!}{n_i! \, (g_i - 1)!}$$

Therefore, the total number X of different ways there are to arrange $n_1, n_2, \ldots$ bosons in the energy levels $E_1, E_2, \ldots$ if there are $g_1, g_2, \ldots$ states in each level is

$$X = \frac{(n_1 + g_1 - 1)!}{n_1!(g_1 - 1)!} \; \frac{(n_2 + g_2 - 1)!}{n_2!(g_2 - 1)!} \; \cdots \qquad\qquad (1)$$

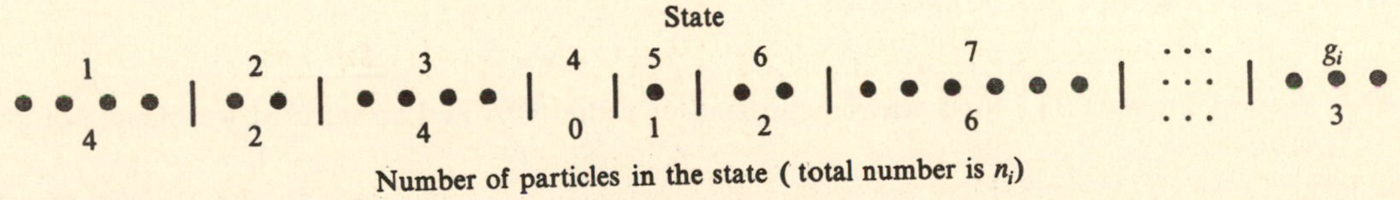

State

Number of particles in the state (total number is n_i)

Fig. 38-5

38.37. For Problem 38.36 find the most probable distribution if the number of particles and total energy are held constant.

Proceeding exactly as in Problems 37.16 and 38.35, we maximize the function X from Problem 38.36 —or, more conveniently, the function

$$\ln \{ [(g_1 - 1)! \, (g_2 - 1)! \cdots] X \}$$

—subject to the constraints

$$\sum_i n_i = N = \text{constant} \qquad\qquad \sum_i E_i n_i = E_T = \text{constant}$$

The conditions for a maximum are found to be

$$\frac{n_i}{n_i + g_i - 1} = e^{-\alpha}e^{-\beta E_i}$$

If we assume $n_i + g_i \gg 1$ we have

$$\frac{n_i}{n_i + g_i} = e^{-\alpha}e^{-\beta E_i} \qquad \text{or} \qquad n_i = \frac{g_i}{e^{\alpha}e^{\beta E_i} - 1}$$

Since n_i is the number of bosons with the energy E_i, and g_i is the number of states with the energy E_i, n_i/g_i is the average number of bosons in a state at energy E_i; that is, n_i/g_i is the Bose-Einstein distribution function, F_{BE}:

$$F_{BE} = \frac{1}{e^{\alpha}e^{\beta E_i} - 1}$$

38.38. Show that a system of bosons which does not have a fixed number of particles (e.g. a system of photons in a blackbody cavity) has a Bose-Einstein distribution with $\alpha = 0$.

In the derivation given in Problem 38.37, setting $\alpha = 0$ is equivalent to dropping the constraint

$$\sum_i n_i = N = \text{constant}$$

Supplementary Problems

BLACKBODY RADIATION

38.39. What is the density of photon states in the energy interval between E and $E + dE$ in a cubical cavity of volume $V = l^3$? *Ans.* $8\pi V E^2/(hc)^3$

38.40. What is the amount of energy per wavelength interval emitted by a blackbody?

Ans. $\dfrac{8\pi V hc}{\lambda^5} \dfrac{1}{e^{hc/\lambda kT} - 1}$

38.41. Replot Fig. 38-1 as $F(\lambda)$ versus λ.

38.42. Repeat Problem 38.5 for a wavelength band of width 50 Å and an infrared wavelength of 25,000 Å. *Ans.* 33.8

FREE ELECTRON THEORY OF METALS

38.43. What is the order of degeneracy of the $57E_0$ level of a three-dimensional, cubical infinite well? *Ans.* 6

38.44. For a cubical, three-dimensional infinite well, what is the density of states in the momentum interval between p and $p + dp$? *Ans.* $4\pi l^3 p^2/h^3$

38.45. At $T = 0$ K, what is the rms speed, in terms of the Fermi energy E_{f0}, in an electron gas in a metal? *Ans.* $v(\text{m/s}) = (4.59 \times 10^5)\sqrt{E_{f0}(\text{eV})}$

38.46. What is the next term in the expansion of (1), Problem 38.12?

Ans. $\dfrac{7}{640}\left(\dfrac{\pi kT}{E_f}\right)^4$

38.47. At $T = 300$ K it is found that the Fermi energy of a certain material is reduced from its value at 0 K by 1.2×10^{-4} eV. What is the Fermi energy of this material at 0 K?   Ans.   4.58 eV

38.48. The average electrostatic potential energy is 11.2 eV for a material whose work function is 3.4 eV.   What is the Fermi energy of the material?   Ans.   7.8 eV.

38.49. At $T = 0$ K, what is the ratio of the maximum speed to the rms speed in an electron gas in a metal? Ans.   1.29

SPECIFIC HEATS OF CRYSTALLINE SOLIDS

38.50. In a crystalline solid, what is the density of states in the wavelength interval between λ and $\lambda + d\lambda$?

Ans. $4\pi V\left(\dfrac{2}{v_t^3} + \dfrac{1}{v_l^3}\right)\dfrac{c^3}{\lambda^4}$

38.51. Refer to Problem 38.23.   For silver find the temperature at which the electronic molar specific heat, C_{Ve}, is 5% of the lattice molar specific heat, C_V.   Ans.   7.83 K

38.52. In a quantum mechanical two-state system with energy levels $E_1 = 0$ and $E_2 = \epsilon$ the probability of finding a particle in a state of energy E is proportional to the Boltzmann factor $e^{-E/kT}$.   Find the total energy of a system of N particles.

Ans. $E_T = \dfrac{N\epsilon e^{-\epsilon/kT}}{1 + e^{-\epsilon/kT}}$

38.53. For the system of Problem 38.52 find the specific heat at constant volume.

Ans. $C_V = \dfrac{Nk(\epsilon/kT)^2 e^{-\epsilon/kT}}{(1 + e^{-\epsilon/kT})^2}$

38.54. Find the low- and high-temperature limits of C_V in Problem 38.53.   Ans.   0 and 0

THE QUANTUM MECHANICAL IDEAL GAS

38.55. What is the rms speed in the Bose gas of Problem 38.25?

Ans. $\left(\dfrac{3kT}{m}\sum\limits_{n=1}^{\infty}\dfrac{e^{-n\alpha}}{n^{5/2}} \Big/ \sum\limits_{n=1}^{\infty}\dfrac{e^{-n\alpha}}{n^{3/2}}\right)^{1/2}$

38.56. Find the temperature T_0 (see Problem 38.26) for liquid ^{4}He, whose density is 0.146×10^3 kg/m^3.
Ans.   3.16 K (Although liquid helium is not a true noninteracting Bose gas, the approximation gives a result quite close to 2.2 K, the temperature at which a rapid increase in the ground state population is observed, corresponding to the change from a normal fluid to a superfluid.)

38.57.   What is the rms speed in the Fermi gas of Problem 38.28?   Assume that $\alpha \geqslant 0$.

$Ans.$ $\left[\dfrac{3kT}{m} \sum\limits_{n=1}^{\infty} (-1)^{n+1} \dfrac{e^{-n\alpha}}{n^{5/2}} \Big/ \sum\limits_{n=1}^{\infty} (-1)^{n+1} \dfrac{e^{-n\alpha}}{n^{3/2}} \right]^{1/2}$

38.58.   Find the zero-point pressure in the electron gas in silver ($E_{f0} = 5.5$ eV; density $= 10.5 \times 10^3$ kg/m^3).
        $Ans.$ 2.06×10^{10} N/m$^2 \approx 2 \times 10^5$ atm

Chapter 39

The Band Theory of Solids

The free electron theory of metals, discussed in Chapter 38, does not explain why some materials are good conductors of electricity, while others, *insulators*, are poor conductors; and why some materials, called *semiconductors*, have conducting properties somewhere in between the previous two. A more successful theory, called the *band theory* of solids, explains why the ratios of resistivities of solids may be as large as 10^{30}.

A solid can be thought of as being formed by bringing together isolated single atoms. Any single atom will possess a large number of discrete energy levels that can be occupied by the electrons of the atom. Normally the electrons exist in the ground state, occupying only the lowest-lying energy levels. It is, of course, possible to excite the electrons into higher energy levels. Usually only the highest-energy, or valence, electrons will participate in these excitations. Consider first the combination of two atoms. If there were no interaction between the two atoms, the value of each energy level would be the same as for each isolated atom, with the number of levels at a particular energy simply being doubled, as shown in Fig. 39-1(*b*). Because of interactions, however, each previously single energy level is split into two levels, as shown in Fig. 39-1(*c*). In a similar fashion, if more atoms were brought together there would be a larger number of splittings of each energy level, one split for each additional atom. Figure 39-1(*d*) shows the splitting of energy levels when five atoms are brought together.

Because there are of the order of 10^{23} atoms/cm^3 in solids, each previously single energy level of an isolated atom will be split into an enormous number of parts. Since the values of the energy levels remain approximately the same, the net effect of assembling a large number of interacting atoms is to form bands of practically continuous energy levels, separated by gaps where no electron states exist, as illustrated in Fig. 39-1(*e*).

The way the electrons occupy the available bands is governed by the Pauli exclusion principle. The bands will fill with electrons in the same manner as the electron states were filled in many-electron atoms (Chapter 24). As an example, $_{11}$Na has all its energy levels filled up to the $3s$ level, which has one electron, and its electron configuration is $1s^2\, 2s^2\, 2p^6\, 3s^1$. Since the $3s$ level can accommodate two electrons, it is only half-filled.

In a similar fashion, the bands in solids may be filled, partially filled or empty, as shown in Fig. 39-2. The highest energy band occupied by the valence electrons and the unoccupied band directly above it determine the conduction properties of a crystalline solid. If the band containing the valence electrons is filled, it will be referred to as the *valence band*, and the next-higher band will be referred to as the *conduction band*; if the band containing the valence electrons is not filled, *it* will be called the conduction band. A good conductor has a conduction band that is approximately half-filled [Fig. 39-2(*a*)], or else the conduction band overlaps the next-higher band [Fig. 39-2(*b*)]. In this situation it is very easy to raise a valence electron to a higher energy level, so these electrons can easily acquire energy from an electric field to participate in electrical conduction.

An insulating material has a filled valence band, and the gap to the conduction band is large [Fig. 39-2(*c*)]. As a result, electrons cannot easily acquire energy from an electric field, so they cannot participate in electrical conduction.

Some materials have a filled valence band, like an insulator, but a small gap to the conduction band [Fig. 39-2(*d*)]. At $T = 0$ K the valence band is completely filled and the conduction band is empty, so the material behaves like an insulator. At room temperature, however, some of the electrons acquire sufficient thermal energy to be found in the conduction band, where they can participate in electrical conduction. In addition, these electrons leave behind unfilled "holes" into

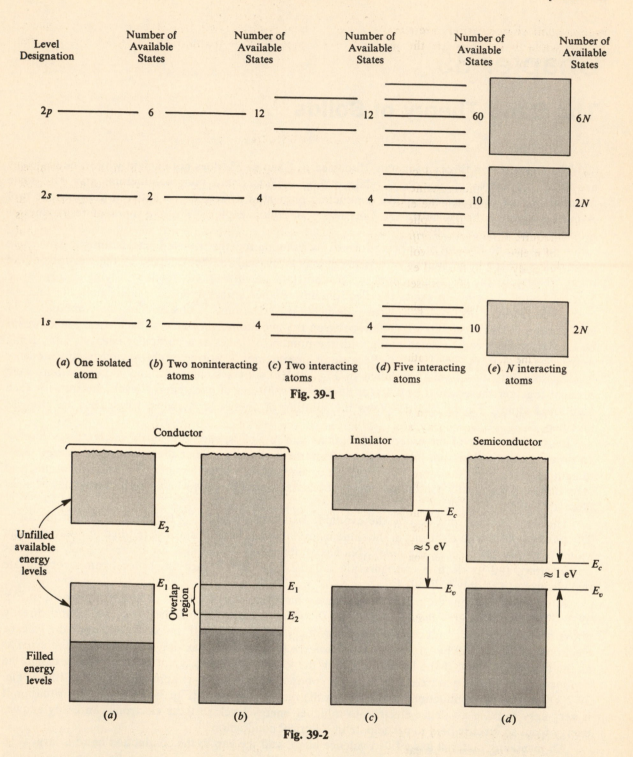

Level Designation	Number of Available States	Number of Available States	Number of Available States	Number of Available States	Number of Available States
2p	6	12	12	60	6N
2s	2	4	4	10	2N
1s	2	4	4	10	2N

(a) One isolated atom   (b) Two noninteracting atoms   (c) Two interacting atoms   (d) Five interacting atoms   (e) N interacting atoms

Fig. 39-1

Fig. 39-2

which other electrons in the valence band can move during electrical conduction. The excitation of electrons into these holes has the net effect of positive charge carriers supporting electrical conduction.

The semiconductors just described are called *intrinsic* semiconductors. It is possible, however, by introducing the proper impurities into a material, to control whether the electrical conduction will be primarily by electron (negative) or hole (positive) charge carriers. Such "doped" semiconductors are called *extrinsic* semiconductors, and serve as the foundation for semiconductor devices. If the predominant charge carriers are electrons, the material is called an *n-type* (for "negative") semiconductor; while if the holes are the predominant charge carriers, the material is called a *p-type* (for "positive") semiconductor.

Solved Problems

39.1. A model for electrical current conduction in a metal is to treat the metal as a contained electron gas, the container being the metallic lattice. When an electrical potential is applied across the material, the electrons, which are moving in a random manner, are accelerated in the direction of the applied electric field, and after many collisions with the heavy lattice ions acquire a net average *drift velocity*, v_d, which results in a net electrical current. For a material of n electrons per unit volume, where the electron mean free path is taken as λ and the average velocity due to thermal excitation is $\bar{v}$, find the resistivity. For simplicity, assume the volume V to be a box of cross-sectional area A and length l.

If a potential ϕ is applied across the material along the side of length l, the force on an electron is

$$F = eE = e\frac{\phi}{l}$$

and the electron's acceleration is

$$a = \frac{F}{m_e} = \frac{e\phi}{m_e l}$$

The average time between the electron's collisions with the lattice is $t = \lambda/\bar{v}$. The velocity, over and above its random velocity, acquired during this time interval is the drift velocity

$$v_d = at = \frac{e\phi\lambda}{m_e l\bar{v}}$$

(More exactly, $v_d = at'$, where t' is the average time *since the last collision*. However, under the assumption that collisions are independent random events, it can be shown that the intercollision time has an exponential distribution, (*1*) of Problem 29.1. It can then be shown that the expected time since the last collision equals the expected time between collisions, i.e. $t' = t$.)

The current, I, is the rate at which charge is transported through the cross-sectional area A. The random part of the electrons' motion produces no net transport, and so

$$I = (ne)Av_d = \frac{ne^2\lambda A}{m_e l\bar{v}}\phi$$

which is Ohm's law, $I = \phi/R$. Defining the *resistivity* ρ of the material by

$$R = \rho\frac{l}{A}$$

we have

$$\rho = \frac{m_e\bar{v}}{ne^2\lambda}$$

39.2. Refer to Problem 39.1. Estimate the resistivity of silver ($A = 108$), whose density is 10.5×10^3 kg/m^3, assuming each atom contributes one electron (the valence electron) for conduction. For purposes of this estimation take λ equal to 100 times the atomic spacing d, and v_d equal to the velocity corresponding to the Fermi energy, $E_{f0} = 5.5$ eV.

The velocity corresponding to the Fermi energy is given by

$$v_f = \left(\frac{2E_{f0}}{m_e c^2}\right)^{1/2}c = \left(\frac{2 \times 5.5 \text{ eV}}{0.511 \times 10^6 \text{ eV}}\right)^{1/2}(3 \times 10^8 \text{ m/s}) = 1.39 \times 10^6 \text{ m/s}$$

Since each atom contributes one electron, the atomic density equals the electron density:

$$n = \frac{(6.02 \times 10^{26} \text{ atoms/kmol})(10.5 \times 10^3 \text{ kg/m}^3)}{108 \text{ kg/kmol}} \left(\frac{1 \text{ electron}}{\text{atom}} \right) = 5.85 \times 10^{28} \frac{\text{electrons}}{\text{m}^3}$$

If each atom takes up a volume of approximately d^3, we then have

$$\frac{\lambda}{100} = d = \left(\frac{1}{5.85 \times 10^{28} \text{ m}^{-3}} \right)^{1/3} = 2.58 \times 10^{-10} \text{ m} \quad \text{or} \quad \lambda = 2.58 \times 10^{-8} \text{ m}$$

Using the result of Problem 39.1, we have

$$\rho = \frac{m_e v_f}{e^2 n \lambda} = \frac{(9.11 \times 10^{-31} \text{ kg})(1.39 \times 10^6 \text{ m/s})}{(1.6 \times 10^{-19} \text{ C})^2 (5.85 \times 10^{28} \text{ m}^{-3})(2.58 \times 10^{-8} \text{ m})} = 3.29 \times 10^{-8} \ \Omega \cdot \text{m}$$

Experimentally it is found that the resistivity of silver varies from $1.5 \times 10^{-8} \ \Omega \cdot \text{m}$ at 0 °C to $6.87 \times 10^{-8} \ \Omega \cdot \text{m}$ at 800 °C. It is thus seen that our crude approximation gives reasonable agreement with what is observed.

39.3. As in Problem 39.2, estimate the resistivity of silicon ($A = 28$), which has a density of $2.42 \times 10^3 \text{ kg/m}^3$ and a valence of 2. Assume the Fermi energy for silicon is about 5 eV.

Proceeding as in Problem 39.2, we find

$$v_f = \left(\frac{2 E_{f0}}{m_e c^2} \right)^{1/2} c = \left(\frac{2 \times 5 \text{ eV}}{0.511 \times 10^6 \text{ eV}} \right)^{1/2} (3 \times 10^8 \text{ m/s}) = 13.3 \times 10^5 \text{ m/s}$$

$$n = \frac{\left(6.02 \times 10^{26} \frac{\text{atoms}}{\text{kmol}} \right)(2.42 \times 10^3 \text{ kg/m}^3)}{28 \text{ kg/kmol}} \left(\frac{2 \text{ electrons}}{\text{atom}} \right) = 1.04 \times 10^{29} \frac{\text{electrons}}{\text{m}^3}$$

The number of atoms per m^3 is one-half the electron density, and if each atom takes up a volume of approximately d^3, we have

$$\lambda = 100d = 100 \left[\frac{1}{\frac{1}{2}(1.04 \times 10^{29} \text{ m}^{-3})} \right]^{1/3} = 2.68 \times 10^{-8} \text{ m}$$

The resistivity is then

$$\rho = \frac{m_e v_f}{e^2 n \lambda} = \frac{(9.11 \times 10^{-31} \text{ kg})(13.3 \times 10^5 \text{ m/s})}{(1.6 \times 10^{-19} \text{ C})^2 (1.04 \times 10^{29} \text{ m}^{-3})(2.68 \times 10^{-8} \text{ m})} = 1.7 \times 10^{-8} \ \Omega \cdot \text{m}$$

At room temperature the resistivity of silicon is found to be about $10^3 \ \Omega \cdot \text{m}$, much larger than predicted by this calculation. The reason for the discrepancy is that we have assumed the valence electrons will produce conduction. However, silicon is a semiconductor, so that there is a gap between the valence and conduction bands. Only those electrons which have energies sufficient for them to be found in the conduction band will support electrical conduction.

39.4. Show that in an intrinsic semiconductor the number of holes with energies between E and $E + dE$ is given by

$$dn_h = (1 - F_{FD}) g(E) dE$$

where $g(E)$ is the density of states in the valence band.

At $T = 0$ K in an intrinsic semiconductor the valence band is filled and no electrons are in the conduction band. For $T > 0$ K some electrons are excited to the conduction band, and their number equals the number of holes created in the valence band. This is illustrated in Fig. 39-3. The distribution function for the number of holes is the shaded area in Fig. 39-3(b). Therefore, the number of holes with energies between E and $E + dE$ is given by

$$dn_h = (1 - F_{FD}) g(E) dE$$

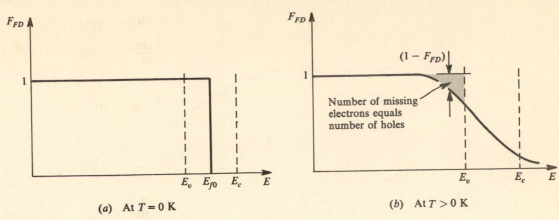

(a) At $T = 0$ K   (b) At $T > 0$ K

Fig. 39-3

39.5. For a semiconductor it can be shown that in the region immediately above the conduction band and below the valence band the density of states functions $g(E)$ are approximately symmetrical about the midpoint of the gap, E_m, as shown in Fig. 39-4(b). (Of course, in the gap itself the density of states will be zero.) Show that the Fermi energy level will lie at the midpoint of the gap.

Because one hole is produced in the valence band for each electron in the conduction band, the total number of electrons, N_e, in the conduction band must be equal to the total number of holes, N_h, in the valence band. The number of electrons, dn_e, with energies between E and $E + dE$ in the conduction band is.

$$dn_e = F_{FD}g(E)\,dE$$

while in the valence band the number of holes, dn_h, with energies between E' and $E' + dE'$ is (see Problem 39.4)

$$dn_h = (1 - F_{FD})\,g(E')\,dE'$$

Therefore,

$$\int_{E_c}^{E_c + E_v} F_{FD}(E)g(E)\,dE = \int_0^{E_v}[1 - F_{FD}(E')]\,g(E')\,dE'$$

We take the upper limit on the left-hand side as $E_c + E_v$ because $g(E') = 0$ for $E' < 0$ and so, by symmetry, $g(E) = 0$ for $E > E_c + E_v$. Changing the variables to

$$x = E - E_m \qquad x = E_m - E'$$

in the left-hand and right-hand integrals, respectively, we obtain

$$\int_{E_c - E_m}^{E_m} F_{FD}(E_m + x)g(E_m + x)\,dx = \int_{E_m - E_v}^{E_m}[1 - F_{FD}(E_m - x)]\,g(E_m - x)\,dx$$

But, by symmetry,

$$E_c - E_m = E_m - E_v \qquad g(E_m + x) = g(E_m - x)$$

and so

$$\int_{E_c - E_m}^{E_m}\{F_{FD}(E_m + x) - [1 - F_{FD}(E_m - x)]\}\,g(E_m + x)\,dx = 0 \qquad (1)$$

Using the explicit expression for F_{FD}, we have for the quantity in braces:

$$\frac{1}{e^{(E_m + x - E_f)/kT} + 1} - \left[1 - \frac{1}{e^{(E_m - x - E_f)/kt} + 1}\right] = \frac{1}{e^{(E_m + x - E_f)kT} + 1} - \frac{e^{(E_m - x - E_f)/kT}}{e^{(E_m - x - E_f)/kT} + 1}$$

$$= \frac{1 - e^{2(E_m - E_f)/kT}}{[e^{(E_m + x - E_f)/kT} + 1][e^{(E_m - x - E_f)/kT} + 1]}$$

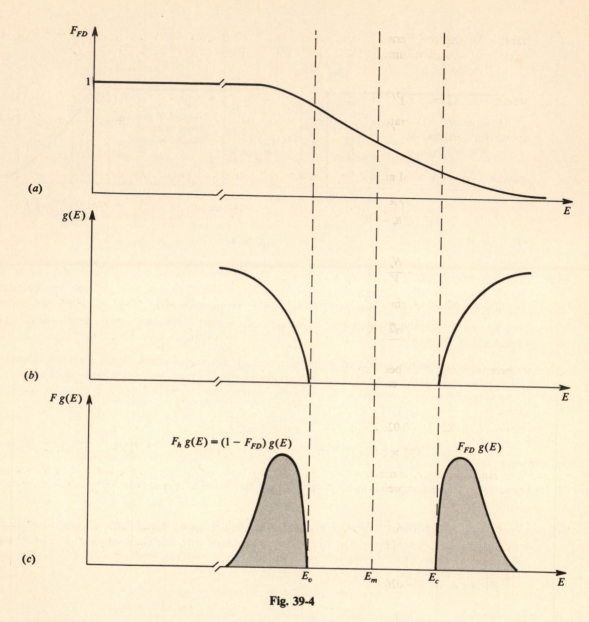

Fig. 39-4

Now, over the entire range of integration in (1), $g(E_m + x)$ is positive. It is seen that the quantity in braces would be everywhere positive if $E_f > E_m$ and everywhere negative if $E_f < E_m$. In either case the integrand in (1) would be one-signed, and the integral would be nonzero. Hence, $E_f = E_m$.

A more refined analysis shows that the density of states is not exactly symmetrical about E_m, so the Fermi energy will not be precisely at the midpoint of the gap. However, the error introduced by our assumption is quite small.

39.6. In Problem 38.9 the density of states in the free electron theory of metals was found to be

$$g(E) = \frac{8\sqrt{2}\ \pi m_e^{3/2} V}{h^3}\ E^{1/2}$$

Assume that this same expression holds for electrons in the conduction band, but with E on the right side replaced by $E - E_c$, where E_c is the energy at the bottom of the conduction band.   Taking the Fermi energy at the center of the gap, show that the number of valence electrons per unit volume in the conduction band at a temperature T varies like

$$n_e = Ce^{-E_g/2kT}$$

where $C = 2(2\pi m_e kT)^{3/2}/h^3$ and $E_g = E_c - E_v$ is the size of the energy gap.

At ordinary temperatures $kT \approx 0.026$ eV, so that in the conduction band $E - E_f \gg kT$, and the approximate expression

$$F_{FD} = e^{-(E-E_f)/kT}$$

may be used.   The total number, N_e, of electrons in the conduction band is then

$$N_e = \int_{E_c}^{\infty} F_{FD} g(E)\, dE = \frac{8\sqrt{2}\,\pi m_e^{3/2} V}{h^3} \int_{E_c}^{\infty} (E - E_c)^{1/2} e^{-(E-E_f)/kT}\, dE$$

or

$$n_e = \frac{N_e}{V} = \frac{8\sqrt{2}\,\pi m_e^{3/2}}{h^3} e^{-(E_c - E_f)/kT} \int_{E_c}^{\infty} (E - E_c)^{1/2} e^{-(E-E_c)/kT}\, dE$$

Using the result (Problem 39.5) that $E_c - E_f = E_g/2$ and substituting $u^2 = E - E_c$, we obtain

$$n_e = \frac{8\sqrt{2}\,\pi m_e^{3/2}}{h^3} e^{-E_g/2kT} 2 \int_0^{\infty} u^2 e^{-u^2/kT}\, du = 2\left(\frac{2\pi m_e kT}{h^2}\right)^{3/2} e^{-E_g/2kT}$$

where the integral has been evaluated by use of Table 37-1 ($n = 2$).

If the values of the constants are substituted in the expression for n_e, one obtains

$$n_e = (4.83 \times 10^{21}) T^{3/2} e^{-E_g/2kT} \text{ electrons/m}^3$$

For $T = 300$ K, $kT = 0.026$ eV and we have

$$n_e|_{300\,K} = (4.83 \times 10^{21})(300)^{3/2} e^{-E_g/2(0.026\ eV)} = (2.51 \times 10^{25}) e^{-E_g/(0.052\ eV)} \text{ electrons/m}^3$$

The expression for n_e is not strictly correct, because the Fermi energy is not exactly at the center of the gap.   However, the expression is sufficiently accurate for reasonable order-of-magnitude estimates.

39.7.   Refer to Problem 39.6.   Estimate the ratio of the electron densities in the conduction bands of the insulator carbon ($E_g = 5.33$ eV) and the semiconductor silicon ($E_g = 1.14$ eV) at room temperature (300 K).

At 300 K, $kT = 0.026$ eV.

$$\frac{n_C}{n_{Si}} = e^{-(5.33\ eV - 1.14\ eV)/(0.052\ eV)} \approx 10^{-35}$$

Thus the conduction population for the insulator is much much smaller than that for the semiconductor, which shows why insulators have such enormously high resistivities, even compared to semiconductors.

39.8.   The *mobility* of charge carriers, defined as $\mu = v_d/E$, where v_d is the drift velocity resulting from an applied electric field E, is a measure of the ability of the charge carriers to move through a material when an electric field is applied.   From Problem 39.1 it is seen that we may also write $\mu = e\lambda/m\bar{v}$.   In silicon ($E_g = 1.1$ eV) at room temperature the mobility of electrons is $\mu_n = 0.13$ m²/V·s and the mobility of holes is $\mu_p = 0.05$ m²/V·s.   Find the *conductivity*, $\sigma = 1/\rho$, of silicon, where ρ is the resistivity (see Problem 39.1).

The conductivity σ has two contributions, one from the electrons and the other from the holes.   Therefore

$$\sigma = \frac{1}{\rho_n} + \frac{1}{\rho_p} = \frac{e^2 n_n \lambda_n}{m_e \bar{v}_e} + \frac{e^2 n_p \lambda_p}{m_p \bar{v}_p}$$

Using $\mu = e\lambda/m\bar{v}$, we may write

$$\sigma = n_n e\mu_n + n_p e\mu_p$$

From the result of Problem 39.6,

$$n_n = n_p = (2.51 \times 10^{25})e^{-(1.1\ eV)/(0.052\ eV)} = 1.6 \times 10^{16}\ m^{-3}$$

whence

$$\sigma = (1.6 \times 10^{16}\ m^{-3})(1.6 \times 10^{-19}\ C)(0.13\ m^2/V \cdot s + 0.05\ m^2/V \cdot s) = 4.6 \times 10^{-4}\ (\Omega \cdot m)^{-1}$$

The overall resistivity is then

$$\rho = \frac{1}{\sigma} = 2.2 \times 10^3\ \Omega \cdot m$$

39.9. Refer to Problem 39.8. An impurity atom which donates one extra electron to the conduction band is added to silicon in the ratio of one impurity atom to 10^{10} silicon atoms. Determine the conductivity of the doped silicon, assuming that the mobility of the donated electrons is the same as the mobility of the host electrons and that the donor density is much greater than the intrinsic density of the host electrons and holes.

From Problem 39.3 the density of the silicon atoms is

$$n_{Si} = 5.2 \times 10^{28}\ atoms/m^3$$

so the density of the donor atoms is 10^{-10} times this value, or

$$n_d = 5.2 \times 10^{18}\ atoms/m^3$$

(In Problem 39.8 it was found that the density of the intrinsic charge carriers was $\approx 10^{16}\ m^{-3}$, which is much smaller than the donor density n_d.) The general expression for the conductivity is

$$\sigma = n_n e\mu_n + n_p e\mu_p$$

where n_n and n_p are the densities of the negative and positive charge carriers. Since the negatively charged donors are predominant in density, we may write

$$\sigma \approx n_d e\mu_n = (5.2 \times 10^{18}\ m^{-3})(1.6 \times 10^{-19}\ C)(0.13\ m^2/V \cdot s) = 0.11\ (\Omega \cdot m)^{-1}$$

This result, when compared with that of Problem 39.8, shows that a very small doping can have a significant effect on the conduction properties of a semiconductor. In practice, the doping ratio is in the range 10^{-5} to 10^{-10}.

39.10. If a thin strip of material carrying a constant current is placed in a magnetic field B which is directed perpendicular to the strip, it is found that a measurable potential difference appears across the strip (this is called the *Hall effect*). Show that the relationship between the electric field E induced in the material, the current density j, the number of charge carriers per unit volume n, and the magnetic field B is

$$E = \frac{jB}{qn}$$

where q is the charge of the charge carriers.

A picture of the equilibrium situation for positive charge carriers is shown in Fig. 39-5. When the magnetic field **B** is initially turned on, the force $\mathbf{F}_B$ causes the left side of the material to assume a net positive charge with respect to the right side. The electric field **E** created by this charge separation will exert a force $\mathbf{F}_E$ on the charge carriers to produce an equilibrium situation where the carriers drift neither to the right nor left. Therefore, with $\mathbf{v}_d$ perpendicular to **B**, we have

$$F_E = F_B \quad\quad \text{or} \quad\quad qE = qv_dB \quad\quad \text{or} \quad\quad E = v_dB$$

and since $j = qnv_d$,

$$E = \frac{jB}{qn}$$

Measuring E, j and B allows us to determine $1/qn$, called the *Hall coefficient*.

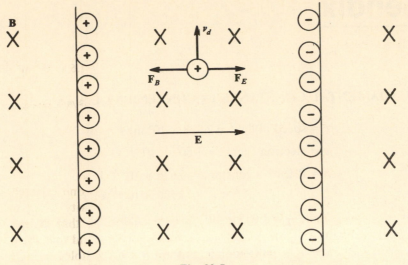

Fig. 39-5

Since the direction of the electric field and the corresponding polarity of the potential difference will depend upon the sign of the charge carriers, it is possible to determine this sign from the Hall effect. It is interesting that even though only electrons are free to move in a material, one does find Hall coefficients that indicate conduction by positively charged carriers. It is the holes, or absence of electrons in the valence band, which behave like positive charge carriers.

Supplementary Problems

39.11. In *thermionic emission* electrons which escape from a metal must have energy of at least $E = E_f + \phi$, with E_f the Fermi energy and ϕ the work function. Assuming that $\phi \gg kT$, find the energy distribution of the emitted electrons. *Ans.* $Ae^{-E/kT}$

39.12. For gold ($A = 197$) the resistivity is 2.04×10^{-8} $\Omega \cdot$ m, the Fermi energy is 5.54 eV, and the density 19.3×10^3 kg/m^3. Find the ratio of the mean free path λ to the interatomic spacing d for the conduction electrons if each gold atom contributes one free electron. *Ans.* $\lambda/d = 161$

39.13. Estimate the ratio of the electron densities in the conduction bands of silicon ($E_g = 1.1$ eV) and germanium ($E_g = 0.7$ eV) at 300 K. *Ans.* $n_{Ge}/n_{Si} = 2.2 \times 10^3$

39.14. An impurity atom which donates one extra hole to the valence band is added to silicon in the ratio of one impurity atom to 10^9 silicon atoms. Determine the conductivity of the doped silicon, assuming that the mobility of the donated holes is the same as the mobility of the intrinsic holes ($\mu_p = 0.05$ m^2/V $\cdot$ s). *Ans.* 0.42 $(\Omega \cdot$ m$)^{-1}$

39.15. After addition of an impurity atom which donates one extra electron to the conduction band of silicon ($\mu_n = 0.13$ m^2/V $\cdot$ s), the conductivity of the doped silicon is measured as 0.54 $(\Omega \cdot$ m$)^{-1}$. Determine the doping ratio. *Ans.* 2 parts in 10^9

39.16. It is convenient to describe the motion of an electron (or a hole) in a band by giving it an *effective mass*, m^*, defined by

$$\frac{1}{m^*} \equiv \frac{1}{\hbar^2} \frac{d^2E}{dk^2}$$

where k is the wave number ($k = 2\pi/\lambda$). For a free electron ($p = \hbar k$), show that $m^* = m$.

Appendix

SOME FUNDAMENTAL CONSTANTS IN CONVENIENT UNITS

$$c = \text{speed of light} = 2.998 \times 10^8 \text{ m/s}$$

$$e = \text{electron charge} = 1.602 \times 10^{-19} \text{ C}$$

$$h = \text{Planck's constant} = 6.626 \times 10^{-34} \text{ J} \cdot \text{s}$$
$$= 4.136 \times 10^{-15} \text{ eV} \cdot \text{s}$$

$$\hbar = \frac{h}{2\pi} = 1.055 \times 10^{-34} \text{ J} \cdot \text{s} = 0.658 \times 10^{-15} \text{ eV} \cdot \text{s}$$

$$k = \frac{1}{4\pi\epsilon_0} = \text{Coulomb constant} = 8.988 \times 10^9 \text{ N} \cdot \text{m}^2/\text{C}^2$$

$$k = \frac{R}{N} = \text{Boltzmann's constant} = 1.38 \times 10^{-23} \text{ J/K}$$
$$= 8.617 \times 10^{-5} \text{ eV/K}$$

SOME USEFUL CONVERSIONS AND COMBINATIONS

$$1 \text{ eV} = 1.602 \times 10^{-19} \text{ J}$$

$$1 \text{ Å} = 10^{-10} \text{ m} = 10^5 \text{ fm}$$

$$hc = 19.865 \times 10^{-26} \text{ J} \cdot \text{m} = 12.41 \times 10^3 \text{ eV} \cdot \text{Å} = 1241 \text{ MeV} \cdot \text{fm}$$

$$\hbar c = 3.165 \times 10^{-26} \text{ J} \cdot \text{m} = 1973 \text{ eV} \cdot \text{Å} = 197.3 \text{ MeV} \cdot \text{fm}$$

$$ke^2 = 1.44 \text{ MeV} \cdot \text{fm}$$

$$\frac{ke^2}{\hbar c} = \text{fine structure constant} = \frac{1}{137}$$

$$\frac{e\hbar}{2m_e} = \text{Bohr magneton} = 9.27 \times 10^{-24} \text{ J/T}$$
$$= 5.79 \times 10^{-5} \text{ eV/T}$$

MASSES OF SOME ELEMENTARY PARTICLES

Particle	Rest Mass, m_0 (kg)	$m_0 c^2$ (MeV)
Electron	9.109×10^{-31}	0.511
Proton	1.673×10^{-27}	938.3
Neutron	1.675×10^{-27}	939.6
Atomic mass unit (1 u)	1.661×10^{-27}	931.5

MASSES OF NEUTRAL ATOMS

In the fifth column of the table an asterisk on the mass number indicates a radioactive isotope, the half-life of which is given in the seventh column.

Z	Element	Symbol	Chemical Atomic Weight	A	Mass (u)	$T_{1/2}$
0	(Neutron)	n		1*	1.008665	12 min
1	Hydrogen	H	1.0079	1	1.007825	
	Deuterium	D		2	2.014102	
	Tritium	T		3*	3.016050	12.26 y
2	Helium	He	4.0026	3	3.016030	
				4	4.002603	
				6*	6.018892	0.802 s
3	Lithium	Li	6.939	6	6.015125	
				7	7.016004	
4	Beryllium	Be	9.0122	7*	7.016929	53.4 d
				9	9.012186	
				10*	10.013534	2.7×10^6 y
5	Boron	B	10.811	10	10.012939	
				11	11.009305	
6	Carbon	C	12.01115	12	12.000000	
				13	13.003354	
				14*	14.003242	5730 y
7	Nitrogen	N	14.0067	14	14.003074	
				15	15.000108	
8	Oxygen	O	15.9994	15*	15.003070	122 s
				16	15.994915	
				17	16.999133	
				18	17.999160	
9	Fluorine	F	18.9984	19	18.998405	
10	Neon	Ne	20.183	20	19.992440	
				21	20.993849	
				22	21.991385	
11	Sodium	Na	22.9898	22*	21.994437	2.60 y
				23	22.989771	
12	Magnesium	Mg	24.312	23*	22.994125	12 s
				24	23.985042	
				25	24.986809	
				26	25.982593	
13	Aluminum	Al	26.9815	26*	25.986892	7.4×10^5 y
				27	26.981539	
14	Silicon	Si	28.086	28	27.976929	
				29	28.976496	
				30	29.973763	
				32*	31.974020	≈ 700 y
15	Phosphorus	P	30.9738	31	30.973765	
16	Sulfur	S	32.064	32	31.972074	
				33	32.971462	
				34	33.967865	
				36	35.967089	
17	Chlorine	Cl	35.453	35	34.968851	
				36*	35.968309	3×10^5 y
				37	36.965898	
18	Argon	A	39.948	36	35.967544	
				38	37.962728	
				39*	38.964317	270 y
				40	39.962384	
				42*	41.963048	33 y
19	Potassium	K	39.102	39	38.963710	
				40*	39.964000	1.3×10^9 y

Z	Element	Symbol	Chemical Atomic Weight	A	Mass (u)	$T_{1/2}$
(19)	(Potassium)			41	40.961832	
20	Calcium	Ca	40.08	39*	38.970691	0.877 s
				40	39.962589	
				41*	40.962275	7.7×10^4 y
				42	41.958625	
				43	42.958780	
				44	43.955492	
				46	45.953689	
				48	47.952531	
21	Scandium	Sc	44.956	45	44.955920	
				50*	49.951730	1.73 min
22	Titanium	Ti	47.90	44*	43.959572	47 y
				46	45.952632	
				47	46.951768	
				48	47.947950	
				49	48.947870	
				50	49.944786	
23	Vanadium	V	50.942	50*	49.947164	$\approx 6 \times 10^{15}$ y
				51	50.943961	
24	Chromium	Cr	51.996	50	49.946055	
				52	51.940513	
				53	52.940653	
				54	53.938882	
25	Manganese	Mn	54.9380	50*	49.954215	0.29 s
				55	54.938050	
26	Iron	Fe	55.847	54	53.939616	
				55*	54.938299	2.4 y
				56	55.939395	
				57	56.935398	
				58	57.933282	
				60*	59.933964	$\approx 10^5$ y
27	Cobalt	Co	58.9332	59	58.933189	
				60*	59.933813	5.24 y
28	Nickel	Ni	58.71	58	57.935342	
				59*	58.934342	8×10^4 y
				60	59.930787	
				61	60.931056	
				62	61.928342	
				63*	62.929664	92 y
				64	61.927958	
29	Copper	Cu	63.54	63	62.929592	
				65	64.927786	
30	Zinc	Zn	65.37	64	63.929145	
				66	65.926052	
				67	66.927145	
				68	67.924857	
				70	69.925334	
31	Gallium	Ga	69.72	69	68.925574	
				71	70.924706	
32	Germanium	Ge	72.59	70	69.924252	
				72	71.922082	
				73	72.923462	
				74	73.921181	
				76	75.921405	
33	Arsenic	As	74.9216	75	74.921596	
34	Selenium	Se	78.96	74	73.922476	
				76	75.919207	
				77	76.919911	
				78	77.917314	
				79*	78.918494	7×10^4 y

Z	Element	Symbol	Chemical Atomic Weight	A	Mass (u)	$T_{1/2}$
(34)	(Selenium)			80	79.916527	
				82	81.916707	
35	Bromine	Br	79.909	79	78.918329	
				81	80.916292	
36	Krypton	Kr	83.80	78	77.920403	
				80	79.916380	
				81*	80.916610	2.1×10^5 y
				82	81.913482	
				83	82.914131	
				84	83.911503	
				85*	84.912523	10.76 y
				86	85.910616	
37	Rubidium	Rb	85.47	85	84.911800	
				87*	86.909186	5.2×10^{10} y
38	Strontium	Sr	87.62	84	83.913430	
				86	85.909285	
				87	86.908892	
				88	87.905641	
				90*	89.907747	28.8 y
39	Yttrium	Y	88.905	89	88.905872	
40	Zirconium	Zr	91.22	90	89.904700	
				91	90.905642	
				92	91.905031	
				93*	92.906450	9.5×10^5 y
				94	93.906313	
				96	95.908286	
41	Niobium	Nb	92.906	91*	90.906860	(long)
				92*	91.907211	$\approx 10^7$ y
				93	92.906382	
				94*	93.907303	2×10^4 y
42	Molybdenum	Mo	95.94	92	91.906810	
				93*	92.906830	$\approx 10^4$ y
				94	93.905090	
				95	94.905839	
				96	95.904674	
				97	96.906021	
				98	97.905409	
				100	99.907475	
43	Technetium	Tc		97*	96.906340	2.6×10^6 y
				98*	97.907110	1.5×10^6 y
				99*	98.906249	2.1×10^5 y
44	Ruthenium	Ru	101.07	96	95.907598	
				98	97.905289	
				99	98.905936	
				100	99.904218	
				101	100.905577	
				102	101.904348	
				104	103.905430	
45	Rhodium	Rh	102.905	103	102.905511	
46	Palladium	Pd	106.4	102	101.905609	
				104	103.904011	
				105	104.905064	
				106	105.903479	
				107*	106.905132	7×10^6 y
				108	107.903891	
				110	109.905164	
47	Silver	Ag	107.870	107	106.905094	
				109	108.904756	
48	Cadmium	Cd	112.40	106	105.906463	
				108	107.904187	

Z	Element	Symbol	Chemical Atomic Weight	A	Mass (u)	$T_{1/2}$
(48)	(Cadmium)			109*	108.904928	453 d
				110	109.903012	
				111	110.904188	
				112	111.902762	
				113	112.904408	
				114	113.903360	
				116	115.904762	
49	Indium	In	114.82	113	112.904089	
				115*	114.903871	6×10^{14} y
50	Tin	Sn	118.69	112	111.904835	
				114	113.902773	
				115	114.903346	
				116	115.901745	
				117	116.902958	
				118	117.901606	
				119	118.903313	
				120	119.902198	
				121*	120.904227	25 y
				122	121.903441	
				124	123.905272	
51	Antimony	Sb	121.75	121	120.903816	
				123	122.904213	
				125*	124.905232	2.7 y
52	Tellurium	Te	127.60	120	119.904023	
				122	121.903064	
				123*	122.904277	1.2×10^{13} y
				124	123.902842	
				125	124.904418	
				126	125.903322	
				128	127.904476	
				130	129.906238	
53	Iodine	I	126.9044	127	126.904070	
				129*	128.904987	1.6×10^{7} y
54	Xenon	Xe	131.30	124	123.906120	
				126	125.904288	
				128	127.903540	
				129	128.904784	
				130	129.903509	
				131	130.905085	
				132	131.904161	
				134	133.905815	
				136	135.907221	
55	Cesium	Cs	132.905	133	132.905355	
				134*	133.906823	2.1 y
				135*	134.905770	2×10^{6} y
56	Barium	Ba	137.34	137*	133.906770	30 y
				130	129.906245	
				132	131.905120	
				133*	132.905879	7.2 y
				134	133.904612	
				135	134.905550	
				136	135.904300	
				137	136.905500	
				138	137.905000	
57	Lanthanum	La	138.91	137*	136.906040	6×10^{4} y
				138*	137.906910	1.1×10^{11} y
				139	138.906140	
58	Cerium	Ce	140.12	136	135.907100	
				138	137.905830	
				140	139.905392	

Z	Element	Symbol	Chemical Atomic Weight	A	Mass (u)	$T_{1/2}$
(58)	(Cerium)			142*	141.909140	5×10^{15} y
59	Praseodymium	Pr	140.907	141	140.907596	
60	Neodymium	Nd	144.24	142	141.907663	
				143	142.909779	
				144*	143.910039	2.1×10^{15} y
				145	144.912538	
				146	145.913086	
				148	147.916869	
				150	149.920960	
61	Promethium	Pm		145*	144.912691	18 y
				146*	145.914632	1600 d
				147*	146.915108	2.6 y
62	Samarium	Sm	150.35	144	143.911989	
				146*	145.912992	1.2×10^{8} y
				147*	146.914867	1.08×10^{11} y
				148*	147.914791	1.2×10^{13} y
				149*	148.917180	4×10^{14} y
				150	149.917276	
				151*	150.919919	90 y
				152	151.919756	
				154	153.922282	
63	Europium	Eu	151.96	151	150.919838	
				152*	151.921749	12.4 y
				153	152.921242	
				154*	153.923053	16 y
				155*	154.922930	1.8 y
64	Gadolinium	Gd	157.25	148*	147.918101	85 y
				150*	149.918605	1.8×10^{6} y
				152*	151.919794	1.1×10^{14} y
				154	153.920929	
				155	154.922664	
				156	155.922175	
				157	156.924025	
				158	157.924178	
				160	159.927115	
65	Terbium	Tb	158.925	159	158.925351	
66	Dysprosium	Dy	162.50	156*	155.923930	2×10^{14} y
				158	157.924449	
				160	159.925202	
				161	160.926945	
				162	161.926803	
				163	162.928755	
				164	163.929200	
67	Holmium	Ho	164.930	165	164.930421	
				166*	165.932289	1.2×10^{3} y
68	Erbium	Er	167.26	162	161.928740	
				164	163.929287	
				166	165.930307	
				167	166.932060	
				168	167.932383	
				170	169.935560	
69	Thulium	Tm	168.934	169	168.934245	
				171*	170.936530	1.9 y
70	Ytterbium	Yb	173.04	168	167.934160	
				170	169.935020	
				171	170.936430	
				172	171.936360	
				173	172.938060	
				174	173.938740	
				176	175.942680	

Z	Element	Symbol	Chemical Atomic Weight	A	Mass (u)	$T_{1/2}$
71	Lutecium	Lu	174.97	173*	172.938800	1.4 y
				175	174.940640	
				176*	175.942660	2.2×10^{10} y
72	Hafnium	Hf	178.49	174*	173.940360	2.0×10^{15} y
				176	175.941570	
				177	176.943400	
				178	177.943880	
				179	178.946030	
				180	179.946820	
73	Tantalum	Ta	180.948	180	179.947544	
				181	180.948007	
74	Wolfram (Tungsten)	W	183.85	180	179.947000	
				182	181.948301	
				183	182.950324	
				184	183.951025	
				186	185.954440	
75	Rhenium	Re	186.2	185	184.953059	
				187*	186.955833	5×10^{10} y
76	Osmium	Os	190.2	184	183.952750	
				186	185.953870	
				187	186.955832	
				188	187.956081	
				189	188.958300	
				190	189.958630	
				192	191.961450	
				194*	193.965229	6.0 y
77	Iridium	Ir	192.2	191	190.960640	
				193	192.963012	
78	Platinum	Pt	195.09	190*	189.959950	7×10^{11} y
				192	191.961150	
				194	193.962725	
				195	194.964813	
				196	195.964967	
				198	197.967895	
79	Gold	Au	196.967	197	196.966541	
80	Mercury	Hg	200.59	196	195.965820	
				198	197.966756	
				199	198.968279	
				200	199.968327	
				201	200.970308	
				202	201.970642	
				204	203.973495	
81	Thallium	Tl	204.19	203	202.972353	
				204*	203.973865	3.75 y
				205	204.974442	
		Ra E″		206*	205.976104	4.3 min
		Ac C″		207*	206.977450	4.78 min
		Th C″		208*	207.982013	3.1 min
		Ra C″		210*	209.990054	1.3 min
82	Lead	Pb	207.19	202*	201.927997	3×10^5 y
				204*	203.973044	1.4×10^{17} y
				205*	204.974480	3×10^7 y
				206	205.974468	
				207	206.975903	
				208	207.976650	
		Ra D		210*	209.984187	22 y
		Ac B		211*	210.988742	36.1 min
		Th B		212*	211.991905	10.64 h
		Ra B		214*	213.999764	26.8 min
83	Bismuth	Bi	209.980	207*	206.978438	30 y

Z	Element	Symbol	Chemical Atomic Weight	A	Mass (u)	$T_{1/2}$
(83)	(Bismuth)			208*	207.979731	3.7×10^5 y
				209	208.980394	
		Ra E		210*	209.984121	5.1 d
		Th C		211*	210.987300	2.15 min
				212*	211.991876	60.6 min
		Ra C		214*	213.998686	19.7 min
				215*	215.001830	8 min
84	Polonium	Po		209*	208.982426	103 y
		Ra F		210*	209.982876	138.4 d
		Ac C′		211*	210.986657	0.52 s
		Th C′		212*	211.989629	0.30 μs
		Ra C′		214*	213.995201	164 μs
		Ac A		215*	214.999423	0.0018 s
		Th A		216*	216.001790	0.15 s
		Ra A		218*	218.008930	3.05 min
85	Astatine	At		215*	214.998663	$\approx$ 100 μs
				218*	218.008607	1.3 s
				219*	219.011290	0.9 min
86	Radon	Rn				
		An		219*	219.009481	4.0 s
		Tn		220*	220.011401	56 s
		Rn		222*	222.017531	3.823 d
87	Francium	Fr				
		Ac K		223*	223.019736	22 min
88	Radium	Ra	226.05			
		Ac X		223*	223.018501	11.4 d
		Th X		224*	224.020218	3.64 d
		Ra		226*	226.025360	1620 y
		Ms Th₁		228*	228.031139	5.7 y
89	Actinium	Ac		227*	227.027753	21.2 y
		Ms Th₂		228*	228.031080	6.13 h
90	Thorium	Th	232.038			
		Rd Ac		227*	227.027706	18.17 d
		Rd Th		228*	228.028750	1.91 y
				229*	229.031652	7300 y
		Io		230*	230.033087	76000 y
		UY		231*	231.036291	25.6 h
		Th		232*	232.038124	1.39×10^{10} y
		UX₁		234*	234.043583	24.1 d
91	Protoac-tinium	Pa	231.0359	231*	231.035877	32480 y
		UZ		234*	234.043298	6.66 h
92	Uranium	U	238.03	230*	230.033937	20.8 d
				231*	231.036264	4.3 d
				232*	232.037168	72 y
				233*	233.039522	1.62×10^5 y
				234*	234.040904	2.48×10^5 y
		Ac U		235*	235.043915	7.13×10^8 y
				236*	236.045637	2.39×10^7 y
		UI		238*	238.048608	4.51×10^9 y
93	Neptunium	Np	237.0480	235*	235.044049	410 d
				236*	236.046624	5000 y
				237*	237.048056	2.14×10^6 y
94	Plutonium	Pu	239.0522	236*	236.046071	2.85 y
				238*	238.049511	89 y
				239*	239.052146	24360 y
				240*	240.053882	6700 y
				241*	241.056737	13 y
				242*	242.058725	3.79×10^5 y
				244*	244.064100	7.6×10^7 y

Index

Absolute space, 7
Absorption
  coefficient, 74
  edge, 152
Activity, 185
Affinity (*see* Electron affinity)
Alkali metal, 140, 147, 227
Alpha
  decay, 185
  particle, 184, 185
Angular momentum, classical, 112
  electron, 100, 114
  intrinsic, 120
  spin, 120
  total, 124
Anomalous Zeeman effect, 115, 135, 144
Antiparticle, 68, 187, 213, 215
Atmospheres, law of, 243
Atom, hydrogen, 99
  hydrogenic, 103
  many-electron, 133
  mu-mesic, 103, 109, 111
  pi-mesic, 103
Atomic
  excited state, 135
  mass unit, 163
  number, 168
  shell model, 133
  state, 133
Attenuation law, 74
Auger effect, 152
Average lifetime, 184, 190
Avogadro's number, 241

Balmer series, 99, 105, 110
Band, conduction, 287
  valence, 287
Band theory, 287
Barn, 202
Baryon, 215
  number, 217
Beta
  decay, 186
  particle, 184, 186
Bethe cycle, 210, 213
Binding energy, 173
Blackbody radiation, 266, 268
Bohr, N., 92
  atom, 99
  correspondence principle, 107
  model, 99
  orbit, 99
  theory, 99, 100

Boltzmann, L., 269
  constant, 242
  distribution, 251, 256
Bonding, covalent, 227
  ionic, 227
  metallic, 228
  van der Waals, 228
Bose-Einstein
  condensation, 278
  statistics, 266, 278
Bosons, 215, 266
Box, particle in a, 90, 94, 128, 270
Brackett series, 99
Bragg
  diffraction, 82
  law, 82
  plane, 82
Bremsstrahlung, 149
Brownian motion, 245

Carbon cycle, 210, 213
Center-of-mass system, 101, 200, 203, 208
Characteristic X-rays (*see* X-ray spectra)
Chadwick, J., 206
Clock synchronization, 8
Complementarity, principle of, 92
Compound nucleus, 200, 202
Compton
  effect, 62
  equation, 62
  scattering, 62
  wavelength, 62, 65, 81, 111
Conduction band, 287
Conductivity, electrical, 293
  heat, 274
Conductor, 287
Configuration, electron, 133
Conjugate variables, 92
Conservation laws
  angular momentum, 216
  baryon number, 217
  charge, 216
  isotopic spin, 217
  lepton number, 217
  linear momentum, 216
  mass-energy, 216
  parity, 219
  spin, 216
  strangeness, 218
Contraction, Lorentz-Fitzgerald, 17
Coordinate transformation, Galilean, 1
  Lorentz, 12
Correspondence principle, 107

Coulomb
    energy, 169, 174
    force, 168
Coupling constant, 216
Covalent bonding, 227
Cowan, C. L., 186
Critchfield cycle, 210, 214
Cross section, 201, 202, 205, 206
Curie, 185, 188
Cycle, Bethe, 210, 213
    carbon, 210, 213
    Critchfield, 210, 214
    proton-proton, 210, 214

Daughter nuclei, 184
Davisson and Germer experiment, 83, 85
De Broglie
    hypothesis, 77, 82, 99
    waves, 77, 82, 88, 99
Debye, P., 229, 274
    frequency, 275
    temperature, 276, 277
    theory, 274
    unit, 229
Decay, alpha, 185
    beta, 185
    gamma, 185
Decay constant, 184
Decay law, 184
Degeneracy, 271, 282
Density of states, 250, 254, 255, 256, 266, 276
Deuterium, 107
Deuteron, 163, 164, 199
Diatomic molecules, 232, 259
Diffraction, electron, 83
    particle, 91
    X-ray, 82
Dipole moment, electric, 135, 229
    magnetic, 112
Dirac, P., 121, 266, 282
Disintegration
    constant, 184
    energy, 186, 187
Dissociation energy, 227, 228
Distribution, Boltzmann, 251
    Bose-Einstein, 266, 282
    Fermi-Dirac, 266, 282
    frequency, 249
    Maxwell-Boltzmann, 256, 260
Distribution function, 249, 250
Doping, 288
Doppler effect, 50
Dulong-Petit law, 259, 274, 276

Edge, X-ray absorption, 152
Einstein, A., 264, 274
    postulates, 7

Electric
    dipole transitions, 115, 135, 229
    quadrupole moment, 164, 165
Electrical
    conduction, 270, 287
    conductivity, 293
Electron, free, 270
    valence, 270
Electron
    affinity, 227
    capture, 186, 187
    configuration, 133
    diffraction, 83
    spin, 120, 124
Electron theory of metals, 270
Electron volt, 40
Elementary particles
    antiparticles, 215
    families, 215
    interactions, 216
    isotopic spin, 217
    mass, 215
    mean lifetime, 215
    spin, 215
Endoergic (endothermic) reaction, 201, 203
Energy, Coulomb, 169, 174
    electron affinity, 227
    magnetic dipole, 113
    neutron-proton excess, 174
    nuclear binding, 169, 173
    pairing, 174
    relativistic, 40
    rotational, 232
    surface, 173
    uncertainty in, 92
    vibrational, 232
Energy level diagrams, 101
Equipartition theorem, 259
Ether, the, 7
Event, 1
Exchange particle, 225
Exclusion principle, 128, 133, 266
Exoergic (exothermic) reaction, 201, 203
Experiment, Fizeau, 37
    Franck-Hertz, 161
    Michelson-Morley, 7, 9, 10, 11
    Stern-Gerlach, 120
    Zeeman, 112
Exponential distribution, 188
Extrinsic semiconductor, 288

Fermi-Dirac statistics, 266, 278
Fermi energy, 266, 272, 293
Fine structure, 124
    constant, 111
Fission, 209
Fizeau experiment, 37

Fluorescence, 153
Franck-Hertz experiment, 161
Free electron theory, 270
Fusion, 210

Galilean transformation, 1
Gamma
   decay, 185
   function, 267
   ray, 184, 185
Goudsmit, S. A., 120
Gerlach, W., 120
Germer, L. H., 85
Gyromagnetic ratio, 121

Half-life, 184
Half-thickness, 74
Hall
   coefficient, 294
   effect, 294
Halogen, 140, 147, 227
Harmonic oscillator, 175, 232
Heat capacity, 274
Heisenberg uncertainty principle, 91
Hertz, G., 161
Hole, 287
Hydrogen
   atom, 99
   spectrum, 99

Ideal gas, classical, 241
   quantum mechanical, 278
Impurity atom, 294
Independent particle model, 133
Indeterminacy (see Uncertainty principle)
Insulator, 287
Interaction, electromagnetic, 216
   gravitational, 216
   strong, 216
   weak, 216
Interferometer, 9
Intrinsic semiconductor, 288
Invariance, 2, 4, 5, 12, 30, 31
Ionic bonding, 227
Ionization
   energy, 100
   potential, 104
Isobar, 168, 179
Isomers, 185
Isotone, 168
Isotope, 168
Isotopic spin, 217

K shell, 150
Kinetic theory, 241

L shell, 150
Lagrange multipliers, 263, 283
Landé g-factor, 143
Larmor precession, 116
Laue, M. von, 82
Law
   atmospheres, 243
   Dulong-Petit, 259, 274, 276
   ideal gas, 241
   Stefan-Boltzmann, 269
   Wien displacement, 269
Length, proper, 17
   rest, 17
Length
   contraction, 17
Lepton, 215
   number, 217
Light, speed of, 7, 8, 12, 34
Liquid drop model, 173
Lorentz
   coordinate transformations, 12
   velocity transformations, 34
Lyman series, 99, 103, 111

M shell, 150
Magic numbers, 175, 182
Magnetic dipole moment, 112, 135
Magneton, Bohr, 114
   nuclear, 163
Many-electron atom, 133
Mass, atomic, 169
   nuclear, 169
   relativistic, 39
Mass-energy relation, 40
Mass excess, 183, 203, 208
Mass formula, semiempirical, 173
Mass number, 168
Maxwell-Boltzmann distribution, 251, 256, 260
Maxwell's equations, 5, 12
Mean free path, 244, 245
Mean lifetime, 184, 215
Measurement
   length (relativistic), 17
   location, 91
   space-time (relativistic), 24
   time (relativistic), 20
Meson, 215
Metallic bonding, 228
Metals, free electron theory, 270
Michelson-Morley experiment, 7, 9, 10, 11
Mirror nuclei, 178
Mobility, 293
Model, atomic shell, 133
   independent particle, 133
   liquid drop, 173
   nuclear shell, 175

Molecular
  energy levels, 233
  rotation, 232
  vibration, 232
Momentum, relativistic, 39, 91
  uncertainty in, 91
Momentum change, 91
Morley, E., 7, 9, 10, 11
Morse potential, 230
Moseley relation, 150
Mu-mesic atom, 103, 109, 111

Natural line width, 93
Neutrino, 186, 216
Neutron, 163, 168, 199, 206
Newton's second law, 39
Noble gas, 140, 147
Normal Zeeman effect (*see* Zeeman effect)
Normalization condition, 249, 257, 273, 279
Nuclear
  binding energy, 169
  fission, 209
  fusion, 210
  magnetic moment, 163
  magneton, 163
  matter, 169
  models, 173
  radius, 169
  reactions, 199
  shell model, 175
  spin, 163
Nucleon, 163

One-dimensional box, 90, 94, 128
Orbit, 133, 177
Orbital, 177
Orbital angular momentum, 114

Pair
  annihilation, 68
  production, 68
Pairing, 174, 177, 180, 182
Parent nucleus, 184
Parity, 219
Particles (*see* Elementary particles)
Paschen series, 99, 103, 110
Pauli, W., 186
  exclusion principle, 128, 133, 266
Periodic table, 133
Perrin, J., 246
Phonon, 275
Photoelectric effect, 56
Photofission, 211
Photon, 53, 270
  emission, 100
  energy, 53
  momentum, 53

Pickup reaction, 199
Pi-mesic atoms, 103
Planck, M., 53, 268
  constant, 53
Position, uncertainty in, 92
Positronium, 108
Potential, harmonic oscillator, 175
  Morse, 230
Probability interpretation (*see* De Broglie waves)
Proper
  length, 17
  time, 20
Propagation vector, 68
Proton, 163, 168, 199

Q-value, 201
Quadrupole moment, 164
Quanta, 53
Quantum number
  magnetic, 114
  orbital angular momentum, 114, 175
  principal, 90, 100, 128, 270
  spin, 121, 177
  total angular momentum, 125, 177
Quantum statistics, 266

Radiation
  atomic, 100
  blackbody, 268
Radioactive decay, 184
Radioastronomy, 122
Radius
  Bohr, 100
  nuclear, 169
Random walk, 244
Range, 198
Reactions
  compound nucleus, 200
  D–D, 210
  D–T, 210
  direct, 200
  pickup, 199
  stripping, 199
Reduced mass, 101, 111
Reines, F., 186
Relativity, principle of, 7
Resistivity, 289
Resonance, 215, 219
Riemann zeta function, 267
Roentgen, W., 149
Root-mean-square speed, 241, 258
Rotation, molecular, 232
Russell-Saunders coupling, 135
Rutherford, E., 199
Rydberg formula, 99

Scattering, elastic, 199
  inelastic, 199, 201

Selection rules, 135, 233
Semiconductors, 287
  doped, 288
  extrinsic, 288
  intrinsic, 288
  n-type, 289
  p-type, 289
Semiempirical mass formula, 173
Shell, atomic, 133
  nuclear, 175
Shell closing, 175
Shell model, 175
Simultaneity, 9, 13, 15
Singlet, 227
Special Theory of Relativity, 1
Specific heat, 274
Spectra, absorption, 238
  hydrogen, 99
  rotational, 233, 234, 235
  vibrational, 233
  X-ray, 150
Spectroscopic notation, 134, 175
Speed, average, 232
  of light, 8, 12
  most probable, 258
  root-mean-square, 241, 258
Spin, electron, 120, 124
Spin-orbit coupling, 124, 135, 177
Square well (*see* Box)
State, ground, 101
  molecular, 250
  quantum, 128
  singlet, 227
  triplet, 227
  vibrational, 275
Stefan-Boltzmann law, 269, 283
Stern-Gerlach experiment, 120
Stirling's formula, 261, 283
Stopping potential, 56
Strangeness, 218
Stripping reaction, 199
Strong interaction, 216
Subshell, 133
Synchronization, 8
System, center-of-mass, 200
  laboratory, 200

Thermal neutron, 211
Thermionic emission, 295
Thomson, G. P., 83

Threshold
  energy, 201, 204
  wavelength, 56, 69, 72
Time, proper, 20
  uncertainty in, 92
Time dilation, 20
Transformation, Galilean acceleration, 2
  Galilean coordinate, 1
  Galilean velocity, 1
  Lorentz coordinate, 12
  Lorentz velocity, 34
  mass-energy, 40
  momentum-energy, 40
Transition, electric dipole, 115
Transition element, 141, 147
Triplet, 227
Triton, 199
Tunneling, 205
Twin effect, 27

Uhlenbeck, G. E., 120
Uncertainty, angular momentum and angle, 94
  energy and time, 92
  position and momentum, 92
Uncertainty principle, 91

Valence band, 287
van der Waals bonding, 228
Vector model, 124, 142
Velocity transformation, 34
Vibration, molecular, 232

Wave, de Broglie, 77
  matter, 91
  probability interpretation of, 88
Wave equation, 5
Wave-particle duality, 77
Weak interaction, 216
Weizsäcker, C. v., 173
Wien displacement law, 269

X-ray, 82, 83, 149, 187
  absorption edge, 152
  diffraction, 82
  fluorescence, 153
  spectra, 150

Yukawa, H., 216

Zeeman effect, anomalous, 115, 135, 144
  normal, 112, 114, 124

Catalog

If you are interested in a list of SCHAUM'S
OUTLINE SERIES send your name
and address, requesting your free catalog, to:

SCHAUM'S OUTLINE SERIES, Dept. C
McGRAW-HILL BOOK COMPANY
1221 Avenue of Americas
New York, N.Y. 10020